PETROLEUM
RADIATION
PROCESSING

PETROLEUM RADIATION PROCESSING

Yuriy Zaikin
Raissa Zaikina

CRC Press
Taylor & Francis Group
Boca Raton London New York

CRC Press is an imprint of the
Taylor & Francis Group, an **informa** business

CRC Press
Taylor & Francis Group
6000 Broken Sound Parkway NW, Suite 300
Boca Raton, FL 33487-2742

First issued in paperback 2019

© 2014 by Taylor & Francis Group, LLC
CRC Press is an imprint of Taylor & Francis Group, an Informa business

No claim to original U.S. Government works

ISBN-13: 978-1-4665-9310-7 (hbk)
ISBN-13: 978-0-367-37915-5 (pbk)

Library of Congress Cataloging-in-Publication Data

Zaikin, Yuriy.
    Petroleum radiation processing / Yuriy Zaikin, Raissa Zaikina.
        pages cm
    Summary: "This book is a summary of the research progress in the field of petroleum radiation processing, giving rise to technology that offers the highest energy savings and lowest capital and operational expenses compared with any existing methods for oil refining. The theoretical part of this book sets out important problems in self-sustaining cracking reactions in hydrocarbons. The next part provides a systematic description of the most important experiments on radiation cracking. The concluding part summarizes progress in the development of new radiation technologies for oil upgrading and deep processing"-- Provided by publisher.
    Includes bibliographical references and index.
    ISBN 978-1-4665-9310-7 (hardback)
    1. Cracking process. 2. Radiation chemistry. I. Zaikina, Raissa. II. Title.

TP690.4.Z35 2014
665.5'33--dc23
                                           2013036987

Visit the Taylor & Francis Web site at
http://www.taylorandfrancis.com

and the CRC Press Web site at
http://www.crcpress.com

Contents

Authors

Yuriy Alexandrovich Zaikin currently serves as chief technology officer of PetroBeam Inc., a US company developing a new radiation technology for heavy oil upgrading based on his and Dr. Zaikina's invention. He earned his MS in 1971 from Kazakh State University (Almaty, Kazakhstan) and his PhD in physics of condensed matter in 1976 from A.A. Baikov Institute of Metallurgy (Moscow, Russia). He earned his DrSci in physics of condensed matter in 1996 from the Institute of Physics and Technology (Almaty, Kazakhstan). From 1976 to 2006, he worked as a professor and head of the radiation physics lab in Al Farabi Kazakh National University (Almaty, Kazakhstan). Dr. Zaikin has published more than 250 articles in the field of radiation physics and chemistry, petroleum radiation processing, radiation methods for the development of materials with enhanced properties, and physics of condensed matter, including detailed studies of diffusion and internal friction in irradiated solids. Since 1991, he has been developing radiation methods for petroleum processing. In 2006, together with Dr. Raissa Zaikina, he described and experimentally proved the phenomenon of self-sustaining chain cracking reactions in liquid hydrocarbons at lowered temperatures. This technology was developed by PetroBeam, Inc., United States, in 2007–2013. From 2006 to 2010, he served as an expert consultant on oil radiation processing for IAEA in Saudi Arabia.

Raissa Fuatovna Zaikina currently serves as chief research scientist in PetroBeam, Inc. (Durham, North Carolina). She earned her MS in 1972 from Kazakh State University and her PhD in physics of condensed matter in 1991 from the Institute of Physics and Technology (Almaty, Kazakhstan). She has performed research and contributed to more than 100 publications and 7 patents in the fields of radiation processing, studies of radiation effects in semiconductor compounds, and radiation technologies for petroleum processing. Since 1991, she has been conducting systematic research on radiation-induced conversion of petroleum and developing basic experimental approaches to radiation processing of heavy and high-paraffin oils.

Introduction

Radiation methods for petroleum processing have attracted the attention of researchers since the early 1960s when the discovery of the phenomenon of radiation-thermal cracking presented an opportunity of using ionizing irradiation for high-rate oil processing accompanied by profound changes in oil fractional contents and chemical composition.

New technologies for high-viscous and heavy oil processing were developed with technical advances in the 1990s. These technologies are now ready to be scaled up for industrial applications. Radiation-thermal cracking of oil feedstock represents a solution to overcoming many acute problems of the oil industry. However, processes based on radiation-thermal cracking require heightened temperatures although they are usually about 40% lower than those characteristic for thermocatalytic cracking. This is acceptable for many refinery operations; however, other applications, such as oil upgrading near the sites of its extraction, require radical reduction of the process temperature.

Observation of radiation-induced chain cracking reactions in hydrocarbons at lowered temperatures initiated the development of improved technological approaches, combining the advantages of radiation-thermal cracking and low-temperature feedstock processing.

Progress in radiation technologies for oil processing is associated with more detailed elaboration of the theory of thermally and radiation-induced self-sustaining cracking reactions. Researchers still face serious difficulties in the practical application of the theory to experimental data interpretation. A kinetic description of chain cracking reactions available in the literature is often in contradiction to the thermodynamic requirements of endothermic chain reactions. It presents difficulties in determining the correct starting cracking temperature and in predicting its dependence on the dose rate in the case of radiation-thermal cracking. Methods for determining the number of chain propagation steps and their dependence on experimental conditions are not sufficiently developed. The reactions and the nature of the molecular states responsible for chain propagation in the case of low-temperature radiation cracking require further research.

The objective of this book is to fill this theoretical gap. The book provides systematic descriptions of the fundamentals of radiation-induced cracking reactions in hydrocarbons and analyzes the basic experiments that have given rise to the rapid development of radiation technology for petroleum radiation processing during the last decades. It also provides a detailed introduction to radiation methods based on radiation-thermal and low-temperature cracking of hydrocarbons, with an emphasis on high-viscous oil feedstock that are difficult to process by conventional methods, such as heavy and high-paraffinic crude oil, fuel oil, and bitumen. The application of promising radiation methods for solving such pressing environmental issues as oil desulfurization and regeneration of used lubricants and other used oil products also receives special attention.

The research in the last 50 years since the discovery of the phenomenon of radiation-thermal cracking of hydrocarbons has resulted in impressive fundamental findings and promising technological approaches to the most acute issues in the oil industry. The principal achievements in this field of radiation science are summarized in this book.

This book will be of interest to chemical technologists and researchers working on technological applications of methods that use ionizing irradiation for oil upgrading, refining, and desulfurization. It can also be useful for students specializing in the fields of chemical and petroleum engineering, physical chemistry, radiation physics, and chemistry.

1 Theory of Radiation-Induced Cracking Reactions in Hydrocarbons

Practical application of the chain reaction theory for the interpretation of data on thermal and radiation cracking of hydrocarbons and calculations of the basic cracking parameters faces certain difficulties associated with the following problems of the theory:

1. The kinetic description of the chain cracking reactions often comes into contradiction with the process of thermodynamics. In the case of thermal cracking (TC), it does not allow a distinct determination of the cracking start temperature, while in the case of radiation-thermal cracking (RTC), it does not allow the determination of the dependence of the cracking start temperature on the irradiation dose rate.
2. The methods for the calculation of the reaction chain length on the base of the process kinetic characteristics are not sufficiently developed. In the case of radiation cracking, dependence of the chain length on temperature and dose rate remains practically unknown.
3. A correct description of the dependence of the cracking rate on temperature and dose rate (in the case of radiation cracking) requires a special theoretical consideration.
4. The nature of the reactions and molecular states, as well as mechanisms of chain propagation in low-temperature radiation cracking (LTRC), still remains unclear.

This chapter deals with the consideration of these problems as applicable to TC, RTC, and LTRC of hydrocarbons.

1.1 THERMAL CRACKING: NUMBER OF PROPAGATION STEPS, CRACKING START TEMPERATURE, REACTION RATE

By definition, the chain length is the number of elementary events in a chain reaction initiated by a single chain carrier. The chain length is often interpreted as the

1

ratio of the chain reaction propagation rate to its termination rate (or the rate of chain initiation to the rate of chain termination).

The commonly used equations for chain initiation, propagation, and termination are

$$K_i = k_i e^{-E_i/k_B T} \tag{1.1}$$

$$K_p = k_p e^{-Ep/k_B T} \tag{1.2}$$

$$K_t = k_t e^{-E_t/kT} \tag{1.3}$$

To calculate the chain length, the number of elementary reactions initiated by all chain carriers in the unit of time should be divided by the number of carriers generated per unit of time. In the case of the radical mechanism (Talrose 1974),

$$v = \frac{K_p R}{K_i} = \frac{k_p R}{k_i} e^{(E_i - E_p)/kT} \tag{1.4}$$

where R is concentration of the chain-initiating radicals.

Let us consider the case of a quadratic chain termination that prevails in the reactions of hydrocarbon cracking.

In the dynamical equilibrium,

$$k_i e^{-E_i/k_B T} = k_t R^2 \tag{1.5}$$

This implies a well-known equation for the radical concentration:

$$R = \left(\frac{k_i}{k_t}\right)^{1/2} e^{-E_i/2kT} \tag{1.6}$$

The number of propagation steps (chain length) can be evaluated as

$$v = \frac{W_p}{K_i} = \frac{\sqrt{k_i/k_t}\, e^{-E_i/2kT}\, k_p\, e^{-E_p/kT}}{k_i\, e^{-E_i/kT}}$$

or

$$v = \frac{k_p}{\sqrt{k_i k_t}} e^{(E_i/2 - E_p)/kT} \tag{1.7}$$

Here, $E_i/2 - E_p \approx 20$ kcal/mol.

Thus we come to an absurd expression inapplicable to a chain endothermic reaction.

Wu et al. (1997b) suggested that the reaction mechanism and rate constants of various reactions in the case of RTC should be the same as those for pure TC. They represented a summary of the reaction mechanisms for liquid-phase cracking of long-chain paraffins proposed by many authors (Voge and Good 1949, Ford 1986, Khorashev and Gray 1993, Song et al. 1994, Wu et al. 1997a) to account for the product distributions.

It was noted that liquid-phase cracking obeys a one-step mechanism, while gas-phase cracking follows a two-step or multistep decomposition model. The following basic reactions were proposed for hydrocarbon cracking in the liquid phase:

Initiation

$$M \xrightarrow{ki} R_j + R_j^* \quad \text{thermal initiation} \tag{1.8}$$

$$M \rightsquigarrow R^* \xrightarrow{P} R_j^* + R_j^* \quad \text{and} \quad H + R^* \quad \text{radiation initiation} \tag{1.9}$$

Propagation

$$R' + M \xrightarrow{k'_H} H + M^* \tag{1.10a}$$

$$M^* \xrightarrow{k_p} R' + 1 - alkene \tag{1.10b}$$

Radical addition

$$M^* + 1 - alkene \xrightarrow{k'_{ad}} {}^{\bullet}C_{18}^{+} \tag{1.11}$$

Termination

$$M^* + M^* \xrightarrow{k_t} product \; \rho_1 \tag{1.12a}$$

$$R' + M^* \xrightarrow{k_t} product \; \rho_2 \tag{1.12b}$$

$$R' + R' \xrightarrow{k_t} product \; \rho_3 \tag{1.12c}$$

$R' = {}^{\bullet}C_1 - {}^{\bullet}C_{15}$ in the case of hexadecane.

In Equations 1.8 through 1.12, M indicates the n-hexadecane molecule and R^* the parent radical; $\rho_1 - \rho_3$ denotes the probabilities of different terminations in the liquid phase and gas phase, respectively. Isomerization of primary to secondary radicals was not included in this scheme. The authors (Wu et al. 1997b) explained it referring to the assumption in the paper (Kossiakoff and Rice 1943) that, prior to H abstraction and β-scission, larger radicals isomerize instantaneously through internal hydrogen abstractions with ring formation. It should be pointed out that this statement is not applicable to alkane isomerization. Isomerization is not an instantaneous process and not necessarily resulting in ring formation.

Kossiakoff and Rice (1943) suggested that isomerization of primary to secondary radicals is possible through the six-member transition complex. This assumption allowed obtaining much better agreement between calculated and experimental data on the composition of high-molecular alkane cracking. In the study of the mechanism of high-molecular alkane cracking (Lavrovskiy et al. 1973), the experimentally measured composition of gaseous and liquid products of the TC of n-dodecane was compared with the calculated compositions of the primary decay products. The calculations were performed using different versions of the radical chain scheme.

Version I (Table 1.1) represents the composition of cracking products calculated according to the original Rice's scheme without taking into account reactions of radical isomerization.

Comparison of calculations using Rice's scheme with experimental data had shown their considerable disagreement with the gas composition. However, the calculated yields of α-olefins, although some understated, were rather close to experimental values.

TABLE 1.1

Calculated Compositions of the Products of n-Dodecane Cracking at 680°C (Moles per 100 Moles of Decomposed n-Dodecane)

	Calculation Version		
	I[a]	II	III
Methane	77.1	47.4	71
Ethane	4.7	52.6	28.1
Ethylene	122.4	254.3	98.9
Propane	1.1	0	1
Propylene	45.9	8.7	30.2
Butane-1	15.2	8.7	23.9
Bivinyl	4.1	8.7	0
Pentene-1	11.4	8.7	16.8
Hexane-1	14.7	8.7	34.4
Heptane-1	12.5	8.7	9.5
Octene-1	13.2	8.7	8.6
Nonene-1	13.1	8.7	11.1
Decene-1	12.1	8.7	11.1
Undecylene	5.6	4.3	4.3

Source: Lavrovskiy, K.P. et al., *Neftekhimiya* (*Oil Chem.*), 13, 422, 1973.)

[a]   I—Experimentally measured composition of the products of n-$C_{12}H_{26}$ cracking (conversion 14.2%). Hydrogen yield −55.2 (moles per 100 moles of decomposed n-$C_{12}H_{26}$).

Version II represents the composition of the products of $n\text{-}C_{12}H_{26}$ cracking calculated according to Kossiakoff–Rice model. It was assumed that primary radicals isomerized by the following scheme:

Secondary radicals further isomerized to other secondary radicals also through the six-member complexes. In equilibrium, secondary radicals were more stable than primary radicals by about 4 kcal/mol (Kossiakoff and Rice 1943). It was assumed that each large alkyl radical forming in the cracking process isomerized to an equilibrium state through the six-member complex. For example, a primary radical $C_{12}H_{25}$ isomerized forming equilibrium systems with the radicals whose free valence is located at the fifth or ninth carbon atom.

The probability of a chain break of any C–C bonds located in β-position with respect to the free valence was considered equally probable with the exception of the case when such bond breaks caused a methyl radical formation. According to Kossiakoff and Rice (1943), the hydrocarbon chain breaks leading to the formation of heavier free radicals proceed three times faster than the chain breaks leading to the formation of a methyl radical. It was also assumed that the activation energy for taking away a primary hydrogen atom is higher by 1200 cal/mol than that required for the detachment of a secondary hydrogen atom. As a result of the substitution reaction, all ethyl and methyl radicals transform to ethane and methane, correspondingly.

The values of methane, ethane and ethylene (in total), propylene, and butylene yields calculated in Version II are in a much better agreement with experimental data than calculations according to the original Rice's model. This analysis shows the importance of taking into account isomerization reactions both in the calculations of the cracking product compositions and in the kinetics of product accumulation. It is still more important as a number of experiments show predominant formation of branched paraffins as a result of the RTC of liquid hydrocarbons.

Equations 1.8 through 1.12 show three paths of cracking termination according to the three opportunities of parent and derivative radical recombination. The following reaction rate equations corresponding to these three cases were derived:

$$W_p = k_p \sqrt{\frac{k_i}{2k_t}}, \quad E = E_p + \frac{1}{2}E_i = 60 - 70 \text{ kcal/mol} \tag{1.13}$$

if termination occurs according to Equation 1.12a;

$$W_p = \sqrt{\frac{k_i k_H k_p [M]}{2k_t}}, \quad E = \frac{1}{2}(E_i + E_H + E_p) = 50 - 80 \text{ kcal/mol} \tag{1.14}$$

if termination occurs according to Equation 1.12b; and

$$W_p = k'_H \sqrt{\frac{k_i}{2k_t}}, \quad E = E_H + \frac{1}{2}E_i = 40 - 50 \text{ kcal/mol} \tag{1.15}$$

if termination occurs according to Equation 1.12c.

Activation energy for RTC E is smaller than that for TC by a factor of $\frac{1}{2}E_i$ because irradiation provides a much larger initiation rate than $k_i [M]$ (see Section 1.2). Roughly estimated E_H, E_p, and E_i were assumed to be 10, 30, and 60–80 kcal/mol, respectively. It was suggested that the recombination or disproportionation of parent radicals, reaction (1.12a), is the main termination in the liquid phase because of the much higher concentration of parent radicals relative to smaller radicals.

The analysis of the paper (Wu et al. 1997b) shows that the cracking reaction rate noticeably depends on the termination pathway. However, all of the pathways considered are the second-order radical recombination. Therefore, none of the equations derived works in the vicinity of the cracking start temperature, and none of them can be used for chain length calculation.

Talrose (1974) had shown that a reasonable expression for the chain length can be obtained only in the case when a linear termination of a chain cracking reaction is taken into account together with the quadratic termination. At $\nu \gg 1$, contribution of the linear termination mechanism is relatively small, and radical concentration can be approximately calculated using Equation 1.6. However, this contribution is principally important in calculations of the number of propagation steps and cracking start temperature.

Calculation of the chain length subject to quadratic and linear chain termination is given as follows. We have

$$K_p R = k_p R e^{-E_p/k_B T} \tag{1.16}$$

$$K_i = k_i e^{-E_i/k_B T}$$

$$K_t = k_{t1} R + k_{t2} R^2$$

In dynamic equilibrium,

$$k_i e^{-E_i/k_B T} = k_{t1} R + k_{t2} R^2 \tag{1.17}$$

It implies

$$R = \frac{k_{t1}}{2k_{t2}} \left[\sqrt{1 + \frac{4k_i k_{t2}}{k_{t1}^2} e^{-E_i/kT}} - 1 \right] \tag{1.18}$$

Therefore, we obtain the following equation for the chain length:

$$v = \frac{(k_p k_{t1}/2k_{t2})\left[\sqrt{1+(4k_i k_{t2}/k_{t1}^2)e^{-E_i/kT}} - 1\right]e^{-E_p/kT}}{k_i e^{-E_i/kT}}$$

$$= \frac{(2k_p/k_{t1})\left[\sqrt{1+(4k_i k_{t2}/k_{t1}^2)e^{-E_i/kT}} - 1\right]e^{-E_p/kT}}{\left(1+(4k_i k_{t2}/k_{t1}^2)e^{-E_i/kT}\right) - 1}$$

or

$$v = \frac{2\left(k_p/k_{t1}\right)e^{-E_p/kT}}{1+\sqrt{1+\left(4k_i k_{t2}/k_{t1}^2\right)e^{-E_i/kT}}} \tag{1.19}$$

Expression (1.19) has the same form as that derived in the work by Talrose (1974). According to Equation 1.21, $v \to 0$ at $T \to 0$, and the chain length increases when $T \to \infty$. This asymptotic behavior of the function $v(T)$ meets the conditions of an endothermic chain process.

As temperature approaches the cracking start temperature, the second term under the square root in Equation 1.19 becomes negligibly small compared with unity. In these conditions, cracking stoppage is determined by the rate of linear chain termination with respect to a reactive center. In this case, the expression under the square root can be expanded to a series accurate within a linear expansion term. As a result, we will have

$$v \approx \frac{k_p}{k_{t1}} e^{-E_p/kT} \tag{1.20}$$

Substitution of the following condition to Equation 1.20

$$v = 1 \quad \text{at} \quad T = T_c$$

will result in

$$T_c = \frac{E_p}{k \ln\left(k_p/k_{t1}\right)} \tag{1.21}$$

Equation 1.21 allows the evaluation of the cracking termination rate k_{t1} from the cracking start temperature. Taking the temperature of TC beginning equal to 550 K (277°C) and activation energy for chain propagation, E_p, equal to 20 kcal/mol, we obtain

$$\frac{k_p}{k_{t1}} = 7.9 \cdot 10^7 \tag{1.22}$$

For estimate, let us take $k_p \approx 10^{12}$ s^{-1}. Then, it follows from Equation 1.21 that $k_{t1} \approx 12,700$ c^{-1}. It implies that

$$\frac{4k_i k_{t2}}{k_{t1}^2} e^{-E_i/kT} \approx \frac{4 \cdot 10^{13} \cdot 10^9}{1.27^2 \cdot 10^8} e^{-80000/2 \cdot 550} = 6.4 \cdot 10^{-18} \ll 1$$

that is, the condition of the smallness of this term at temperatures close to the cracking start temperature, used in the derivation of Equations 1.20 and 1.21, is valid.

The reaction of chain termination, which determines the cracking start, can be represented as the recombination of a pair of closely located radicals formed as a result of hydrocarbon molecule decomposition. The formation of such radical pairs can be associated with the cage effect whose models are discussed in Section 1.6.

Constant k_{t1} is the frequency of radical recombination on the condition that all radicals annihilate in the first-order reactions. It can be represented as a difference of two terms:

$$k_{t1} = k'_{t1} - k''_{t1} \tag{1.23}$$

Here, the first term is the recombination rate of close radical pairs at the given temperature. It is determined by the rates of isomerization and disproportionation processes leading to the stabilization of such pairs and their subsequent recombination. The second term is the rate of the decomposition of the coupled radical pairs due to thermal excitation. The lifetime τ of the radicals that contribute to the propagation of the chain reaction is determined by the first term:

$$\tau = \frac{1}{k'_{t1}} = \frac{1}{k_{t1} + k''_1} \tag{1.24}$$

As temperature decreases, $k'_{t1} \to k_{t1}$. At the temperature of cracking start, $\tau = (1/k_{t1}) \approx 7.9 \cdot 10^{-5}$ s, that is close to the estimate of a lifetime of the radiation-excited unstable molecular states according to the data of the work (Zaikin and Zaikina 2008a). A large value of a radical pair lifetime can be explained by the reactions of disproportionation and isomerization, which lead to stabilization of its electronic structure. Until these processes are completed, such a radical pair is unstable.

This derivation is a fundamental proof of the existence of the long-living unstable states of hydrocarbon molecules. Thermal generation of such states requires a high activation energy of about 80 kcal/mol, that is, they can appear only at high temperatures. However, thermal excitation at high temperatures impedes radical pair stabilization, and the rate of their recombination becomes a negligibly small quantity.

At the same time, application of high dose rates of ionizing irradiation allows the generation of high concentrations of such unstable states at lowered temperatures. These molecular states can react with radical chain carriers. It will be shown in the following text that these interactions provide chain propagation in the LTRC of hydrocarbons.

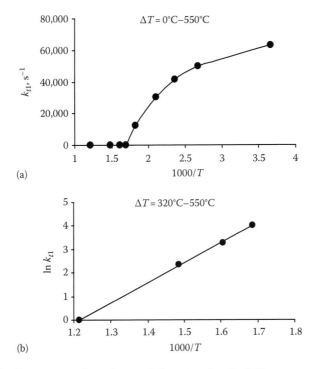

(a)

(b)

FIGURE 1.1 Temperature dependence of the rate of radical linear recombination k_{rl}. (a) $T=0°C–550°C$, (b) $T=320°C–550°C$. (From Brodskiy, A.M. et al., *Neftekhimiya* (*Oil Chem.*), 3(3), 370, 1961.)

Figure 1.1 shows temperature dependence of the quantity k_{rl} calculated on the basis of the estimated contribution of radiation-excited radical pairs to the propagation rate of RTC (see Section 1.2) and experimental data on the critical dose rate above which LTRC in high-viscous oil becomes possible at the given temperature.

At temperatures higher than the cracking start temperature, the rate of radical pair recombination in the first-order reactions rapidly tends to zero with increase in temperature because of the thermal excitation and decoupling of the radical pairs (Figure 1.1b). In the temperature range below the cracking start temperature, the quantity k_{rl} increases with a decreasing rate as temperature goes down. At low temperatures, k_{rl} tends to its higher limit, which corresponds to the rate of electron structure stabilization in radical pairs in a completely "frozen" state (Figure 1.1a).

Temperature dependence of the quantity k_{rl} shown in Figure 1.1 will be used in the following approximate calculations. However, it should be noted that parameters of this dependence may considerably change depending on the type and structure state of a hydrocarbon feedstock. In particular, oil's tendency to polymerize and to form low-temperature structures leads to a more considerable increase in k_{rl} values at low temperatures.

Temperature dependence of the number of propagation steps calculated using Equation 1.19 is shown in Figure 1.2. In the calculations, it was supposed that the quantity k_{rl} is also a function of temperature as shown in Figure 1.1.

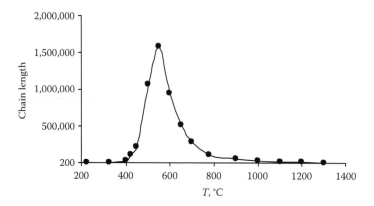

FIGURE 1.2   Temperature dependence of chain length for TC of hydrocarbons.

The maximal chain length ($\nu_{max} \approx 1.6 \cdot 10^6$) was observed at the temperature of 550°C. At higher temperatures, chain length decreases because of the rapid increase in the rate of the reaction initiation and, therefore, the rate of its termination.

At temperatures below 450°C, the second term under the square root in Equation 1.19 is much smaller than unity, and chain length can be calculated using the formula

$$V = \frac{k_p}{k_{t1}} e^{-E_p/kT} \tag{1.25}$$

At temperatures higher than the temperature of the maximum chain length, the quantity k_{t1} rapidly tends to zero, and the expression for the chain length (1.19) takes the form

$$V = \frac{k_p}{\sqrt{k_i\, k_{t2}}} e^{(E_i - 2E_p)/2kT} \tag{1.26}$$

Only those interactions of radicals with hydrocarbon molecules that have a chain length $\nu > 1$ contribute to a chain cracking reaction. If $\nu \leq 1$, then big radicals appearing due to such interactions do not disintegrate and, ultimately, disappear in the reactions of disproportionation, isomerization, and recombination. Therefore, the rate of RTC is equal to the production of the rate of reaction initiated by the chain carriers, $K_p R$, and the probability, $(\nu - 1)/\nu$, that this reaction does not occur only once. Thus the rate of RTC is determined by the equation

$$W = \begin{cases} \dfrac{\nu - 1}{\nu} Rk_p\, e^{-E_p/kT}, & \nu \geq 1 \\[2mm] 0, & \nu < 1 \end{cases} \tag{1.27}$$

For RTC chain length, $\nu \gg 1$,

$$W_p \approx Rk_p\, e^{-E_p/k_B T} \tag{1.28}$$

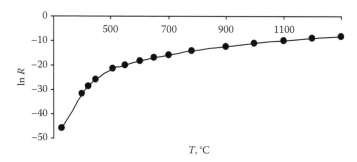

FIGURE 1.3 Temperature dependence of radical chain carrier concentration in the case of TC.

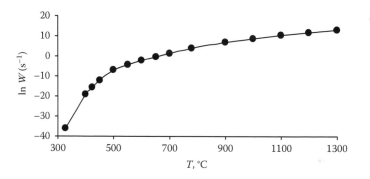

FIGURE 1.4 Dependence of logarithm TC rate on temperature.

Dependence of the concentration of thermally generated radicals on temperature is shown in Figure 1.3. At temperatures higher than the cracking start temperature by 30°C–40°C, the quadratic chain termination is a dominating process, and radical concentration can be calculated using Equation 1.6 instead of a more general expression (1.18).

Temperature dependence of the cracking rate calculated using Equations 1.19 and 1.27 with the same set of constants is shown in Figure 1.4. At temperatures above 320°C, the cracking rate can be calculated using a simplified formula (1.28). Equation 1.27 introduces considerable amendments to the calculated cracking rate only in the vicinity of the cracking start temperature where radical recombination in the first-order reactions becomes prevailing.

Temperatures below the temperature of intense disintegration of the unstable radical pairs (500°C–550°C in this example) are characteristic for the combination of a relatively small chain length with a low concentration of thermally generated radicals and, therefore, a low cracking rate. In the next section, it will be shown that the application of ionizing irradiation allows a considerable increase in the cracking rate at lowered temperatures.

1.2   RADIATION-THERMAL CRACKING

As a rule, RTC proceeds in the temperature range of 350°C–450°C in the conditions when initiation of a chain reaction mainly arises from radiation generation of radicals while chain propagation occurs mainly by thermal activation of the excited molecular states.

In the case of RTC, a member responsible for radiation generation of radicals appears in the expression for the rate of cracking initiation:

$$K_i = k_i e^{-E_i/k_B T} + GP$$

Here

G is the radiation-chemical yield of radical chain carriers

P is the irradiation dose rate

Subject to this additional member, the expressions for radical concentration and chain length will take the form

$$R = \frac{k_{t1}}{2k_{t2}} \left[\sqrt{1 + \frac{4k_{t2}}{k_{t1}^2}(k_i e^{-E_i/kT} + GP)} - 1 \right] \tag{1.29}$$

$$v = \frac{2\left(k_p/k_{t1}\right)e^{-E_p/kT}}{1 + \sqrt{1 + \left(4k_{t2}/k_{t1}^2\right)(k_i e^{-E_i/kT} + GP)}} \tag{1.30}$$

In the temperature range of RTC, $k_i e^{-E_i/kT} \ll GP$. Therefore,

$$R \approx \frac{k_{t1}}{2k_{t2}} \left[\sqrt{1 + \frac{4k_{t2}}{k_{t1}^2}GP} - 1 \right] \tag{1.31}$$

$$v \approx \frac{2\left(k_p/k_{t1}\right)e^{-E_p/kT}}{1 + \sqrt{1 + \left(4k_{t2}/k_{t1}^2\right)GP}} \tag{1.32}$$

For $v \gg 1$,

$$W_p \approx R k_p e^{-E_p/kT} \tag{1.33}$$

If $\left(4k_{t2}/k_{t1}^2\right)GP \gg 1$, then the chain length and the radical concentration do not depend on k_{t1}:

$$v = \frac{k_p}{\sqrt{k_{t2}GP}} \tag{1.34}$$

$$R = \sqrt{\frac{GP}{k_{t2}}} \qquad (1.35)$$

which corresponds to the dependences predicted by RTC theory in supposition of the quadratic chain termination.

At $P \to 0$ in the vicinity of the cracking start temperature, chain carrier concentration is proportional to the dose rate:

$$R \approx \frac{GP}{k_{t1}} \qquad (1.36)$$

Substituting the condition $\nu(T_c) = 1$ to Equation 1.32, we will obtain the following equation for the temperature of RTC start:

$$T_c = \frac{E_p}{k \ln \dfrac{2k_p/k_{t1}}{1 + \sqrt{1 + \left(4k_{t2}/k_{t1}^2\right)GP}}} \qquad (1.37)$$

Equation 1.37 shows that the cracking start temperature generally depends on the dose rate of ionizing irradiation. The rate of radical linear recombination k_{t1} depends on temperature (Figure 1.1). Therefore, Equation 1.37 represents the cracking start temperature as an implicit function of the absorbed dose rate. However, calculations show that this dependence is very weak for the earlier-selected set of the reaction constants. To increase the cracking start temperature by 1°C, a dose rate higher than 200 kGy/s would be needed that is usually unachievable in an ordinary experiment.

It should be noted that estimations above are sensitive to the selection of the rate of cracking propagation. As an example, Figure 1.5 illustrates calculation of T_c dependence on the dose rate using Equation 1.37 for the case when the rate of cracking propagation $k_p = 10^{10}$ s⁻¹, that is, two orders of magnitude lower than a value used in the previous calculations. It was supposed that $G = 5$ radicals/100 eV $\approx 10^{-7}$ kg/J.

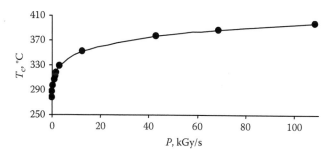

FIGURE 1.5   Dependence of RTC start temperature on dose rate ($k_p = 10^{10}$ s⁻¹).

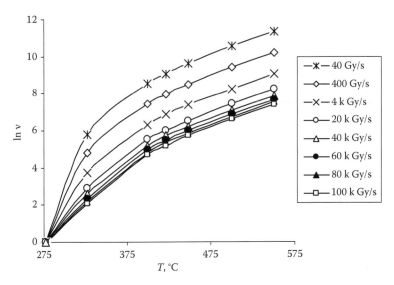

FIGURE 1.6 Temperature dependence of RTC chain length for different dose of ionizing irradiation.

Figure 1.5 shows that, in this case, the cracking start temperature considerably increases as the dose rate goes up. Increase in the dose rate from 0 to 100 kGy/s leads to T_c increase by more than 100°C.

Dependences of the chain length, radical concentration, and the rate of RTC on temperature and dose rate calculated using Equations 1.27, 1.28, and 1.32 are shown in the following text.

Comparison of Figures 1.3 and 1.6 shows that RTC chain length increases with temperature slower than its temperature increase in the case of TC. RTC is characteristic for shorter chains of radical interactions with hydrocarbon molecules, especially at heightened dose rates. At high temperatures (above 400°C in the example shown in Figure 1.6), the temperature dependence of chain length is exponential.

Figure 1.7 demonstrates a monotonous decrease in the RTC chain length as the dose rate of ionizing irradiation increases. At temperatures above 320°C, the chain length is inversely proportional to the square root of dose rate even at low P values.

Thus RTC chain length is considerably lower than that in TC in a wide range of dose rates characteristic of the conventionally used radiation sources. However, even with a lower chain length, a high rate of RTC reaction is provided by a higher concentration of chain carriers: hydrogen atoms and light alkyl radicals. Radical concentration increases as dose rate increases, as shown in Figure 1.8.

For the selected set of parameters, radical concentration increases according to Equation 1.35 proportionally to the square root of dose rate. A deviation from this dependence near the cracking start temperature is due to the increase in the fraction of radical linear recombination at lowered temperatures and dose rates. Generally, TC is characteristic of the long chains of radical interactions with hydrocarbon molecules and relatively low radical concentrations; RTC is characteristic of short interaction chains and high radical concentrations.

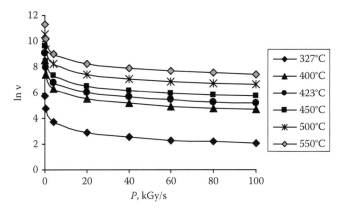

FIGURE 1.7   Dependence of RTC chain length on dose rate of ionizing irradiation for different reaction temperatures.

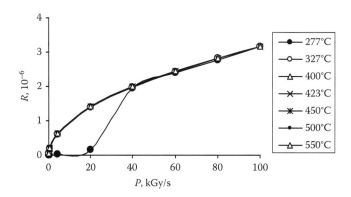

FIGURE 1.8   Dependence of radiation-generated radical concentration on dose rate.

Temperature dependence of the RTC rate for different dose rates is shown in Figure 1.9. At higher dose rates, the rate of RTC increases more rapidly reaching higher values as temperature increases.

At a given temperature, an increase in the dose rate leads to a decrease in the chain length with simultaneous increase in the chain carrier concentration. As a result, the cracking rate increases slower in the range of high dose rates (Figure 1.10). These calculations are sensitive to the value of chain propagation rate k_p. Increase in k_p values may cause appearance of maxima in the cracking rate dependence on dose rate. In this case, such maxima determine the highest possible rate of RTC and the optimal dose rate of the feedstock processing at lowered temperatures.

Generally, all calculations given earlier are sensitive both to the selection of the rate constant of chain propagation k_p and to the form of temperature dependence of the rate of radical linear recombination, k_{t1}. Therefore, they may considerably differ for different hydrocarbon compositions.

The RTC theory as interpreted in this chapter follows the scheme of the well-known RTC studies (Topchiev and Polak 1962, Lavrovskiy 1976) and does not take

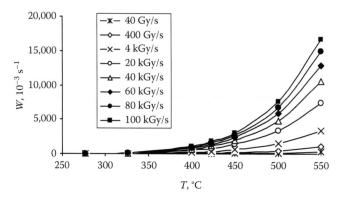

FIGURE 1.9 Dependence of RTC rate on temperature for different dose rates of ionizing irradiation.

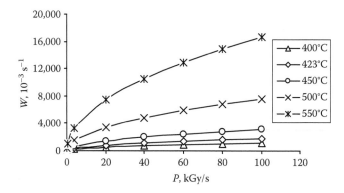

FIGURE 1.10 Dependence of RTC rate on dose rate of ionizing irradiation for different reaction temperatures.

into account the contribution of radiation-generated molecular states to propagation of the chain reaction. It will be shown in Section 1.2 that this contribution considerably increases the calculated chain length and RTC rate at the given temperature.

The calculations of the basic parameters of cracking chain reactions subject to the contribution of radiation to chain propagation are useful not only for better agreement with the available experimental data, but more importantly they allow determination of the conditions when radiation cracking can be observed at lowered temperatures far beyond the conventional temperature range of RTC.

Nevertheless, the RTC theory considered above (with the amendments to low-temperature effects) is useful for experimental data interpretation. It is valid for high RTC temperatures when contribution of radiation to chain reaction propagation is negligibly small. On the other hand, it describes the extreme case practically important for heavy oils and bitumen when low-temperature cracking requires too high, practically unachievable dose rates of ionizing irradiation. Radiation cracking of such types of oil feedstock, more effective at higher temperatures, obeys to the basic laws of the conventional RTC theory.

1.3 OTHER ESTIMATIONS OF CHAIN LENGTH AND CRACKING TEMPERATURE

A reasonable evaluation of the cracking start temperature was made by Brodskiy and coworkers (1961). It was noted that, in a formal kinetic description, oil radiolysis can be reduced to two processes: (1) formation of excited molecules and stable products of their disintegration, and (2) formation of hydrocarbon radicals and products of the radical reactions.

It was stated that, in the case of radiolysis of liquid alkanes, dependence of ln G on $1/T$ (G is the number of molecules converted per 100 eV of absorbed energy) has a characteristic form shown in Figure 1.11. Below the critical temperature T_c (about 550–600 K), G values change relatively slowly as temperature increases (mode I, Figure 1.11). Starting from the critical temperature T_c, G values rapidly increase with the activation energy of 21 ± 5 kcal/mol (mode II). The transition from mode I to mode II is very sharp. As temperature further increases, mode II smoothly turns to TC (mode III).

Brodskiy et al. (1961) indicated that the degree of oil decomposition in mode III depends not only on the total absorbed dose of radiation energy, as characteristic for modes I and II, but also on the dose rate of ionizing irradiation. However, further, more detailed studies of radiation cracking in a wider range of dose rates (Zaikin 2008, 2013a, Zaikin and Zaikina 2008a) have shown that the cracking rate in mode I depends on dose rate stronger than in mode II. Moreover, in mode I, the rate of a chain cracking reaction, unlike that in slow radiolysis reactions, is mainly determined by the dose rate (see Section 1.4).

The existence of mode II was first stated in the work by Topchiev et al. (1960), where the term "radiation-thermal cracking" was introduced for mode II specification.

Földiàk (1981) found that there is a region in the temperature range from 250°C to 310°C having an activation energy of 58.7 ± 14.8 kJ/mol, while the RTC region is characteristic of a higher activation energy ($E \approx 85$–126 kJ/mol). These observations indicate three specific types of radiation-chemical conversion of hydrocarbons.

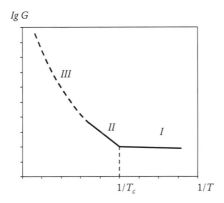

FIGURE 1.11 Schematic dependence of ln G on $1/T$. I, II, III—modes of "pure" radiolysis, RTC, and TC, respectively; T_c—cracking start temperature. (From Brodskiy, A.M. et al., *Neftekhimiya (Oil Chem.)*, 3(3), 370, 1961.)

The first-type process is observed at low temperatures. The radicals formed under ionizing irradiation enter various reactions, such as recombination, disproportionation, and interaction with hydrocarbon molecules. Disintegration of radicals to smaller fragments does not occur in this temperature range. The second-type process includes radical disintegration. Interaction of radicals having a small number of carbon atoms with the starting hydrocarbon molecules is a well-pronounced chain process initiated by radiation-generated radicals. The third-type process is RTC. It was supposed that the second-type process in the intermediate temperature range could not be observed at high dose rates of irradiation.

It is important that all these observations relate to hydrocarbon radiation processing at relatively low dose rates (below 10 kGy/s). As shown in the next section, application of high dose rates substantially changes the behavior of cracking reactions and allows the observation of chain reactions of hydrocarbon cracking at much lower temperatures.

The behavior of radicals is very different in the low-temperature region (mode 1) and RTC region (mode II). At relatively low dose rates in mode I, radicals enter reactions of recombination and disproportionation, while in mode II, they disintegrate forming olefin molecules and radicals of a smaller molecular mass. The latter participate in further reactions of substitution and decomposition. Starting from the critical temperature T_c, mode I turns to mode II. It can be considered that RTC (mode II) starts when the rate of radical decomposition, $K_p[R]$, becomes equal to the reaction rate in mode I. Therefore, at the transition temperature T_c, it can be written that

$$K_p\left[\dot{R}\right] \cong aGP \tag{1.38}$$

where
$K = k_p \exp(-E_p/kT)$ is the averaged rate constant of radical decomposition
$[\dot{R}]$ is the concentration of high-molecular radicals
G is the radiation-chemical yield of products
P is the dose rate
a is the numeric factor

The balance equation for the high-molecular radicals was written in the form

$$GP - K_p\left[\dot{R}\right] - k_{t2}\left[\dot{R}\right]\left[\bar{R}\right] = 0 \tag{1.39}$$

where $[\bar{R}]$ is the total radical concentration according to the equation

$$GP = \bar{k}_{t2}\left[\bar{R}\right]^2 = 0 \tag{1.40}$$

In Equation 1.40, $\bar{k}_{t2}$ is responsible for both recombination and disproportionation reactions and G is the radiation-chemical yield of radicals $\dot{R}$. It was suggested that constants k_{t2} and $\bar{k}_{t2}$ were temperature independent. In a further estimate, it was assumed that $k_{t2} \approx \bar{k}_{t2}$. By substitution of expressions (1.39) and (1.40) into Equation 1.38, the following equation for T_c was obtained:

$$\frac{k_p e^{-E_p/kT_c} G}{k_p + \left(k_{t2} G_{\bar{R}} P\right)^{0.5}} \cong aG_R \qquad (1.41)$$

Here, $G_{\bar{R}}$ is the total radiation-chemical yield of radicals.

From here,

$$T_c = \frac{E_p}{k} \frac{1}{\ln\left(k_p b/(k_{t2} G_{\bar{R}} P)^{1/2}\right)} \qquad (1.42)$$

where b is a constant close to unity. It does not depend on temperature and can be omitted. Substitution of the values $G_{\bar{R}} = 5$ radicals/100 eV, $k_{t2} = 10^{-11}$ cm^3/mol·s, $k_p = 10^{13}$ s^{-1}, $E = 105$–126 kJ/mol, and $P = 1.6 \cdot 10^2$ Gy/s into Equation 1.42 results in $T_c = 600$ K, which is in a satisfactory agreement with experimental observations.

In this derivation, similar to our analysis in the previous section, the quantity T_c is proportional to E_p and changes very slowly with the dose rate P. Equation 1.38 assumes a linear chain termination, that is, radical recombination in the first-order reaction. The same assumption was made in the derivation of Equation 1.37. Therefore, Equation 1.42 gives an estimate of the cracking start temperature close to the estimate we made in the previous section. Actually, it can be easily shown that at sufficiently high dose rates ($P \gg 1$ Gy/s), Equation 1.37 can be written in the form

$$T_c = \frac{E_P}{k} \frac{1}{\ln\left(k_p/(k_{t2} GP)^{1/2}\right)}$$

which coincides with Equation 1.42.

In the book by Saraeva (1986) with a reference to the work (Gabsatarova and Kabakchi 1969), it was noted that the yield of RTC products of n-heptane is inverse the dose rate. To explain this effect, a model based on the cage effect was used. However, the conclusion that RTC yields decreased with the dose rate is inaccurate. Experimental data by Gabsatarova and Kabakchi (1969) really indicate some decrease in the product yields as dose rate increases in the case when the fixed parameter is irradiation dose but not irradiation time. The analysis of the product yield dependence on irradiation time and dose rate shows that the cracking rate at a fixed irradiation time increases as dose rate increases in full agreement with RTC theory. A similar dependence of the cracking rate on the dose rate was observed in other hydrocarbon systems (Zaikin and Zaikina 2008a).

In the paper by Wu et al. (1997b), the estimates were made for RTC chain length in n-hexadecane. It was noted that the chain length can be calculated as a ratio of the cracking propagation rate and chain termination rate. However, referring to the difficulties of such calculations (see Section 1.2), Wu et al. (1997b) recommended to calculate the reaction chain length using the data on radiation-chemical yields of radicals and decomposed feedstock.

Radiation-chemical yield of radicals $G(R)$ is defined as the number of radicals produced per 100 eV of absorbed radiation energy. $G(-M)$ was defined as the number of converted molecules of the starting material per 100 eV of absorbed energy of radiation. Then, the quantity

$$Y = \frac{G(-M)}{G(R)} \qquad (1.43)$$

can be considered as a direct measure of RTC chain length in the temperature range where contribution of the thermal component to chain propagation is negligibly small.

A value $G(R) = 5.5$ radicals/100 eV was taken for the radiation-chemical yield of radicals in n-hexadecane. The quantity $G(-M)$ was determined experimentally from the dose dependence of n-hexadecane conversion under gamma irradiation at the dose rate of 0.13 Gy/s. Calculations have shown that the reaction chain length rapidly increases with temperature varying from about 190 at 330°C to 3800 at 400°C.

In the temperature range where contribution of the thermal component to chain propagation is noticeable, the true reaction chain length can be calculated using the equation

$$Y = v \left[1 + \frac{W_{th}}{W*} - \left(\frac{W_{th}}{W*} \right)^{1/2} \left[1 + \frac{W_{th}}{W*} \right]^{1/2} \right] \qquad (1.44)$$

where $W_{th}/W*$ is the ratio of the thermal component to the total cracking rate.

It should be noted that Equation 1.43 can be used for the estimation of the reaction chain length subjected to thermal effects if the quantity $G(-M)$ is defined as the full number of converted molecules and $G(R)$ is interpreted as the full number of radical chain carriers per unit of absorbed (thermal and radiation) energy.

Actually, the quantity $G(-M)$ can be represented in the form

$$G(-M) = -\frac{dC}{dD}\bigg|_{D=0} = -\frac{1}{P}\frac{dC}{dt}\bigg|_{t=0} = \frac{W}{P}$$

where
  C is the concentration of the reacting material
  D is the absorbed dose
  t is the irradiation time
  P is the dose rate
  W is the initial cracking rate

In this formula, $G(-M)$ is expressed in kg/J.

It implies the following expression for the reaction chain length:

$$v = \frac{W}{PG_r} = \frac{k_p e^{-E_p/kT} R}{PG(R)}$$

Substituting R from Equation 1.31, we come again to Equation 1.30 for the reaction chain length, which is applicable for both RTC and TC.

$$v = \frac{2\left(k_p/k_{t1}\right)e^{-E_p/kT}}{1+\sqrt{1+\left(4k_{t2}/k_{t1}^2\right)\left(k_i e^{-E_i/kT}+GP\right)}}$$

Thus Equations 1.29 through 1.37 of the previous section provide a simple method for the evaluation of the reaction chain length, cracking start temperature, and cracking rate from the general positions of kinetic theory.

1.4   LOW-TEMPERATURE RADIATION CRACKING

The concept of LTRC comes from the probability of radiation generation of unstable molecular state capable of chain reaction propagation at lowered temperatures. Concentration of these molecular states C^* is proportional to the dose rate of ionizing irradiation. In dynamic equilibrium,

$$G^*P = K^*C^* = (k_{t1}+k_{t1}'')C^* = \frac{1}{\tau}C^* \tag{1.45}$$

where
   G^* is the radiation-chemical yield of radiation-generated unstable molecular states
   k_{t1} is the recombination rate of closely located radical pairs
   k_{t1}'' is the temperature-dependent rate of their thermal decomposition

The ratio of the rates of radiation and thermal chain propagation is

$$\frac{W_{p\,rad}}{W_{p\,RTC}} = G^*P\tau\,e^{E_p/kT} \tag{1.46}$$

The total cracking rate can be written as

$$W = W_{p\,rad} + W_{p\,RTC}$$

If the reaction chain length $v \gg 1$, then, at RTC temperatures, the radiation-thermal component $W_{p\,RTC}$ (responsible only for radiation contribution to cracking initiation) is proportional to $P^{1/2}$, while the "pure" radiation component (responsible for radiation contribution to both chain initiation and chain propagation) is proportional to $P^{3/2}$.

A hydrocarbon molecule can react with a light alkyl radical if the molecule is electronically excited under ionizing irradiation up to the level necessary for the reaction. In this case, concentration of thermally excited molecules $\exp(-(E_p/kT))$ in formula (1.46) should be increased by the addition of concentration c^* of the molecules that have received electron excitation up to the reaction level. For this reason, the rate of RTC in the works (Zaikin 2008, Zaikin and Zaikina 2008a) was represented in the form

$$W = K_p\left(\frac{G}{k_{t2}}\right)^{1/2}\left[\exp\left(-\frac{E_p+\Delta E}{kT}\right)+c^*\exp\left(-\frac{\Delta E}{kT}\right)\right]P^{1/2} \tag{1.47}$$

where ΔE was interpreted as activation energy for light radical diffusion in a hydrocarbon mixture. According to the estimate of the paper (Zaikin and Zaikina 2008a), $\Delta E \approx 8570$ J/mol. Equation 1.47 can be rewritten as follows:

$$W \cdot \exp\left(\frac{E}{kT}\right) P^{-1/2} = A + B(T)P \qquad (1.48)$$

where

$$A = K_p \left(\frac{G}{k_{t2}}\right)^{1/2}, \quad B(T) = K_p \left(\frac{G}{k_{t2}}\right)^{1/2} \frac{G^*}{K^*} \exp\left(\frac{E_p}{kT}\right) \qquad (1.49)$$

Equations 1.48 and 1.49 were verified using reprocessed experimental data on RTC available in the literature and special experiments on radiation thermal processing of different types of oil feedstock. Some of the recalculated data of different works represented in Figures 1.12 through 1.14 demonstrate the validity of Equation 1.48.

Similar results were obtained by processing our experimental data on RTC of different types of heavy hydrocarbon feedstock (Figures 1.15 through 1.18).

In Figure 1.19, the yields of light fractions, $Y-Y_0$, from a heavy oil fraction with a high content of high-molecular paraffins obtained in the study (Ponomarev 2010) are plotted in coordinates $Y-Y_0 - P^{1/2}$. It can be seen from Figure 1.19 that this dependence is far from $Y-Y_0$ proportionality to $\sqrt{P}$ predicted by the conventional RTC theory.

Figure 1.20, where the same dependence is plotted in coordinates $(Y-Y_0)/P^{1/2} - P$, shows that the quantity $Y-Y_0$ is nearly proportional to $P^{3/2}$. According to Equation 1.48, it means that, for the given type of feedstock and given experimental conditions, the radiation component makes a major contribution to propagation of the chain RTC reaction.

The estimated value of the rate constant K^* in the papers (Zaikin 2008, Zaikin and Zaikina 2008a) was $2 \cdot 10^6$ s^{-1}. It corresponded to the lifetime of excited molecular states of about $5 \cdot 10^{-7}$ s. However, a more reliable evaluation of the lifetimes of such states should take into account the dependence of coefficients A and B in

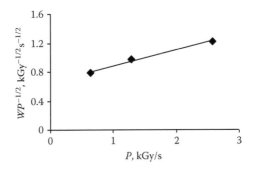

FIGURE 1.12 Dependence of the rate of RTC of pentadecane at 500°C on dose rate in coordinates $WP^{-1/2} - P$. (Recalculated from Mustafaev, I.I., *Khimiya Vysokikh Energiy (High Energy Chem.)*, 24, 22, 1990.)

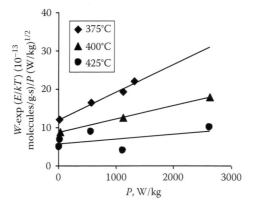

FIGURE 1.13   Dependence of the rate of hydrogen formation on radiation dose rate during RTC of pentadecane for different temperatures of RTC in coordinates $W \cdot \exp(E/kT)$ $P^{-1/2} - P$. (Recalculated from Mustafaev, I.I., *Khimiya Vysokikh Energiy, (High Energy Chem.)*, 24, 22, 1990.)

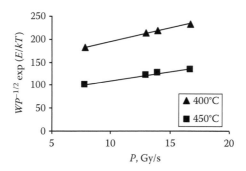

FIGURE 1.14   Dependence of the rate of *n*-hexadecane RTC at different temperatures on dose rate in coordinates $W \cdot \exp(E/kT)P^{-1/2} - P$. (Recalculated from Panchenkov, G.M. et al., *Khimiya Vysokikh Energiy, (High Energy Chem.)*, 15, 426, 1991.)

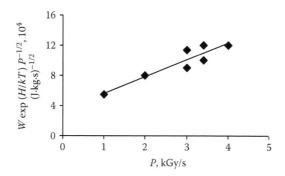

FIGURE 1.15   Dependence of the rate of RTC for Karazhanbas crude oil in the temperature range of 370°C–400°C on the dose rate of electron irradiation in coordinates $W \cdot \exp(E/kT)$ $P^{-1/2} - P$. (From Zaikin, Y.A. et al., *Radiat. Phys. Chem.*, 60, 211, 2001.)

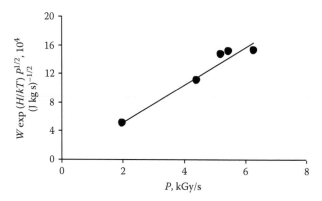

FIGURE 1.16 Dependence of the rate of RTC for Bugulma crude oil (Tatarstan, Russia) in the temperature range of 380°C–400°C on the dose rate of electron irradiation in coordinates $W \cdot \exp(E/kT)P^{-1/2} - P$. (From Zaikin, Y.A. and Zaikina, R.F. New trends in the radiation processing of petroleum. In: *Radiation Research Progress*. New York: Nova Science Publishers, pp. 17–103, 2008a.)

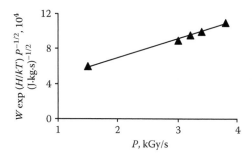

FIGURE 1.17 Dependence of the rate of RTC for fuel oil produced by Zuzeeevneft (Tatarstan, Russia) in the temperature range of 400°C–410°C on the dose rate of electron irradiation in coordinates $W \cdot \exp(E/kT)P^{-1/2} - P$. (From Zaikin, Y.A. and Zaikina, R.F., Development of the methods for processing of oil products using complex radiation-thermal treatment and radiation ozonolysis. Final Report on IAEA Project (Research Contract # 11837/RO), Almaty, Kazakhstan, 34pp, 2004a.)

Equation 1.48 on temperature and dose rate subject to the fact that these dependences can be different for different types of feedstock. Combining Equations 1.46 and 1.48, we shall obtain the following expression for coefficients A and B:

$$\frac{B}{A} = \frac{W_{prad}}{W_{pRTC} \cdot P} = G^* \tau \, e^{E_p/kT} \tag{1.50}$$

Application of Equation 1.50 to experimental data shown in Figures 1.12 through 1.18 results in much higher values of the lifetimes of radiation-excited molecular states (close radical pairs), from $5 \cdot 10^{-4}$ s to $1.5 \cdot 10^{-3}$ s for different hydrocarbon compositions in the ranges of temperature and dose rate under consideration (Zaikin 2013b).

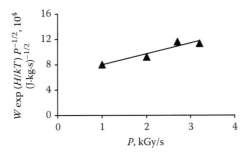

FIGURE 1.18 Dependence of the rate of RTC for fuel oil produced by Atyrau refinery (Kazakhstan) in the temperature range of 390°C–400°C on the dose rate of electron irradiation in coordinates $W \cdot \exp(E/kT)P^{-1/2} - P$. (From Zaikin, Y.A. and Zaikina, R.F., Development of experimental facility for processing hydrocarbon components of oil bitumen, Technical Report on ISTC project K-930. Almaty, Kazakhstan, 44pp, 2006.)

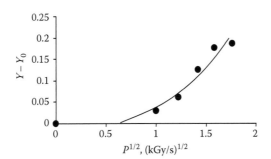

FIGURE 1.19 Dependence of the yields of light fractions from heavy paraffinic fraction on dose rate of electron irradiation: RTC temperature $-410°C$; absorbed dose—710 kGy; $Y - Y_0$—difference of light fraction concentrations in product and feedstock.

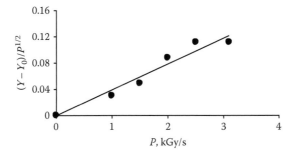

FIGURE 1.20 Dependence of the yields of light fractions from heavy paraffinic fraction on dose rate of electron irradiation in coordinates $(Y - Y_0)P^{-1/2} - P$: RTC temperature $-410°C$; absorbed dose—710 kGy.

According to Equation 1.45, the lifetime of close radical pairs can be expressed through the constant of radical linear recombination, k_{rl}:

$$\tau = \frac{1}{k_{rl} + k_{rl}''} \tag{1.51}$$

As temperature decreases, $k_{rl}'' \to 0$ and $\tau \to 1/k_{rl}$.

In calculations represented in the previous section, a temperature dependence of the quantity k_{rl} was used as shown in Figure 1.1.

As before, at any temperatures and dose rates of irradiation, concentration of radical chain carriers will be determined by formula (1.29):

$$R = \frac{k_{rl}}{2k_{r2}}\left[\sqrt{1 + \frac{4k_{r2}}{k_{rl}^2}(k_i\, e^{-E_i/kT} + GP)} - 1\right]$$

In Equation 1.30 for the reaction chain length, concentration of radiation-excited molecular states capable of chain propagation should be added to the concentration of thermally activated molecules:

$$v = \frac{\left(2k_p/k_{rl}\right)\left(e^{-E_p/kT} + G*P\tau\right)}{1 + \sqrt{1 + \left(4k_{r2}/k_{rl}^2\right) k_i e^{-E_i/kT} + GP}} \tag{1.52}$$

At low cracking temperatures, the lifetime of radiation-excited states τ is approximately equal to $1/k_{rl}$. At heightened temperatures, characteristic of RTC, radiation-excited radical pairs do not completely annihilate. An indirect indication to their considerable concentration is an extremely high level of alkane isomerization in the RTC process. Moreover, the analysis of RTC rate dependence on dose rate in the works (Zaikin 2008, 2013a, Zaikin and Zaikina 2008a) shows that contribution of radiation-excited states remains comparable to the thermal contribution up to the temperatures of 450°C–500°C. This condition can be approximately satisfied with a supposition that $\tau \approx 1/k_{rl}$ in the whole temperature range considered. Then expression (1.52) can be written in the form

$$v \approx \frac{\left(2k_p/k_{rl}\right)\left(e^{-E_p/kT} + \left(G*P/k_{rl}\right)\right)}{1 + \sqrt{1 + \left(4k_{r2}/k_{rl}^2\right)\left(k_i e^{-E_i/kT} + GP\right)}} \tag{1.53}$$

In the extreme case of low temperatures,

$$v = \frac{\left(2k_p/k_{rl}\right)G*P\tau}{1 + \sqrt{1 + \left(4k_{r2}/k_{rl}^2\right)GP}} \tag{1.54}$$

The equation for the cracking start temperature will also change and take the form

$$T_c = \frac{E_p}{k \ln \dfrac{2k_p/k_{t1}}{1 - 2\left(k_p/k_{t1}^2\right)G * P + \sqrt{1 + \left(4k_{t2}/k_{t1}^2\right)GP}}} \qquad (1.55)$$

Equation 1.55 shows that $T_c \to 0$ as $P \to P_c$ where the critical dose rate P_c is determined by the equation

$$1 - 2\frac{k_p}{k_{t1}^2(T_c)}G * P_c + \sqrt{1 + \frac{4k_{t2}}{k_{t1}^2(T_c)}GP_c} = 0$$

At dose rates $P \geq P_c$, cracking becomes possible at any temperature $T \geq T_c$. This condition (the rate of chain propagation is greater than the rate of its termination) can be written in the form of inequality:

$$\frac{2k_p}{k_{t1}^2(T_c)}G * P > 1 + \sqrt{1 + \frac{4k_{t2}}{k_{t1}^2(T_c)}GP} \qquad (1.56)$$

or

$$P > P_c = \frac{k_{t1}^2(T_c)}{k_p G *}\left(1 + \frac{k_{t2}G}{k_p G *}\right) \qquad (1.57)$$

As temperature decreases, the lifetime of radiation-generated unstable molecular states decreases and the critical dose rate increases.

Figure 1.21 shows the dependence of the critical dose rate, that is, the minimal dose rate necessary for the "cold" cracking start, on temperature. This dependence was plotted using experimental data on radiation cracking of heavy oil feedstock at low temperatures and used for the determination of the temperature dependence of the constant of radical linear recombination k_{t1} (Figure 1.1).

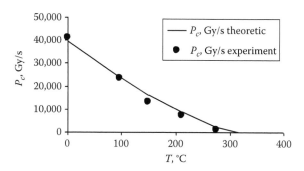

FIGURE 1.21   Dependence of cracking start temperature on critical dose rate (at the dose rate equal to P_c radiation cracking can proceed at any temperatures higher than T_c).

It follows from Figure 1.21 that LTRC can be observed only at high dose rates of ionizing irradiation. This requirement can be satisfied by the application of electron accelerators with a high beam current. Such accelerators are currently widely used in various technological applications demonstrating high reliability of operation in the industrial environment.

For the given type of feedstock, the critical dose rate for the cracking start is about 40 kGy at a temperature of 0°C. As temperature increases to 200°C, the critical dose rate drops down to about 9 kGy/s. It will be shown in the following text that, even in this case, a noticeable rate of "cold" cracking can be observed at dose rates higher than 20 kGy/s. It is important that the lifetime and radiation-chemical yields of radiation-excited molecular states depend on both temperature and the structure state of oil. In the case of heavy petroleum feedstock, they may considerably differ from the earlier-used values.

Brodskiy et al. (1961) have shown that the chain RTC reaction becomes impossible at low temperatures because big radicals formed as a result of radical interactions with hydrocarbon molecules do not decompose at the temperatures below the critical temperature and, therefore, the chain reaction cannot be propagated. To provide chain propagation at low temperatures, it is necessary to generate long-living, unstable molecular states that could be easily decomposed in the interactions with hydrogen atoms and light alkyl radicals. The formation of such molecular states requires energy compared to activation energy for chain initiation. Therefore, thermal activation of such molecular states is less probable. The phenomenon of LTRC indicates a considerable probability of formation of molecular states capable of chain reaction propagation under ionizing irradiation. As shown in the work (Zaikin 2013b), appearance of the long-living, unstable molecular states is associated with the intense radiation-induced isomerization of alkanes at the high dose rates of ionizing irradiation.

Dependences of the reaction chain length on temperature and irradiation dose rate subject to the earlier-described radiation and thermal effects are shown in Figure 1.22.

Comparison of Figure 1.22a and b shows that account of radiation contribution to propagation of the chain cracking reaction leads to the shift of the cracking start temperature at the given dose rate to the lower temperatures with simultaneous considerable increase in the absolute values of the reaction chain length at the given temperature.

Dependence of the chain length on dose rates shown in Figure 1.22b has radically changed compared with that calculated for RTC in the previous section. Taking into account radiation contribution to the chain reaction propagation leads to its increase with the dose rate in contrast to decrease shown in Figure 1.7. However, it should be noted that contribution of the radiation component to chain propagation is determined by the critical dose rate P_c, the minimal dose rate that allows radiation-induced chain cracking reactions at the given temperature. As P_c increases, the dependence shown in Figure 1.22b gradually turns to the dependence shown in Figure 1.7.

Generally, LTRC is characteristic of short interaction chains responsible for cracking propagation, and a high rate of the cracking reaction can be provided only by a great number of such interaction chains.

Subject to the contribution of radiation-excited molecular states, Equation 1.27 for the cracking rate will take the form

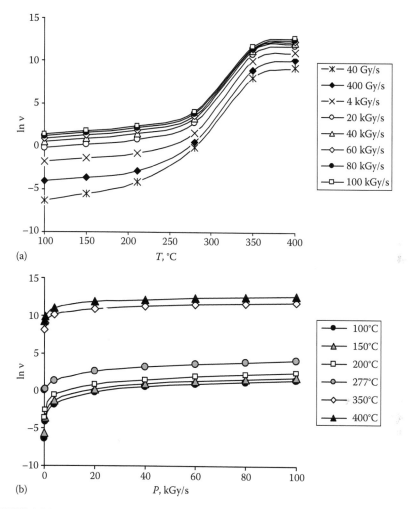

FIGURE 1.22 Dependence of reaction chain length on (a) temperature and (b) dose rate subject to radiation generation of unstable molecular states.

$$W = \begin{cases} \dfrac{v-1}{v} Rk_p(e^{-E_p/kT} + G*P\tau), & v \geq 1 \\ 0, & v < 1 \end{cases} \tag{1.58}$$

or

$$W \approx \begin{cases} \dfrac{v-1}{v} Rk_p\left(e^{-E_p/kT} + \dfrac{GP}{k_{t1}}\right), & v \geq 1 \\ 0, & v < 1 \end{cases}$$

Dependence of the rate of radiation cracking on temperature is shown in Figure 1.23.

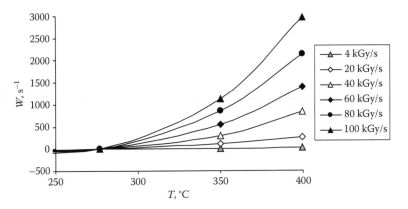

FIGURE 1.23   Dependence of radiation cracking rate on temperature.

Account of contribution of the radiation-excited molecular states to chain propagation leads to considerable increase in the calculated cracking rate at the given temperature. At low dose rates, RTC can be observed only at heightened temperatures (Figure 1.24b). As dose rate increases, a noticeable rate of radiation cracking can be reached at lowered temperatures (down to room temperature).

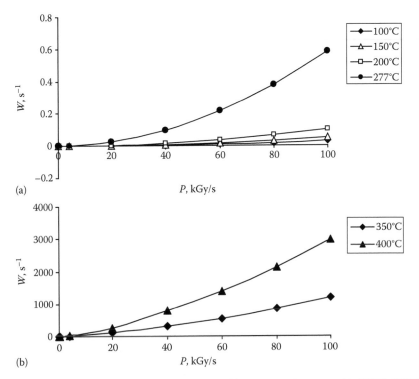

FIGURE 1.24   Dependence of the radiation cracking rate on dose rate. (a) 100°C–277°C, (b) 350°C and 400°C.

The experiments on radiation processing of heavy oil and bitumen (Zakin and Zaikina 2004a,b, 2008b) have shown that formation of low-temperature structures in this kind of feedstock may cause a considerable increase in the critical dose rate of the cracking start and decrease in the cracking rate at the given dose rate by more than an order of magnitude.

At very high values of the critical dose rate for the cracking start, "cold" cracking becomes practically impossible and RTC behavior is determined by the equations given in Section 1.2. At low values of the rate constant of chain propagation k_p, realization of RTC in such feedstock at relatively low temperatures (250°C–350°C) may require lower dose rates (1–5 kGy/s) and increased irradiation doses (see Section 1.2).

1.5 THERMODYNAMIC ASPECTS OF RADIATION CRACKING

TC is an example of a reaction whose energetic characteristics are dominated by entropy (ΔS) rather than by enthalpy (ΔH) in the Gibbs free energy equation

$$\Delta G = \Delta H - T\Delta S \tag{1.59}$$

where ΔG, ΔH, and ΔS are differences in Gibbs free energy, enthalpy, and entropy of the reaction product and the original matter, respectively.

A chain reaction may proceed only if the inequality $\Delta G < 0$ is satisfied. In the case of an endothermic reaction ($\Delta H > 0$, $\Delta S > 0$), it implies the availability of a critical temperature T_c above which the chain reaction is impossible.

Although the dissociation energy for a carbon–carbon single bond is relatively high (about 375 kJ/mol) and cracking is highly endothermic, the large positive entropy change resulting from the fragmentation of one large molecule into several smaller pieces, together with the extremely high temperature, makes the $T\Delta S°$ term larger than the $\Delta H°$ term, thereby favoring the cracking reaction.

Thus, the self-sustainable chain reaction of TC can proceed only at temperatures higher than

$$T_c = \frac{\Delta H}{\Delta S} \tag{1.60}$$

The start of a chain cracking reaction is defined by the equation

$$e^{-\Delta G/kT} = e^{\Delta S/k}e^{-\Delta H/kT_c} = 1 \tag{1.61}$$

From the positions of kinetic theory, this condition can be written in the form

$$v = \frac{W(propagation)}{W(initiation)} = \frac{W(propagation)}{W(term)} = 1 \tag{1.62}$$

where
  W is the reaction rate
  v is the chain length

In the case of an endothermic chain reaction, the following conditions should be satisfied:

$$\nu > 1 \quad \text{at} \quad T > T_c$$
$$\nu \leq 1 \quad \text{at} \quad T \leq T_c \tag{1.63}$$

Fulfillment of the conditions (1.60)–(1.63) can be considered as a criterion of a kinetic model's correctness. Let us note that an approximate equality (1.33)

$$W \approx R k_p e^{-E_p / k_B T}$$

often used for the estimation of the cracking rate does not satisfy these conditions. At $T \leq T_c$, this expression still yields finite W values. In the case of a high concentration of the chain carriers generated by radiation in the RTC process, a calculated cracking rate may have large values in the temperature range where it should be equal to zero according to the thermodynamic requirements.

Kinetic equations (1.18), (1.19), (1.21), and (1.27) are in agreement with the thermodynamic criteria. In particular, the reaction chain length in the case of TC, subject to quadratic and linear chain termination, is determined by Equation 1.19:

$$\nu = \frac{2\left(k_p / k_{t1}\right) e^{-E_p / kT}}{1 + \sqrt{1 + \left(4 k_i k_{t2} / k_{t1}^2\right) e^{-E_i / kT}}}$$

As $T \to T_c$, this expression takes the form

$$\frac{k_p}{k_{t1}} e^{-E_p / kT_c} \approx 1$$

Let us compare this equality with the thermodynamic condition (1.61). In the process of cracking, free energy of the system changes as a result of decomposition of hydrocarbon molecules with activation energy, E_p, that is,

$$\Delta H = E_p \approx 20 \text{ kcal/mol} = 84 \text{ kJ/mol} \tag{1.64}$$

Then,

$$\Delta S = k \ln \frac{k_p}{k_{t1}} \approx 36 \text{ cal/mol} \cdot \text{K} \approx 153 \text{ J/mol} \cdot \text{K} \tag{1.65}$$

Note that expressions (1.64) and (1.65) only roughly estimate thermodynamic characteristics of a hydrocarbon cracking reaction.

Generally, limitation on the cracking temperature (1.61) does not apply to highly nonequilibrium processes where concentrations of the reactive particles responsible for the reaction initiation and propagation and, therefore, initiation and propagation rates are not subjected to thermodynamic limitations.

Radiation generation of radicals in the process of RTC introduces substantial corrections into the equations for cracking start temperature, radical concentration, reaction chain length, and rate of a chain reaction. All of these quantities become functions of both temperature and dose rate of ionizing irradiation. As dose rate increases, chain cracking reaction can occur at lower temperatures. The thermodynamic limitations become weaker, and irradiation dose rate acquires the role of the main parameter in the determination of reaction chain characteristics.

Radiation generation of the reactive particles responsible for initiation and propagation of a chain cracking reaction does not obey thermodynamics and requires very low energy consumption compared with a thermal process. As a result, a high rate of radiation cracking can be provided at temperatures much lower than characteristic temperatures of TC.

At the same time, it should be emphasized that reactions with participation of the radiation-generated particles proceed according to the laws of thermodynamics. Therefore, radiation cracking remains an endothermic reaction where product formation requires absorption of the heat energy. To provide a constant reaction rate, the temperature of oil in the reactor should be kept at the given level. If this requirement is not satisfied, the temperature of the system decreases. In turn, it leads to a decrease in the cracking rate.

Let us estimate a value of this effect for RTC of heavy fuel oil. The heat capacity of fuel oils can be evaluated in a wide temperature range using an empiric formula (Grigoryev et al. 1999):

$$C_p \; (\text{kJ/kg} \cdot {}^\circ\text{C}) = 1.68 + 2.31 \cdot 10^{-3}\, T \; ({}^\circ\text{C}) \tag{1.66}$$

The quantity of heat absorbed in the process of fuel oil heating from temperature t_1 to temperature t_2 (or liberated in the process of cooling from t_2 to t_1) can be calculated as follows:

$$Q = \int_{t_1}^{t_2} C_p \, dt$$

or

$$Q = 1.68(T_2 - T_1) + \frac{2.31 \cdot 10^{-3}}{2}(T_2^2 - T_1^2) \tag{1.67}$$

It implies

$$\Delta T = T_2 - T_1 = \frac{Q}{1.68 + 2.31 \cdot 10^{-3}\,\overline{T}} \tag{1.68}$$

where $\overline{T} = (T_1 + T_2)/2$.

In the case of 50% conversion characteristic of radiation processing of fuel oil and irradiation dose rate of 4 kGy/s, the temperature fall due to absorption of the reaction heat at $\overline{T} = 350°\text{C}–400°\text{C}$ will make 30°C–35°C. In this case, a decreased cracking rate at the decreased temperature will be still rather high (see Section 1.2).

At lowered temperatures of "cold" radiation cracking, the role of this effect is more important. For example, at the cracking temperature $\bar{T} = 150°C$ and dose rate of 20 kGy/s, a decrease in temperature by about 50°C can be accompanied by a decrease in the reaction rate down to zero. This effect should be taken into account in low-temperature oil processing.

1.6 MECHANISMS OF RADIATION-EXCITED MOLECULAR STATE FORMATION

1.6.1 LIFETIMES OF EXCITED MOLECULES

The earlier-estimated value of the radical pair lifetime at a temperature of 550 K is about $5 \cdot 10^{-7}$ s that corresponds to the rate of energy loss by a radiation-excited molecular state of about 130 s^{-1}. At the first glance, it is in contradiction with the well-known statement based on experimental observations that the typical rate of molecule excitation energy transfer is $\sim 10^{14}-10^{15}$ s^{-1} (Tophiev and Polak 1962, Lavrovskiy 1976), which approximately corresponds to the time between electron–electron collisions.

A more detailed analysis of the lifetimes of the excited states (Pikaev 1987) shows that two types of excited molecular states can appear in hydrocarbons under the action of ionizing radiation: short-living singlet and long-living triplet states.

The triplet excited states were observed in radiolysis studies of alkane and cyclo-alkane systems both for solute species and the solvent (as a rule, they were observed in aromatic compounds). Such excited states form as a result of ion neutralization. An example of this process is the reaction

$$A^+ + A^- \rightarrow A^* + A \tag{1.69}$$

It was noted (Pikaev 1987) that triplet excited states can appear not only due to collapse of the coupled ion pairs but also in other processes too. One of these processes is the direct excitation of molecules by the electrons. Another process is ion recombination in spurs that contain several ions. In these cases, such processes are possible as, for example, fast neutralization of ions with uncorrelated spins from the neighboring pairs. The remaining charges form a pair in the triplet state.

Lifetimes of excited triplet molecules of aromatic compounds in liquid hydrocarbons at room temperature can amount to tens and even hundreds of microseconds if quenchers are absent. In the paper (Zaikin and Zaikina 2008a), it was suggested that the observed nonlinear dependence of the cracking rate on $\sqrt{P}$ may be caused by the contribution of excited molecules in the triplet state. However, triplet radical pairs are not capable of recombination, and they should be first turned to singlet pairs (Denisov 1971).

The singlet excited molecules usually have a lifetime much lower than that of triplet excited molecules; such states were studied using luminescence measurements by the methods of nano- and picosecond radiolysis. The singlet-excited molecule solvent can be formed, first, due to the direct interaction of ionizing radiation with hydrocarbons

$$RH \wedge\wedge\wedge\wedge \rightarrow RH* \tag{1.70}$$

and, second, according to the reaction

$$RH^+ + e_s^- \rightarrow RH* \tag{1.71}$$

In consideration of their short lifetime, the singlet-excited states cannot contribute directly to the propagation of the cracking chain reaction proceeding by the radical mechanism. However, as a result of inter-combinational inversion, the singlet-excited molecules can pass into long-living triplet states (Pikaev 1987). Disintegration of excited molecules into radicals

$$RH* \rightarrow R' + R'' \tag{1.72}$$

can contribute to the radical chain initiation.

In different conditions of irradiation of hydrocarbon mixtures, the radical mechanism of radiation cracking can be altered and supplemented by other mechanisms of initiation and propagation of the chain reaction. The description provided in the previous sections does not specifically define the cracking mechanism but suggests that the quasi-equilibrium concentration of the chain carriers is controlled by their coupled recombination.

This approach can be justified by observations showing that due to high mobility of light radicals in liquid hydrocarbon mixtures, even at room temperature, about 70% of free radicals diffuse beyond the bounds of spurs (Burns et al. 1966). It diminishes the role of processes in tracks and spurs that cause changes in the order of the reaction of radical recombination and suppress radical reactions as the dose rate increases.

The analysis of cracking chain characteristics in the previous sections shows that contribution of the radiation-excited states to chain propagation is directly connected with radiation generation of long-living radical pairs, which recombine in the first-order reactions.

It is known that the main product of alkane radiolysis is hydrogen formed as a result of monomolecular and bimolecular decomposition of the excited hydrocarbon molecules (Földiàk 1981). A comprehensive study of the bond rupture in the radiolysis of n-alkanes using gel permeation chromatography (Wojnarovits and Schuler 2000) has shown that scavengeable alkyl radicals are produced in a total yield of 5.3 of which 25% represent primary radicals produced by the rupture of the hydrocarbon backbone and most of the remainder by loss of H from the primary and secondary positions. Other radicals are produced only in very low yield, viz. $G < 0.01$. At millimolar concentrations of the iodine scavenger, there was no evidence for skeleton rearrangement or other secondary reactions of the alkyl radicals occurring within the time scale of scavenging, that is, 100 ns.

In the process of low-temperature cracking, chain carriers can interact directly with excited molecular states or react as scavengers of light alkyl radicals and hydrogen atoms providing accumulation of the heavier unstable radicals and subsequent chain propagation. At the heightened dose rates, the concentration and the lifetime of the hydrogen atoms are sufficient for the high rate of the chain cracking reaction.

Thus the $R + R'$ pairs giving rise to chain propagation are most probably $R + H$ pairs. The main mechanism of such pair formation is known as the cage effect.

1.6.2 CAGE EFFECT

Cage effect is an important phenomenon that plays a key role in the formation of long-living radical pairs and, therefore, strongly affects the kinetics of low-temperature cracking. Cage effect (or Franck–Rabinowitch effect) is a general name for the phenomenon characteristic of the reactions in liquid and solid phases when pairs of the reacting particles are confined in a small area (cage) surrounded by molecules of the medium (Kochi 1973). A typical cage effect can be observed in the studies of hydrocarbon decomposition.

In the liquid phase, radicals are surrounded by a cage of the neighboring molecules. To remove radicals to a distance where they would behave as independent particles, an additional barrier equal to activation energy of radical diffusion from the cage should be surmounted:

$$
R_1 + R_2 \underset{k_{-D}}{\overset{k_D}{\rightleftarrows}} (R_1 \cdots R_2)_{cage} \overset{k_t}{\underset{k_d}{\rightarrow}} \begin{array}{l} RH \\ R_1' + R_2' \end{array} \tag{1.73}
$$

where
k_t is the rate constant of radical recombination
k_d is the rate constant of the reaction in the cage
k_D is the rate constant of radical diffusion

Thus cage effect leads to the changes in the stationary radical concentration. In the case of low-temperature radiation cracking, it means that a confined unstable pair of radicals can contribute to propagation of a chain reaction due to interactions with other radiation-generated radicals.

Estimations of the lifetime of radiation-generated radical pairs in a cage and cage dimensions were performed in the work (Brodskiy et al. 1961). In the case of a strong cage effect, radical concentration R was divided into two parts:

$$
\left[\bar{R} \right] = \left[\bar{R}' \right] + \left[\bar{R}'' \right] \tag{1.74}
$$

where
$[\bar{R}']$ is the number of radicals coupled in a cage per unit of volume
$[\bar{R}'']$ is the bulk concentration of radicals that went out of the cage due to diffusion

A balance equation for the radicals can be written in the form

$$G_{\bar{R}'}P - \frac{1}{\tau_t}\left[\bar{R}'\right] - K_p\left[\bar{R}'\right] = 0 \tag{1.75}$$

where

$1/\tau_t$ is the rate of $[\bar{R}']$ decrease due to recombination

$G_{\bar{R}'}$ is the radiation-chemical yield of radicals $\bar{R}'$

K_p is the rate of chain propagation that was assumed to be equal to the rate of radical pair decomposition

In the extreme case under consideration, radical diffusion was neglected in the balance equation (1.75). At temperatures close to the cracking start temperature, a balance equation for radicals $[\bar{R}'']$ takes the form

$$\frac{1}{\tau_D}\left[\bar{R}''\right] - K_p\left[\bar{R}''\right] - k_t\left[\bar{R}''\right] = 0 \tag{1.76}$$

where τ_D is the time of radical diffusion from a cage. Using Equations 1.75 and 1.76, we can find the rate of hydrocarbon decomposition with olefin formation:

$$G(olefin)P = K_p\left[\bar{R}\right] = K_p\left(\left[\bar{R}'\right]+\left[\bar{R}''\right]\right) \cong \left(K_p + \frac{1}{\tau_D}\right)\frac{G_{\bar{R}'}P}{\left(1/\tau_T\right)+K_p} \tag{1.77}$$

The reaction of the bulk recombination is neglected in Equation 1.77 because of its relatively small contribution to the process. From this equation, it follows that at temperatures satisfying the inequality

$$K_p \ll \frac{1}{\tau_D} \ll \frac{1}{\tau_T} \tag{1.78}$$

temperature dependence of hydrocarbon decomposition will be relatively weak. According to (1.77), the transition to RTC occurs at such a temperature when

$$K_p = k_p e^{-E_p/kT_c} \approx \frac{1}{\tau_D} \tag{1.79}$$

If $K_p \gg 1/\tau_D$ in the RTC mode, then temperature dependence of the process is determined by the equation

$$\frac{K_p}{\left(1/\tau_T\right)+K_p} \cong \frac{K_p}{1/\tau_D} \tag{1.80}$$

since $K_p \ll 1/\tau_T$.

Formula (1.79) allows the evaluation of the cage dimensions. In the work (Brodskiy et al. 1961), it was assumed that $E_p \approx 20$ kcal/mol $= 84$ kJ/mol and $k_p = 10^{13}$ s^{-1}. Substituting these values to equality,

$$\frac{1}{\tau_D} = \frac{D}{d^2} \tag{1.81}$$

we shall obtain that the cage size $d \approx 10^{-8}$ m at $T_c = 600$ K, and $D = 10^{-9}$ m^2/s.

Substitution of the values of the parameters used in calculations in Sections 1.1 through 1.4 ($k_p = 10^{12}$ s^{-1}, $E_p = 20$ kcal/mol, and $T_c = 550$ K) into Equation 1.81 yields about three times greater cage size and an order of magnitude greater radical lifetime at the same values of the cracking start temperature, T_c, and diffusion coefficient, D.

The calculations of the work (Brodskiy et al. 1961) provide a reasonable evaluation of the cage size. However, these estimations of the cage size and, especially, the radical lifetime are very rough because of the uncertainty in the values of the diffusion coefficient. A value of the effective diffusion coefficient used in this estimate can be considerably lower than that in the case of free radical diffusion in the bulk of a material. As a result, the radical lifetime can be considerably greater than that estimated in the paper (Brodskiy et al. 1961). In the calculations of the chain reaction characteristics in the previous sections, the selection of reaction parameters was based on experimental data for heavy oils characteristic of high viscosity and low diffusion coefficients.

In the earlier-given estimate, the radical lifetime in a cage is considered as a time of radical exit from a cage due to diffusion without taking into account the processes inside the cage. In a more detailed diffusion-kinetic model of the cage effect (Denisov 1987), three possibilities for a radical pair in a cage were considered: recombination at the rate k_t, exit from the cage with the rate constant K_D, and transformation to a new pair of radicals with the rate constant K_d. In this consideration, the following equation is valid for the yield of the product of intracellular recombination:

$$\frac{1}{y} - 1 = \frac{K_D}{k_t} + \frac{K_d}{k_t} \tag{1.82}$$

K_D is determined by the equation

$$K_D = \sqrt{\frac{2D}{t} \frac{1}{(l-r)^2}} \tag{1.83}$$

where
    r is the effective radical radius
    l is the distance between radicals in a cage

Time t needed for radicals to go out of the cage depends on k_t and K_d:

$$t^{-1} = k_t + K_d \tag{1.84}$$

Therefore, dependence of y on η has the form

$$\frac{1}{y} - 1 = \frac{1}{l-r}\sqrt{\frac{2D(k_t + K_D)}{k_t^2}} + \frac{K_d}{k_t} \qquad (1.85)$$

or, taking into account that diffusion coefficient D is approximately inverse of viscosity η,

$$\frac{1}{y} - 1 = \frac{const}{\sqrt{\eta}} + \frac{K_d}{k_t} \qquad (1.86)$$

Equation 1.86 shows that both the yield of the products of intracellular recombination and the lifetime of radicals in a cage substantially depend on oil viscosity. On the other hand, Equation 1.84 shows that the radical lifetime depends on the rates of two processes: recombination k_t and intercellular transformation K_d. A characteristic feature of radiation cracking is the extremely high level of alkane isomerization and correlation of the isomerization rate with the rate of cracking reaction. Therefore, we suggest that isomerization is the basic intercellular process that determines the radical lifetime.

1.6.3 ROLE OF ISOMERIZATION REACTIONS IN HYDROCARBON CRACKING

Together with the reactions of hydrocarbon decomposition, intense reactions of isomerization under ionizing irradiation were observed at precracking and cracking temperatures. For example, radiation cracking of butane (Matsuoka et al. 1975) resulted in the formation of butane and butylene isomers. The yields of these products were affected by NH_3 addition. Formation of iso-butane and iso-butylene was associated with a chain process. Neither iso-butane nor iso-butylene was formed in the thermal process.

A strong effect of radiation-induced isomerization of alkanes during RTC of heavy oil was observed in the work (Zaikin et al. 2001). Further studies (Zaikin and Zaikina 2004a,c, 2006, 2007, 2008a, 2013) have shown that RTC is always accompanied by intense isomerization especially pronounced in heavy oils and bitumen with high contents of heavy aromatic compounds. The analysis of the data on radiation cracking of hydrocarbons indicates a special role of isomerization processes in the mechanism of chain reaction propagation at lowered temperatures (see Section 1.7).

As shown in Sections 1.1 through 1.4, self-consistent thermodynamic and kinetic interpretation of radiation cracking and a correct evaluation of the reaction chain length are possible if contribution of the first-order reactions to chain termination is taken into account.

In the case of the radical mechanism of cracking, an alkane molecule, which absorbed energy of about 3.5 eV as a result of thermal or radiation action, decomposes into two radicals: a reactive radical (a hydrogen atom or a light alkyl radical)

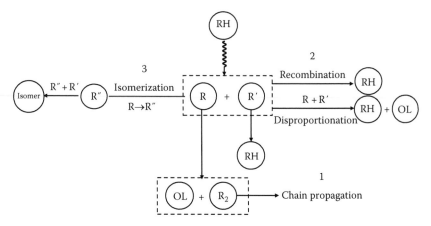

FIGURE 1.25 Schematic primary reactions of hydrocarbon cracking: RH—saturated hydrocarbon molecule; OL—olefin molecule.

and an unstable big residual radical. The fate of the latter can be different. The most probable transformation paths for a big radical formed as a result of hydrocarbon molecule decomposition are shown in Figure 1.25.

The relationship of reactions 1, 2, and 3 (Figure 1.25) depends on temperature and, in the case of radiation cracking, on irradiation dose rate. An increase in temperature and/or irradiation dose rate leads to a higher probability of reaction 1, which results in the decomposition of big alkyl radicals and furthers chain propagation. As temperature and dose rate decrease, the probability of radical stabilization by isomerization and recombination increases. As temperature approaches to the cracking start temperature and concentration of the thermally generated chain carriers considerably decreases, radical recombination in the first-order reactions becomes predominant.

Below the critical temperature, TC is impossible because of the thermodynamic limitations. From the point of the process chemistry, it means that big radicals do not decompose and the chain reaction cannot be propagated. At temperatures above the critical temperature, such radicals disintegrate when receiving energy of about 0.87 eV. Destabilization of alkyl radicals below the critical temperature requires a considerably higher excitation level compared to the energy of primary radical formation resulting from the decomposition of a hydrocarbon molecule (about 3.5 eV).

A high concentration of such excited molecular states can be produced at the high dose rates of ionizing irradiation. However, to provide a high cracking rate, these molecular states must have a sufficiently long lifetime.

Such molecular states can be created if a pair $R + R'$ formed as a result of molecule radiolysis does not recombine but remains unstable as long as electron structure transformation in radical R proceeds until its stabilization.

The most probable processes of the stabilization of radical electron structure are disproportionation and isomerization. Experimental data on kinetics of liquid alkane isomerization are summarized in the paper (Allara 1980). According to these data, activation energy for isomerization is in the range of 10–20 kcal/mol, and inverse

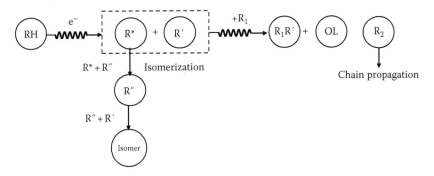

FIGURE 1.26   Schematic chain propagation in LTRC.

isomerization rate at the temperature of 550°C is about 10^{-2}–10^{-1} s. Thus, alkane isomerization is a rather slow process that can determine termination of a chain reaction of TC or RTC at lowered temperatures.

As a result of interaction of an unstable radical pair $R + R'$ with a light alkyl radical, the unstable radical R' can decompose forming a new reactive radical capable of chain propagation. In this case, a reaction of the following type can take place:

$$\left(CH_3 - CH_2 - \dot{C}H_2 + \dot{H}\right) + R \rightarrow RH + CH_2 = CH_2 + \dot{C}H_3$$

Chain propagation proceeds by channel 3 (Figure 1.25), as schematically shown in Figure 1.26.

If no interaction of such unstable molecular states with a reactive radical happens during its lifetime, then a big alkyl radical will stabilize its electron structure and form an isomer by attaching another radical.

Therefore, there are reasons to conclude that alkane isomerization makes a major contribution to both cracking termination at the critical temperature and generation of unstable molecular states responsible for chain propagation in LTRC.

In the paper (Zaikin 2013b), high concentrations of the radiation-generated unstable molecular states necessary for the propagation of the low-temperature chain cracking reactions in the petroleum feedstock were interpreted in association with the earlier-reported effect of the intense radiation-induced isomerization especially characteristic of the high-viscous oils (Zaikin and Zaikina 2004a,c, 2006, 2007, 2008a, 2013).

The theory of LTRC (Zaikin 2008, 2013a, Zaikin and Zaikina 2008a) is based on the concept of the long-living excited states of hydrocarbon molecules. It assumes that in the dynamic equilibrium, a sufficiently high concentration of such states necessary for the propagation of the chain cracking reaction is generated by ionizing radiation at sufficiently high dose rates.

A key to understand the mechanism of radiation-induced generation of the long-living excited molecular states is the phenomenon of the intense isomerization of alkanes, especially pronounced in the hydrocarbon mixtures with a high content of heavy aromatic compounds. In the papers (Zaikin et al. 2001, 2004), this phenomenon was explained with the assumption that big alkyl radicals may transfer their excess energy to the radiation-resistant aromatic molecules. As a result, such a radical does not disintegrate to fragments but realigns its structure and forms an isomer.

This model was modified and used for the interpretation of the formation of the long-living excited molecular states in liquid hydrocarbon mixtures under ionizing irradiation (Zaikin 2013b).

1.6.4 DESCRIPTION OF THE MODEL

According to the commonly accepted theory of RTC of alkanes, in the primary event of cracking initiation, an alkane molecule disintegrates under irradiation into two radicals: a light radical R that plays the role of the chain carrier and a bigger unstable radical R′. Radical R′ disintegrates with the formation of an olefin and a light radical that can also work as a chain carrier. The chain propagation is provided by interaction of the radicals R with the excited states of alkane molecules, which can be generated as a result of thermal (Topchiev and Polak 1962) or radiation (Zaikin and Zaikina 2008a) activation.

As shown earlier, unstable molecular states necessary for the chain propagation in radiation cracking reactions can be considered as the pairs of the bound radicals R–R′. Together with a probability of the molecular state disintegration, there is a probability that radical R′ will stabilize its electron structure and form an isomer by attaching to radical R. In general, isomerization of excited hydrocarbon molecules may proceed not only by a radical but also by an ion mechanism similar to the heterogeneous catalytic isomerization of *n*-alkanes with the participation of carbocations (Egiazarov et al. 1989).

A radical or an ion formed as a result of radiation excitation of a hydrocarbon molecule can interact with the neighboring molecules and induce electron realignment in these molecules. Therefore, isomerization initiated by the appearance of an excited state can cover a group of molecules, whose size depends on excitation energy and conditions of excitation energy transfer to the neighboring molecules.

Although mechanisms of isomerization can differ much for the unstable molecular states of alkanes, it is important that relatively low reaction rates are characteristic of isomerization, and a molecule participating in the process can stay in the unstable state for a relatively long period of time. Interaction of such unstable molecules with radiation-generated radicals R results in their disintegration. Concentration of the unstable molecular states may be high enough to provide a high propagation rate of the chain reaction of hydrocarbon cracking if the dose rate of ionizing irradiation is sufficiently high.

The probability of formation for a long-living unstable state in this process depends on the efficiency of the excess energy transfer to the neighboring molecules. The aromatic molecules, especially heavy aromatic compounds, have a higher capacity of excess energy absorption compared with the paraffin molecules. On the one hand, it causes some decrease in the radiation-chemical yields of radical chain carriers (Topchiev and Polak 1962). On the other hand, the high absorption capacity of the aromatic compounds reduces energy transfer at each interaction with a hydrocarbon molecule and creates shorter interaction chains. With a higher probability, it leads to isomerization rather than to cracking.

Interpretation of the experimental data should take into account that 20%–30% radicals in liquid hydrocarbons recombine in spurs having no time to go out beyond their bounds (Burns et al. 1966). In this model, only the pairs of radicals (or ions) located outside of spurs contribute to isomerization. Isomerization within the spurs was considered to be less probable.

Taking into account that isomerization caused by excitation of the hydrocarbon molecules may cover a group of n molecules, the following equation can be written for the concentration of the long-living unstable molecular states in the dynamic equilibrium:

$$C_{rad}^* = \frac{nG * P}{K*} \tag{1.87}$$

where
  $G*$ is the radiation-chemical yield of the unstable states
  P is the dose rate
  $K*$ is the rate of the molecule electron structure stabilization in the isomerization
     process

At the heightened temperatures (350°C–450°C) of RTC, the main contribution to the propagation of the chain cracking reaction is made by the thermally excited molecular states with excitation energy $E \approx 80$ kJ/mol. Concentration of such states $e^{-E/kT}$ can be high enough, but these states are short-living and they cannot considerably contribute to isomerization. We can suggest that formation of the long-living unstable states requires excitation energy close to the activation energy of cracking initiation (i.e., energy of radical chain carrier formation).

Concentration of these states C_{therm}^* can be estimated by

$$C_{therm}^* = \frac{K_0 e^{-E_0/kT}}{K*} \tag{1.88}$$

where $K_0 e^{-E_0/kT}$ is the rate of thermal generation of the excited states ($K_0 \approx 10^{12}$ s^{-1}; $E_0 \approx 340$ kJ/kg).

Using Equations 1.87 and 1.88, we shall find the concentration ratio of the radiation and thermally excited molecular states that can principally contribute to the alkane isomerization:

$$\frac{C_{rad}^*}{C_{therm}^*} = \frac{nG * P}{K_0 e^{-E_0/kT}} \tag{1.89}$$

Calculation using Equation 1.89 shows that at the dose rate of 1 kGy/s, concentration of the thermally excited long-living molecular states in hydrocarbons with the molecular mass of 300 g/mol becomes comparable with the concentration of such states generated by irradiation only at temperatures above 850°C.

In this model, a high degree of isomerization caused by ionizing irradiation is associated with the direct excitation of the molecular electron system when an unstable molecular state can be considered as a system of two bound reactive particles, one of which demonstrates a high chemical reactivity and another one is prone to isomerization.

1.6.5 Radiation-Excited Unstable States Responsible for Propagation of the Chain Cracking Reaction

In Section 1.4, concentration of the radiation-excited unstable states was estimated using Equation 1.48:

$$W \cdot \exp\left(\frac{E}{kT}\right) P^{-1/2} = A + B(T)P$$

Concentrations of the radiation-excited states in heavy hydrocarbon feedstock, calculated from the experimental values of constants A and $B(T)$, can be represented by an approximate empiric equation:

$$C * e^{-E'/kT} = \frac{B}{A} P \exp\left(-\frac{E}{kT}\right) \approx 5 \cdot 10^{-10} P \qquad (1.90)$$

where P dimension is Gy/s.

We shall assume that disintegration of a radiation-excited molecule in the unstable state additionally requires a minor thermal activation energy E'. Then, radiation-excited states will disintegrate with the probability $e^{-E'/kT}$, while in the absence of thermal activation, isomerization involving n molecules will be observed. Concentration of the unstable radiation-excited states can be represented as a sum of the two terms responsible for the two different types of molecular states:

$$C* = C_c^* + \frac{C_{iso}^*}{n} = C * e^{-E'/kT} + C * \left(1 - e^{-E'/kT}\right) \qquad (1.91)$$

The first term $C_c^* = C * e^{-E'/kT}$ defines contribution of the radiation-excited states to propagation of the chain cracking reaction. The second term is responsible for isomerization. Taking into account that a group of n molecules may be involved into isomerization, this term can be represented in the form

$$\frac{C_{iso}^*}{n} = C * \left(1 - e^{-E'/kT}\right) \qquad (1.92)$$

where C_{iso}^* is the concentration of the isomerization centers.

Both types of states are generated with the same rate GP, but their lifetimes are different and they disappear with different rates:

$$K_c^* = \frac{G*P}{C*} e^{E'/kT} \qquad (1.93)$$

$$K_{iso}^* = \frac{G^*P}{C^*\left(1-e^{-E'/kT}\right)} \approx \frac{nG^*P}{C_{iso}^*} \tag{1.94}$$

Dependence of the quantity C_c^* on the dose rate derived from the analysis of the experimental data is described by Equation 1.90. Combining Equations 1.90 and 1.93, we shall find

$$\frac{G^*}{K_c^*} \approx 5 \cdot 10^{-10} \frac{\text{kg} \cdot \text{s}}{\text{J}} \tag{1.95}$$

The lifetime of the radiation-excited unstable states responsible for propagation of the chain cracking reaction estimated from Equation 1.95 is $1/K_c^* \sim 10$ ms. At a temperature of about 400°C, concentration of such states is comparable with the concentration of molecules thermally excited at the level necessary for chain propagation.

1.6.6 Radiation-Excited Unstable States Responsible for Isomerization

Now we shall compare the earlier-given estimations of the concentrations of the radiation-excited molecular states with the kinetics of alkane isomerization observed in the experiments on hydrocarbon feedstock irradiation.

In the simplified model considered in this section, kinetics of radiation-induced isomerization of paraffins can be described by the following system of equations:

$$\frac{dC_{iso}}{dt} = K_{iso}^* C_{iso}^* \tag{1.96}$$

$$\frac{dC_{iso}^*}{dt} = nG^*P(C_{alk} - C_{iso}) - K_{iso}^* C_{iso}^* \tag{1.97}$$

where
C_{iso}^* is the concentration of the unstable states of n-alkanes
C_{iso} is the concentration of iso-alkanes
C_{alk} is the entire concentration of alkanes of normal and branched structure
K_{iso} is the isomerization rate constant

Taking into account that the steady-state concentration of the unstable molecular states sets in under irradiation in a short period of time and then slowly changes in the course of alkane isomerization, we can assume that $dC_{iso}^*/dt = 0$. Under this assumption, solution of the systems of Equations 1.12 and 1.13 is of the form

$$K_{iso}^* = \frac{nG^*P(C_{alk} - C_{iso})}{C_{iso}^*} \tag{1.98}$$

$$C_{alk} - C_{iso}(t) = \left[C_{alk} - C_{iso}(0)\right]e^{-nG^*Pt} \tag{1.99}$$

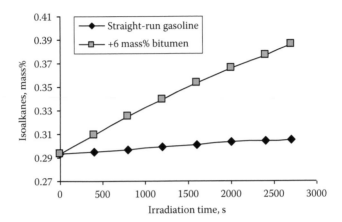

FIGURE 1.27 Kinetics of alkane isomerization under irradiation of straight-run gasoline with bremsstrahlung x-rays from 2 MeV electrons. (From Zaikin, Y.A., *Radiat. Phys. Chem.*, 84, 6, 2013b.)

Kinetics of alkane isomerization was studied in the series of experiments (Zaikin and Zaikina 2004a, 2007, 2008a) on irradiation of the straight-run gasoline with the bremsstrahlung x-rays from 2 MeV electrons. Irradiation was conducted at a temperature of 25°C and a dose rate of 16 Gy/s. After irradiation with the dose of 43.2 kGy, iso-alkane concentration in gasoline increased by 1.2 mass% while *n*-alkane concentration decreased by 1.1 mass%.

Comparison of the calculations using Equation 1.99 with the experimental data is given in Figure 1.27. In calculations, it was assumed that increase in the iso-alkane concentration as a result of isomerization is equal to the decrease in the concentration of normal alkanes.

The best agreement with the experimental data was obtained for the following set of parameters in Equations 1.98 and 1.99: $K_{iso}^* = 5.0 \cdot 10^{-3} \, \text{s}^{-1}$; $nG^* = 1.2 \cdot 10^{-6}$ kg/J ≈ 20 molecules/100 eV. With the estimation of $G^* = 5$ molecules/100 eV, we come to the conclusion that isomerization caused by the primary excitation involves about four molecules.

The energy of thermal activation E' necessary for the disintegration of a radiation-excited molecule calculated using Equation 1.90 is 26.5 kJ/mol. It can be noted that the calculated value of the activation energy E' approximately corresponds to the interaction energy of two single-charged ions separated by 2–3 intermolecular distances.

Using Equation 1.90, we shall also estimate the contribution of radiation-excited unstable states to the propagation of chain cracking reaction in gasoline at a temperature of 25°C and dose rate of bremsstrahlung x-ray irradiation of 16 Gy/s:

$$C_c^* = C^* \, e^{-E''/kT} = 6.4 \cdot 10^{-9}$$

Concentration of the radical chain carriers in the given irradiation conditions can be calculated using the equation $R = \sqrt{G * P/K_r}$, where K_r is the rate of radical recombination ($K_r \approx 10^9$ s). The calculated value is $R \approx 2 \cdot 10^{-8}$.

The rate of "cold" cracking W in these conditions can be estimated using the equation $W = K_0 R C_c^*$, where $K_0 \approx 10^{12}$ s^{-1}. The calculated initial cracking rate makes $1.3 \cdot 10^{-4}$ s^{-1} while the initial rate of isomerization

$$\left. \frac{dC_{iso}}{dt} \right|_{t=0} = nG*P\left(C_{alk} - C_{iso}(0)\right)$$

is $7.8 \cdot 10^{-6}$ s^{-1}. It explains why relatively modest isomerization observed in these conditions was accompanied by a noticeable gas evolution and changes in the hydrocarbon content of the feedstock.

Addition of heavy aromatic leads to a considerable increase in the rate of alkane isomerization. Calculations show that addition of 6% bitumen in the same irradiation conditions results in increase in the G-values of the unstable molecular products by 8.4 times. Thus, the primary radiation excitation of alkanes in the presence of heavy aromatic compounds leads to the isomerization that involves 30–35 molecules. Estimate of the isomerization rate constant results in the same value as that in the absence of the heavy aromatic additive $K_{iso}^* = 5.0 \cdot 10^{-3}$ s^{-1}.

As bitumen concentration increases by 2.5 times (from 6 mass% to 15 mass%), a probability of heavy aromatic contact with the molecules of normal alkanes increases by 2.3 times. At the same time, the initial rate of isomerization increases only by 1.2 times. Therefore, the effect of the increase in the rate of isomerization comes to saturation at a moderate concentration of the heavy aromatic additive. It should be noted that, in these experiments, a bitumen additive contained a considerable amount of asphaltenes. It is known that paraffin constituents in the gasoline fraction further asphaltene coagulation. Therefore, a relatively small increase in the isomerization rate at the high bitumen concentration can be caused, at least partially, by formation of asphaltene aggregates and colloid structures that deteriorated conditions of the contact between molecules of alkanes and heavy aromatic compounds.

A strong dependence of the isomerization rate on the concentration of heavy aromatic compounds was also observed in the experiments where radiation ozonation of heavy oil feedstock was used for the increased rate of RTC (Zaikin and Zaikina 2004d). A foregoing bubbling of the hydrocarbon feedstock with ozone-containing ionized air allowed considerable decrease in the cracking temperature. It was found that this effect reached a maximum at the approximately equal concentrations of paraffin hydrocarbons and heavy aromatic compounds. However, as distinct from the earlier-described experiments on gasoline isomerization, asphaltenes were absent in the feedstock used in the experiments by Zaikin and Zaikina (2004a).

In the same study, it was shown that increase in the cracking rate and decrease in the reaction temperature led to a more intense alkane isomerization. It is an additional indication to the interconnection of the cracking reaction propagation and isomerization process.

Increase in the iso-alkane concentration was accompanied by a noticeable decrease in the olefin concentration in the gasoline fraction. This observation allows the suggestion that olefins also participate in the isomerization process; probably, olefin molecules can stay in the excited unstable state for a relatively long period of time.

The maximal concentration of isomers in the gasoline fraction obtained in the optimal conditions of RTC of heavy oils and bitumen was in the range of 33%–47% while iso-alkane concentration in the paraffin part of gasoline reached 60%–75%.

Thus, the model connecting the radiation generation of unstable molecular states responsible for the propagation of the chain cracking reactions with the phenomenon of radiation-induced isomerization in liquid hydrocarbons finds support in many experimental observations.

1.7 RADIATION CRACKING OF EXTRA-HEAVY OIL AND BITUMEN

Radiation cracking is characteristic of the high yields of light fractions and their fast increase with the dose rate, P, of ionizing irradiation (in the case of cold cracking, the reaction rate is approximately proportional to $P^{3/2}$). Increase in the dose rate is accompanied by a decrease in the irradiation dose corresponding to maximal yield of light products. The practical realization of this process has shown that such behavior is characteristic of the most part of oil feedstock. However, formation of highly radiation-resistant low-temperature colloid structures in extra-heavy oils and bitumen upsets the mechanisms described earlier because of the considerable degree of oil polymerization and adsorption of light fraction by a polymerized heavy residue.

In particular, even at relatively high temperatures, radiation-resistant structures can display themselves in a "delayed" radiation cracking when considerable yields of light products are observed only after absorption of a certain dose of radiation required for the partial destruction of low-temperature structures in thermal or radiation processes (see Chapter 5).

Characteristic features of low-temperature irradiation of strongly polymerized heavy oil are appearance of maximums in the dose dependence of light product yields and more complicated dependence of product yields on dose rate. The specific behavior of extra-heavy oil feedstock under irradiation cannot be described in the frames of theoretical concepts developed for lighter hydrocarbon compositions and processes proceeding at higher temperatures.

1.7.1 Electron Irradiation of Extra-Heavy Oil in Static Conditions

The anomalous behavior of viscosity in heavy hydrocarbon feedstock was very pronounced in a series of our experiments on low-temperature bitumen irradiation in static conditions (Zaikin and Zaikina 2008a, Zaikin 2013a). The experiments have shown that in the presence of the developed low-temperature structures formed by asphaltene aggregates, the surface effects associated with such structures are an additional limiting factor for oil radiation conversion.

Figures 1.28 and 1.29 demonstrate some of these limitations on the example of bitumen samples irradiated by 3 MeV electrons at a temperature of 70°C. A pronounced dependence of bitumen viscosity on the thickness of the irradiated layer was observed for all the dose rates and all the doses of electron irradiation with the same viscosity minimum at 1.5–2.0 mm. For a layer thickness greater than 3.5 mm, no reduction in bitumen viscosity was observed for all irradiation conditions used in the experiment. In this case, irradiation provoked only oil polymerization and increase in

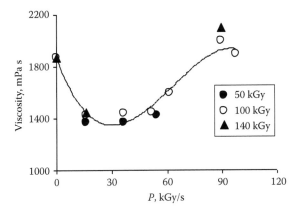

FIGURE 1.28 Dependence of viscosity on dose rate for bitumen irradiated with electrons in the 2 mm layer. (From Zaikin, Y.A., Radiation-induced cracking of hydrocarbons, *Radiation Synthesis of Materials and Compounds*, Kharissov, B.I., Kharissova, O.V., and Mendez, U.O., eds., Taylor & Francis, Boca Raton, FL, pp. 355–379, 2013a.)

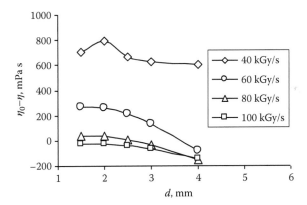

FIGURE 1.29 Dependence of viscosity on the thickness of the bitumen layers irradiated by electrons with the dose of 50 kGy at different dose rates. (From Zaikin, Y.A., Radiation-induced cracking of hydrocarbons, *Radiation Synthesis of Materials and Compounds*, Kharissov, B.I., Kharissova, O.V., and Mendez, U.O., eds., Taylor & Francis, Boca Raton, FL, pp. 355–379, 2013a.)

its viscosity. Even in thin layers (≤ 2 mm), viscosity reduction was observed only in favorable irradiation conditions defined by irradiation dose and dose rate.

Availability of the critical thin layer favorable for radiation processing can be associated with the surface tension caused by the interactions in the system of low-temperature aggregates on bitumen surface. The breaks of the bonds between aggregates lying on the surface create the unbalanced shear that probably prevents dissipation of radiation energy through the system of aggregates.

The depth of the unbalanced shear stress spreading can be roughly estimated from the equation

$$\frac{\sigma}{d} = G\frac{d}{h}$$ (1.100)

where
σ is the surface tension
G is the shear modulus
d is a characteristic size of a low-temperature aggregate
h is the depth of shear stress spreading

The typical value of the surface tension for bitumen at 100°C is about 24 mN m⁻¹ (Drelich et al. 1994) and a characteristic value of the shear modulus is about 0.6 GPa (Batzle and Zhijing 1992). By substituting these values into Equation 1.100, it can be calculated that the depth of shear spreading will be 1–2 mm for a characteristic aggregate size of 2–3 nm. This estimate is in agreement with the experimental data.

In the case of irradiation of thick samples, the shear energy in the surface layer is transferred to the bulk of the sample; the greater the sample thickness, the higher the rate of energy dissipation. The decrease in the reaction rate with the thickness of a sample is accompanied by a decrease in the difference of the feedstock and product viscosities. Dependence of the cracking rate on the thickness of the irradiated layer can be additionally caused by more complete oxidation of thin layers. Radiation-oxidative scission leads to a lower level of polymerization (Ivanov 1988) and increase in the cracking rate.

Figure 1.28 shows that the effect of the dose rate on oil viscosity is much stronger than the effect of the irradiation dose. The points in the graph relate to different doses, but all of them satisfactorily fit the same curve. The viscosity values at the zero dose rate were conventionally taken equal to the viscosity of the original feedstock.

Dependence of the difference in the feedstock and product viscosities on the thickness of the irradiated bitumen layer for the dose rates of electron irradiation in the range of 40 kGy/s–100 kGy/s is shown in Figure 1.29 for the same dose of 50 kGy. The greatest decrease in viscosity was observed after irradiation at the dose rate of 40 kGy/s. Irradiation at the dose rate of 100 kGy/s provokes polymerization for any thickness of the irradiated layer. The higher the layer thickness is, the greater is the increase in bitumen viscosity.

Figures 1.28 and 1.29 show that the rate of bitumen cracking in these processing conditions does not obey the law $W \sim (P)^{3/2}$. They indicate the maximum rate of radiation-induced bitumen decomposition at the dose rate of about 40 kGy/s.

The observed dose rate dependence can be explained in frames of the model of heavy oil radiation cracking developed in the paper (Zaikin and Zaikina 2013).

1.7.2 POLYMERIZATION AS A LIMITING FACTOR FOR LIGHT PRODUCT YIELDS IN RADIATION CRACKING OF HEAVY OIL AND BITUMEN

The kinetic model of RTC and LTRC subject to the specificity of radiation-induced processes and the effect of structure in extremely heavy hydrocarbon feedstock was developed in the paper (Zaikin and Zaikina 2013). In this model, radiation-induced polymerization and chemical adsorption were considered as important factors limiting heavy oil conversion.

For the heaviest types of oil feedstock characterized by the availability of dense radiation-resistant low-temperature colloid structures and high degree of oil polymerization, dose and dose rate dependences of the cracking rate become more complicated, and additional factors limiting oil conversion should be taken into account. To describe the specific features of reactions in the processes of radiation-thermal and cold radiation cracking in anomalously heavy oils and bitumen, it is necessary to take into account the following experimental facts (Zaikin 2008, Zaikin and Zaikina 2008a):

1. In the given conditions of radiation processing, kinetics of the process and yields of light fractions depend on the original structure of the feedstock (Yen 1998). The thixotropic properties of heavy oil and bitumen strongly affect the efficiency of heavy oil and bitumen radiation processing. Decomposition of the low-temperature oil structures under shear stress applied in the stream is accompanied by the drop in viscosity and sharp increase in the cracking rate, sometimes by more than an order of magnitude (Zaikin and Zaikina 2008).
2. A characteristic feature of heavy oil and bitumen cracking is availability of the well-pronounced maximums in the dose dependence of the light product yields or heavy residue conversion. Such maximums are not observed in the light oils where the dose dependence of the product yields comes to saturation at the high irradiation doses (Zaikin and Zaikina 2008a).
3. The maximal yield of the stable light products is limited by the critical irradiation dose when the rate of radiation cracking becomes comparable with the rates of the competing reactions, such as polymerization and adsorption of light fractions by the reactive residue. As a rule, increase in the dose rate allows greater yields of light fractions at lower dose values. The cracking products obtained as a result of heavy oil radiation processing at the doses above the critical value are unstable. Polymerization of the product and adsorption of the light fractions may continue after irradiation (Zaikin et al. 2004, Zaikin 2013a).
4. The repeated irradiation of the products of radiation-induced cracking of heavy oil and bitumen at the high dose rates does not provide considerable yields of light fractions and usually leads to additional polymerization of the product (Zaikin and Zaikina 2008a).
5. A long-term storage of the heavy oil feedstock, especially in the conditions of varying temperature and humidity, leads to the formation of the radiation-resistant structures. At the heightened dose rates, the cracking rate of such a feedstock considerably decreases; the effect of the shear stresses on oil viscosity in a flow becomes less significant. On the one hand, it affects the reproducibility of the data on radiation cracking obtained for the same time of feedstock. On the other hand, the fact of formation of radiation-resistant oil structures in the anomalously heavy oil feedstock necessitates special feedstock preparation for the effective radiation processing (Zaikina and Mamonova 1999).

6. As viscosity and the degree of oil polymerization increase, the dependence of the cracking rate on dose rate becomes weaker. Strong polymerization and formation of radiation-resistant asphaltene aggregates in bitumen may lead to appearance of the surface effects, for example, dependence of oil conversion on the thickness of an irradiated layer and additional peculiarities in the dependence of oil conversion on the dose rate, for example, appearance of maximums in this dependence.

1.7.3 KINETIC MODEL OF RADIATION-INDUCED CRACKING OF HEAVY OIL FEEDSTOCK

In contrast to the relatively light petroleum oils, the kinetics of radiation cracking of heavy oils should take into account the important fact that, in this case, the main cracking subject is not free hydrocarbon molecules but rather aliphatic groups and fragments chemically bound with the anthracene skeleton of the system. In this model, we shall suggest that the original feedstock consists only of such aliphatic groups attached to the radiation-resistant structure formed by the asphaltene aggregates. For simplicity, we shall assume that free hydrocarbon molecules are absent and that any aliphatic group detached from the anthracene skeleton can be considered as a cracking product. Correspondingly, we shall take into account only those cracking reactions that proceed at the contact of the asphaltene and aliphatic components of the feedstock.

The appearance of maximums in the dose dependence of the product yield is explained in this model by the fact that radiation-induced cracking is accompanied by polymerization of all the aliphatic part of the system and adsorption of the light fractions of the cracking product by the heavy residue. In this model, the long-living radicals formed as a result of the detachment of aliphatic groups from the anthracene framework are considered as the centers of polymerization and adsorption.

The earlier-listed specific features of chain cracking reactions in the extremely heavy oils can be described by the following system of equations:

$$\frac{d(C*-C)}{dt} = -AR(C*-C)\left(1-C_0^*\right) - B(C*-C)R' \tag{1.101}$$

$$\frac{dC*}{dt} = -B(C*-C)R' - DCR' \tag{1.102}$$

$$\frac{dR'}{dt} = AR(C*-C)\left(1-C_0^*\right) - BC*R' \tag{1.103}$$

where
  C is the time-dependent concentration of the cracking product (detached aliphatic fragments and hydrocarbon molecules)
  $C*$ is the time-dependent concentration of the nonpolymerized aliphatic component of the feedstock chemically bound with the anthracene skeleton that can be potentially converted to the cracking product

t is the irradiation time

R is the concentration of the chain carriers (hydrogen atoms and short-living light alkyl radicals)

R' is the concentration of polymerization and adsorption centers, that is, radicals formed as a result of the aliphatic fragment detachment from the anthracene skeleton

A is the cracking rate constant

B is the polymerization rate of the aliphatic part of the feedstock

D is the rate of the light cracking product adsorption by the heavy residue

The initial conditions can be written in the form

$$C(0) = 0; \quad R'(0) = 0; \quad C^*(0) = C_0^* \qquad (1.104)$$

The equilibrium concentration of the chain carriers is defined by the equation

$$R = \left(\frac{G_r P}{K_r} \right)^{1/2}$$

where

G_r is the radiation-chemical yield of the chain carriers

K_r is the constant of the short-living radical recombination

P is the dose rate of ionizing irradiation

Equation 1.101 describes the cracking of the aliphatic part of the feedstock, that is, detachment of the aliphatic fragments from the heavy residue. The reactive centers of polymerization/adsorption formed as a result of cracking cause the decrease in the concentration of the aliphatic component C^* that can be potentially converted to the cracking product. This reduction in the potential product yield is described by Equation 1.102. It is assumed that the polymerized (or adsorbed) products are radiation-resistant and do not further participate in the cracking reactions. Equation 1.103 describes accumulation and disappearance of the long-living radicals R'.

The system of Equations 1.101 through 1.103 can be solved numerically. However, it is useful to consider its approximate solution as a clear demonstration of the anomalously heavy oil behavior under ionizing irradiation.

For the simplicity of the analysis, we shall consider the rates of adsorption and polymerization equal to each other ($B=D$). The approximate solution of the system of Equations 1.101 through 1.103 obtained by the simple iteration method with the initial approach of $C^*=$const is of the form

$$C = C^* \left[1 - e^{-AR\left(1-C_0^*\right)t} \right] \qquad (1.105)$$

$$R' = C^* \left[1 - e^{-AR\left(1-C_0^*\right)t} \right] \qquad (1.106)$$

$$C^* = \frac{C_0^*}{1 + BC_0^* t - \dfrac{BC_0^*}{AR\left(1-C_0^*\right)\left[1-e^{-AR\left(1-C_0^*\right)t}\right]}}$$ (1.107)

Expressions (1.105) and (1.107) describe the yields of the cracking products during irradiation depending on the main parameters of radiation processing (dose, dose rate, and temperature).

In the case of RTC (350°C–420°C), radiation generation of the excited molecules is negligible compared with their thermal activation. In this case, the cracking rate constant can be written in the form

$$A = A_0 e^{-E/kT}$$ (1.108)

where $A_0 \approx 4 \cdot 10^{-10}$ cm³/molecules · s ($\approx 10^{12}$ s⁻¹); $E \approx 80$ kJ/mol.

In the following calculations, the following values of radiation-chemical yields of radical chain carriers and constants of their recombination were used: $G = 5$ radicals per 100 eV; $K_r \approx 3 \cdot 10^{-13}$ cm³/molecules · s ($\approx 10^9$ s⁻¹).

At relatively low temperatures (below 350°C), thermal activation of the hydrocarbon molecules becomes negligibly small. In this case, sufficiently high concentration of excited molecules C_{exc} necessary for propagation of the chain reaction can be generated by ionizing irradiation at sufficiently high dose rates. The rate of such "cold" radiation cracking can be written in the form

$$A = A_0 R C_{exc}$$ (1.109)

where C_{exc} is proportional to P.

1.7.4 RADIATION-THERMAL CRACKING

Comparison of the theoretical calculations of RTC kinetics with the experimental data for heavy crude oil (Zaikin and Zaikina 2013) is given in Figure 1.30. The calculations are in good agreement with the experiment at the following set of parameters: $C_0^* = 0.7$; the cracking rate, $ARC_0^*(1 - C_0^*)$, is 1.3 s⁻¹ at 400°C and 2.1 s⁻¹ at 420°C; absorption coefficient B is 4.45 s⁻¹ at 400°C and 1.79 s⁻¹ at 420°C.

Activation energy of 92.0 kJ/mol determined from the comparison of RTC rates at 400°C and 420°C is a characteristic value for chain propagation in hydrocarbons. Decrease in the adsorption coefficient with the increase in the process temperature is associated with decomposition of the low-temperature colloid structures.

Similar kinetic calculations were made for the case of RTC of bitumen (Zaikin and Zaikina 2013) (Figure 1.31). A specific feature of the dose dependence of the light fraction yields in the experiments on RTC of bitumen is the cracking "delay" that corresponds to the absorbed irradiation dose of about 1 kGy. We attribute this

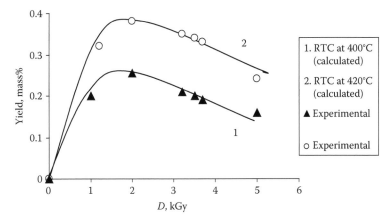

FIGURE 1.30 Yields of light fractions boiling out below 450°C after RTC of heavy crude oil versus the dose of electron irradiation (2 MeV, 3 kGy/s). (From Zaikin, Y.A. and Zaikina, R.F., *Radiat. Phys. Chem.*, 84, 2, 2013.)

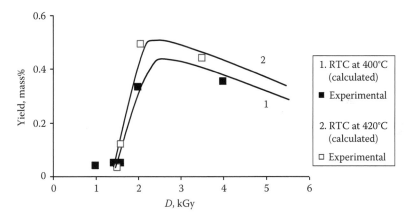

FIGURE 1.31 Yields of light fractions boiling out below 450°C after RTC of bitumen versus the dose of electron irradiation (2 MeV, 4 kGy/s). (From Zaikin, Y.A. and Zaikina, R.F., *Radiat. Phys. Chem.*, 84, 2, 2013.)

peculiarity to radiation-induced structure transformation preceding cracking. The following set of parameters providing the best fit to the experimental data was used in the calculations: $C_0^* = 0.7$; the cracking rate $AR (1 - C_0^*)$ is 3.3 s^{-1} at 400°C and 5.0 s^{-1} at 420°C; adsorption coefficient B is 1.73 s^{-1} at 400°C and 1.21 s^{-1} at 420°C.

Activation energy for propagation of the cracking reaction determined from the temperature dependence of the initial cracking rate is 69.7 kJ/mol. A lower value of the activation energy in the case of RTC of heavier bitumen compared with the lighter crude oil can be explained by the fact that detachment of aliphatic chains from the aromatic framework has a higher specific weight for bitumen. On the other hand, it requires smaller energy than that necessary for the decomposition of paraffin molecules.

1.7.5 Low-Temperature Radiation Cracking

In Figure 1.32, theoretical calculations of the kinetics of LTRC are compared with experimental data on electron irradiation of high-viscous oil in static conditions. Dose dependence of the light fraction yields relates to different values of temperature and dose rate of electron irradiation. They were calculated using a self-consistent set of parameters. The absorption coefficient B was taken as 0.031 s^{-1}, and the effective concentration of the aliphatic part of the feedstock C_0^* was considered to be 0.5. The calculated values of the initial cracking rate for different temperatures and dose rates of electron irradiation are given in Table 1.2.

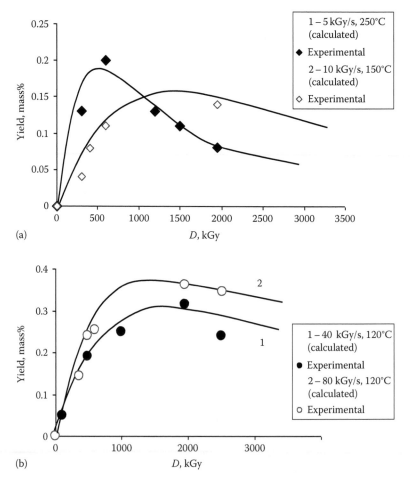

FIGURE 1.32 Yields of light fractions boiling out below 450°C after LTRC of heavy crude oil versus the dose of electron irradiation (3 MeV) at the dose rates of (a) 1–10 kGy/s and (b) 40–80 kGy/s. (From Zaikin, Y.A. and Zaikina, R.F., *Radiat. Phys. Chem.*, 84, 2, 2013.)

TABLE 1.2
Initial Rates of Low-Temperature
Radiation Cracking for
High-Viscous Oil

Dose Rate, kGy/s	Temperature, °C	Initial Cracking Rate, s⁻¹
5	250	0.004
10	150	0.0025
40	120	0.002
80	120	0.113

Source: Zaikin, Y.A. and Zaikina, R.F., *Radiat. Phys. Chem.*, 84, 2, 2013.

In the calculations, it was assumed that the initial cracking rate can be represented as the sum $W_r + W_{rt}$ of two terms: radiation component $W_r \sim P^{3/2}$ and radiation-thermal component $W_{rt} \sim P^{1/2} e^{-E/kT}$ (where P is the dose rate). Activation energy determined from the dependence of the initial cracking rate on dose rate and temperature was made 69.3 kJ/mol; this value is close to the activation energy for RTC of bitumen.

Estimate of the absorption coefficient B leads to much lower values compared with those calculated for the case of RTC. However, polymerization and absorption of light fractions by the heavy residue strongly affect the cracking kinetics at the low values of cracking rate.

At relatively low process temperatures, heavy oil and bitumen demonstrate pronounced thixotropic properties. In flow conditions, the low-temperature oil structures partially fail in shear that is accompanied by the viscosity reduction, considerable increase in the cracking rate, and decrease in the absorption coefficient.

The specific radiation behavior of extra-heavy oils can be explained in the frames of the earlier-described kinetic model. In this model, polymerization and adsorption of light products by the heavy residue increase with a higher rate at higher dose rates of radiation.

The qualitative calculations with conventionally selected parameters (initial cracking rate, adsorption coefficient, and initial concentration of the aliphatic component) indicate increase in the product yield at a dose of 40 kGy, availability of a maximum at a dose of 80 kGy, and decrease in the product yield with the dose rate at a dose of 100 kGy and over (Figure 1.33).

Figure 1.33b qualitatively corresponds to the conditions of the earlier-described experiment in bitumen irradiation in static conditions: the maximum in the dependence of the light cracking product yield on dose rate corresponds to the minimal viscosity of the product.

The model considered earlier provides satisfactory description of the characteristic features of the kinetics of RTC and LTRC of heavy oil and bitumen and can be used for the analysis of the yields and stability of the products of radiation cracking. The critical parameters of the model—initial rate of radiation-induced cracking and

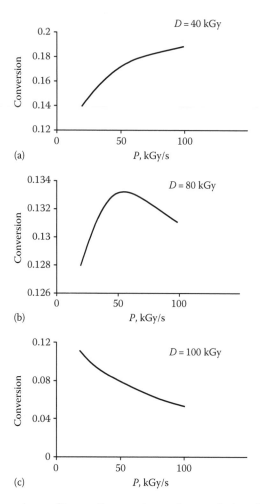

FIGURE 1.33 Dependence of heavy oil conversion on dose rate in the cold cracking process at different irradiation doses: (a) 40 kGy, (b) 80 kGy, and (c) 100 kGy.

absorption coefficient—strongly depend on the structural state of the feedstock. To a great extent, the control and purposeful transformation of heavy oil structure predetermine the efficiency of heavy oil radiation processing.

At a high initial degree of feedstock polymerization, radiation cracking becomes less effective. Polymerization impedes detachment of the aliphatic components from the anthracene skeleton that is accompanied by adsorption of light fraction in the polymerized residue. In addition, a considerable part of chain carriers (hydrogen atoms and light alkyl radicals) is formed and recombined in conditions of their high concentrations in spurs that leads to the suppression of chain cracking reactions at the high dose rates.

One of the important factors that affect the efficiency of radiation cracking is a hydrocarbon content of the original feedstock. The most favorable for radiation processing is a composition that provides the maximal rate of the cracking reaction and, at the same time, prevents formation of asphaltene aggregates. It is known that aromatic hydrocarbons and pitches assist asphaltene peptization while paraffinic–naphtenic hydrocarbons contribute to their coagulation.

The paraffin hydrocarbons are mostly subjected to radiation cracking. They are the main source of chain carriers in the chain reactions of hydrocarbon decomposition while aromatic compounds are remarkable for their high radiation resistance. However, the subject of radiation cracking in heavy oil and bitumen is not so much free alkane molecules as rather alkyl substituents having weaker bonds with the polynuclear aromatic system. A degree of the system polymerization and adsorption of light fractions by the heavy residue determines the product stability and the kinetic characteristics of radiation cracking.

In this connection, application of flow conditions characteristic of high shear stresses together with other special methods for feedstock pretreatment (Zaikin and Zaikina 2008a, Zaikin 2013a) is necessary for the destruction of radiation-resistant low-temperature structures and providing favorable conditions for the efficient radiation processing of extra-heavy oils and bitumen.

Among these methods, such techniques are used as oil mechanical agitation, acoustic treatment, feedstock bubbling with special agents, gas injection to the reaction zone, etc. In many cases, application of flow conditions characteristic of high shear rates is capable of increase in the cracking rate by more than an order of magnitude (Chapter 5).

1.7.6 Effect of Flow Conditions on the Efficiency of High-Viscous Oil Radiation Processing

A considerable effect of flow conditions on the efficiency of low-temperature cracking of high-viscous oils is caused by their thixotropic properties. High-viscous oils and bitumen pertain to the disperse systems whose structure and viscosity depend on the applied shear stress.

This effect is of a special importance in view of the fact that low-temperature coagulation structures in heavy oils are highly radiation-resistant that makes their radiation processing difficult. In high-viscous oils and bitumen, such structures are usually caused by asphaltene aggregation. Knowledge of the rheology of such systems is necessary for the understanding of the changes in oil structure in flow conditions of its radiation processing.

The phenomenon of thixotropy leads to the appearance of hysteresis loops in the dependence of shear stress on shear rate (Mezger 2006). These dependences are schematically shown in Figure 1.34. The area between the upward and the downward curves in Figure 1.34 is known as "hysteresis area" and can be considered as a measure of the thixotropic effect.

The main characteristics of the rheological behavior of heavy oils and bitumen were described in a theoretical work by Mikhailov and Lichtheim (1955).

The flow process of structured systems and associated structure destruction can be represented in the following scheme (Figure 1.35).

In the work (Mikhailov and Lichtheim 1955), it was supposed that each microparticle in oil bitumen forms a node of a volume network where it oscillates similar

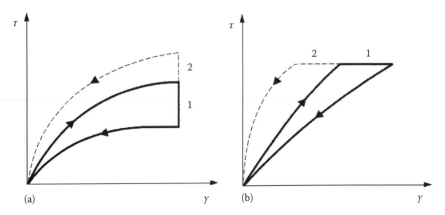

FIGURE 1.34 Dependence of shear stress on shear rate showing a hysteresis area. (a) Using controlled shear rate and (b) using controlled shear stress. (1) With decreasing, and (2) with increasing structural strength.

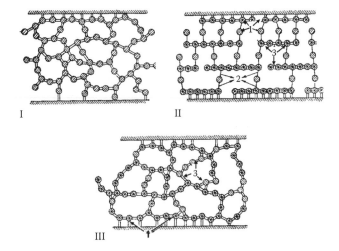

FIGURE 1.35 Schematic flow process of structured liquids: I—before flow, II—in flow process, III—after flow discontinuance. 1—colloidal particles, 2—cross-links, 3—longitudinal links.

to a molecule. A node of the volume network is the equilibrium position of a micro-particle considered as a part of this system. In the absence of an external force, such coagulation structure has equal strength in all directions (Figure 1.35, I).

Application of an external shearing force, first of all, leads to the breaks of weaker bonds in the volume network. The velocity gradient that appears in the flow process causes ordering of the volume network. As a result, the network realigns to more strong longitudinal chains (in the direction of the applied force), which are connected with each other by cross-links formed by separate particles or transverse chains. A scheme of a structure ordered in the flow process is shown in Figure 1.35, II.

At low velocity gradients, almost all the links broken in the flow process restore because the effect of Brownian motion prevails over the effects of velocity gradient. As a result, the liquid flows as a Newtonian one with the initial viscosity value, η_0. Starting from a certain value, $P = P_r'$, the velocity gradient considerably affects the flow process: as it increases, a certain amount of the particles involved in the cross-links pass on to the longitudinal chains. In this transition, a particle surmounts the repulsive forces, and the layers between longitudinal chains get thinner. The longitudinal chains are bound with the attractive forces weakened by the intermediate layer resistance.

Each velocity gradient in the steady flow corresponds to a certain amount of the cross-links decreasing with the velocity gradient increase. All links corresponding to the given gradient break down in the flow process and completely restore themselves. As velocity gradient increases, contraction of the transversal chains becomes more and more considerable. Therefore, a scheme in Figure 1.35, II shows a greater amount of particles in the longitudinal chains and a shorter distance between the chains in the domains with a smaller amount of particle involved in the cross-links.

As velocity gradient increases, the number of transversal links decreases because of the transfer of the particles involved in these links to the longitudinal chains. It is accompanied by the strengthening of the longitudinal chains and, therefore, decreases probability of such transfer. As soon as a certain stress value $P = P_m'$ is reached, the number of cross-links does not decrease anymore because the effect of Brownian motion on cross chain deviation comes down to a minimum and the chain strength becomes so high that no particles from the cross-links can be squeezed into longitudinal chains. Besides, the number of such particles becomes very small and a probability of their transfer to the longitudinal chains abruptly falls down. From this moment, the liquid flows as a Newtonian one but having a viscosity of η_m corresponding to a structure destroyed down to the limit.

As the stress is removed, the layers between the longitudinal chains expand and the cross chains become longer and curved. Oscillations of the particles in such unloaded curved chains lead to the chain winding and their motion in the disperse medium. In the process of such semi-constrained chaotic motion of the particles in the cross chains, separate sections of the neighboring chains approach and interact with each other surmounting the repulsive forces caused by solvate shells at the interaction sites. As a result, they form an irregular-shaped node of the volume network. As the number of such contacts increases, a free length of the chains gets shorter that constrains the Brownian motion of separate particles in the chain. The particles from the cross chains also participate in structure restoration accelerating this process. Kinetics of the contact formation dies down and the system restores in a thixotropic manner.

The process of thixotropic restoration of a destroyed structure is illustrated in Figure 1.35, III. The time needed for restoration of the original volume network depends on viscosity of the disperse medium; the higher the viscosity, the longer is the time required for structure restoration.

The full rheological curve of a structured liquid (dependence of shear rate on shear stress) has a characteristic S-like shape (Figure 1.36). The area limited by the rheological curve $\dot{\varepsilon}(\tau)$ and axis $\dot{\varepsilon}$ is numerically equal to the power needed for maintaining a stationary flow.

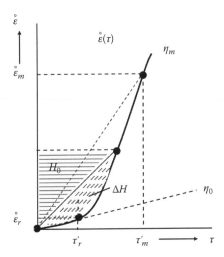

FIGURE 1.36 Diagram of powers per unit of volume required for steady flow maintenance.

This power consists of two parts: power H_0 required for the maintenance of Newtonian flow with a viscosity equal to the instant viscosity value and power ΔH necessary for structure destruction at any time of the process.

Viscosity dependence on the shear rate has the form

$$\eta(\dot{\varepsilon}) = \eta(\tau) = \eta_m + \Delta\eta_s \qquad (1.110)$$

where η_m is the viscosity of a structure destroyed down to the limit. It represents the lowest degree of the volume network development when it restores in a flow even at high shear rates. It implies that the quantity η_m is larger than a viscosity value determined by Einstein's formula (and only in the extreme case of the complete structure destruction equal to it):

$$\eta_m \geq \bar{\eta}(1 + \alpha\varphi) \qquad (1.111)$$

where
  $\bar{\eta}$ is the viscosity of the dispersion medium
  α is the particle shape coefficient
  φ is the volume fraction coefficient

In Equation 1.110, $\Delta\eta_s$ is the increment of "structural viscosity," that is, a viscosity fraction that restores in a thixotropic manner in the flow process and has a certain value for each velocity gradient or shear stress. This increment also depends on the distance between particles of the disperse medium and characterized a degree of their interaction.

Based on the molecular-kinetic concepts developed by Frenkel (1945, 1972), surmounting of the energy barrier was considered in the study (Mikhailov and Lichtheim 1955) as a break of the bonds between replacing particles (Figure 1.37).

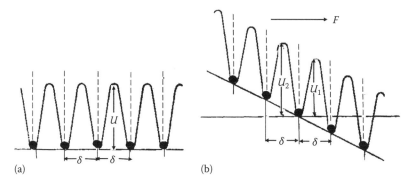

(a)   (b)

FIGURE 1.37 Schematic potential relief of a particle in the (a) absence and (b) presence of external force F.

In this model, dependence of viscosity on shear stress is determined by the equation

$$\eta = \eta_m + (\eta_0 - \eta_m) \cdot \frac{x}{sh\ x} \tag{1.112}$$

where
 x is the dimensionless variable $x = \tau/\tau_0$
 η_0 is the viscosity of the practically undestroyed oil structure ($\eta_0 = \eta|_{\tau=0}$)
 η_m is the viscosity of the completely destroyed structure ($\eta_m = \eta|_{\tau=\infty}$)
 τ_0 is the effective shear static stress that characterizes the strength of the undestroyed structure

Dependence $\eta(\tau)$ is shown in Figure 1.38.
 It implies the following equation for the shear rate:

$$\dot{\varepsilon} = \frac{\tau}{\eta_m + (\eta_0 - \eta_m)(x/sh\ x)} \tag{1.113}$$

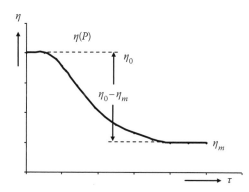

FIGURE 1.38 Schematic dependence $\eta(P)$ according to Equation 1.112.

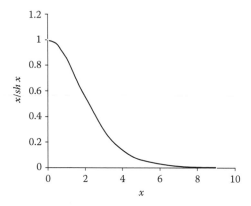

FIGURE 1.39   Dependence $f(x) = x/sh\ x$.

It can be found from Equation 1.112 that

$$\eta(\tau_0) = 0.85\eta_{max} + 0.15\eta_{min} \approx 0.85\eta_{max} \qquad (1.114)$$

Equation 1.114 allows approximate evaluation of the characteristic shear stress τ_0 as that corresponding to the viscosity drop by 15%.

Dependence of $(\eta - \eta_m)/(\eta_0 - \eta_m)$ as a function of the dimensionless variable x is shown in Figure 1.39.

Rheological properties of heavy oils and bitumen and their thixotropic behavior are described in many papers (Rebinder 1978, 1979, Uryev 1980, Pierre et al. 2004, Malkin and Isayev 2007, Rojas et al. 2008, Bazyleva et al. 2010).

Thixotropic behavior of bitumen is well pronounced in the dependences of its dynamic viscosity and shear stress on shear rate (Figure 1.40).

Viscosity changes under the action of external shear stresses are reversible. However, it is important that, in the appropriate flow conditions, they are accompanied by the destruction of the structures unfavorable for efficient radiation processing of heavy oil feedstock. Therefore, heavy oil processing in a flow characteristic of high shear stresses considerably increases the rate of radiation cracking.

Figure 1.41 shows conversion of a heavy oil residue boiling above 350°C in the time of radiation processing in static conditions at a temperature of 70°C. Although low-temperature processing was conducted at a high dose rate of electron irradiation (80 kGy/s), the conversion of 14% was reached only after 6 s of irradiation with a very high dose of 480 kGy.

The results of the same oil processing in flow conditions are shown in Figure 1.42.

Although, in this case, the dose rate was eight times lower, conversion of the heavy residue of 55% was reached only after 2.6 s of irradiation. Thus, application of the flow conditions provided an increase in the process rate by about 10 times and, correspondingly, 10-fold energy savings in oil radiation processing.

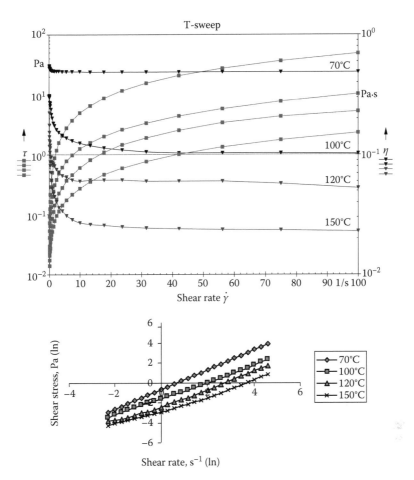

FIGURE 1.40 Dependence of bitumen viscosity, η, and shear stress, τ, on shear rate in normal and double logarithmic scales.

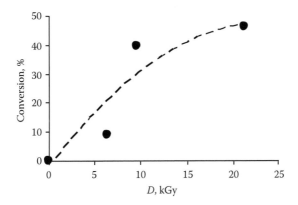

FIGURE 1.41 Conversion of heavy oil residue after irradiation in the continuous mode at the electron dose rate of 80 kGy/s.

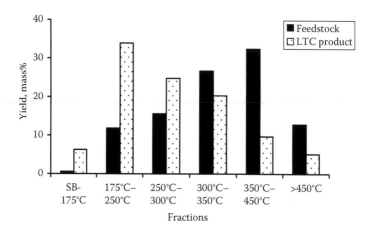

FIGURE 1.42 Fractional contents of heavy crude oil and the product of its LTRC. Electron dose—26 kGy, dose rate—10 kGy/s, process temperature –220°C.

1.8 EFFECT OF ELECTRON BEAM PULSE CHARACTERISTICS ON THE RATE OF RADIATION-THERMAL CRACKING OF PETROLEUM FEEDSTOCK

The rate of radiation cracking in liquid hydrocarbon systems was estimated in the paper (Zaikin et al. 2007b) on the basis of the analytic solutions of the differential equations describing kinetics of accumulation of radicals and excited molecules with an arbitrary order of reaction.

RTC of hydrocarbons by radical mechanism is a self-sustaining chain reaction controlled by radiation generation of chain carriers, that is, light hydrocarbon radicals. In this type of cracking, chain carriers are generated by ionizing radiation; the heightened temperatures of 350°C–450°C are necessary for chain propagation (Section 1.2).

The rate of RTC is proportional to the concentration of light hydrocarbon radicals. The latter is an increasing function of the dose rate. However, dependence of the chain carrier concentration on dose rate is different for different pulse characteristics of the electron irradiation. Therefore, one of the important problems of the electron-beam processing of hydrocarbon feedstock is a proper account of the pulse irradiation mode.

Due to the high diffusion mobility of light radicals, especially at the heightened temperatures of RTC, suppression of radical reactions associated with spur overlapping as the high dose rate in a pulse increases is not as important in liquid hydrocarbons as it is in such systems as solid polymers.

Subject to the high rate of radical recombination in liquid hydrocarbons (about 10^9 s^{-1}), in this case, it is more important to take into account "algebraic" effects that determine radical accumulation and recombination during an electron pulse and in the interim between the two pulses.

This problem is still more important for PetroBeam technology (Zaikin and Zaikina 2012), a method that allows high-rate hydrocarbon processing at lowered temperatures down to room temperature. The PetroBeam process is based on radiation generation of the long-living excited molecular states. In this process, both initiation and propagation of the chain cracking reaction are provided by the action of ionizing irradiation.

In this case, the reaction rate is proportional to the production of concentrations of chain carriers and radiation-excited molecules. Therefore, the process is characterized by a stronger sensitivity to the dose rate compared with RTC.

Estimation of the effect of electron beam pulse characteristics on the reaction rate of the low-temperature cracking requires analysis of accumulation of chain carriers and excited hydrocarbon molecules in different irradiation modes.

1.8.1 ACCUMULATION OF RADICALS IN CONTINUOUS AND PULSE MODES OF ELECTRON IRRADIATION

1.8.1.1 Radical Recombination in Second-Order Reactions

To compare results of radiation cracking of hydrocarbons in different modes of electron irradiation, we shall consider accumulation of free radicals in an arbitrary pulse mode that will be characterized by pulse width Δt, time between two pulses l (or frequency $f_0 = 1/l$), and dose rate P in a pulse. For comparison of radical accumulation in different modes, we shall adduce the results of calculations to the time-averaged dose rate

$$\langle P \rangle = P f_0 \Delta t = P \Delta t / l.$$

Dependence of radical concentrations on irradiation time is schematically shown in Figure 1.43. After a time, the quasi-stationary mode is settled. In this mode, radical concentration changes periodically with time from the minimum value R_∞ to the maximum value R_∞^* (Figure 1.44).

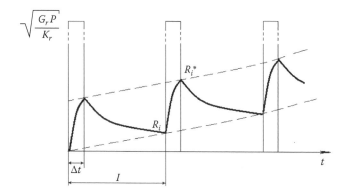

FIGURE 1.43 Schematic dependence of radical concentrations in hydrocarbons on irradiation time. R_i and R_i^* are radical concentrations at the beginning and at the end of the i-pulse; G_r is the G-value for radical generation; K_r is radical recombination rate. (From Zaikin, Y.A. et al., *Radiat. Phys. Chem.*, 76, 1404, 2007a; Zaikin, Y.A. et al., Effect of electron beam characteristics on the rate of radiation-thermal cracking of petroleum feedstock, *Proceedings of the 8th International Topical Meeting on Nuclear Applications and Utilization of Accelerators*, Pocatello, ID, American Nuclear Society, pp. 701–707, 2007b.)

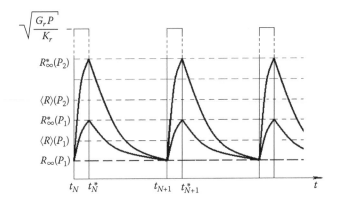

FIGURE 1.44 Dependence of radical concentration on irradiation time in the quasi-stationary mode of electron irradiation. $P_2 > P_1$. (From Zaikin, Y.A. et al., *Radiat. Phys. Chem.*, 76, 1404, 2007a; Zaikin, Y.A. et al., Effect of electron beam characteristics on the rate of radiation-thermal cracking of petroleum feedstock, *Proceedings of the 8th International Topical Meeting on Nuclear Applications and Utilization of Accelerators*, Pocatello, ID, American Nuclear Society, pp. 701–707, 2007b.)

In the paper (Zaikin et al. 2007b), accumulation of radiation-generated radicals and excited molecules was studied by the methods of perturbation theory (Mathews and Walker 1964). Complicated and cumbersome intermediate computations are omitted here.

Assuming that radical recombination is defined by the second-order kinetics, we shall write the following equation that describes radical accumulation in a pulse mode:

$$\frac{dR}{dt} = \begin{cases} G_r P - K_r R^2, & t_N \leq t < t_r^* \\ -K_r R^2, & t_N^* \leq t < t_{N+1} \end{cases} \tag{1.115}$$

with the initial condition

$$R(0) = 0 \tag{1.116}$$

Solution of this equation can be written in the form

$$R(\tau) = \begin{cases} a\,\dfrac{1 - C_N \exp(-2a\tau)}{1 + C_N \exp(-2a\tau)}, & N\Lambda \leq \tau < N\Delta\tau \\[2mm] \dfrac{1}{N\Lambda + \Delta\tau - C_N^*}, & \Delta\tau + N\Lambda \leq \tau < (N+1)\Lambda \end{cases}$$

where

$$C_N = \frac{a - R_N}{a + R_N}\exp(2aN\Lambda), \quad C_N^* = \frac{R_N^*(N\Lambda + \Delta\tau) - 1}{R_N^*}$$

$$a = (G_r P/K_r), \ \tau = K_r t$$

$$\Delta\tau = K_r\Delta t, \quad \Lambda = lK_r, \quad N = \frac{t}{l} \tag{1.117}$$

The approximate expressions for R_N and R_N^* could be written as expansions by the small parameter ε:

$$\varepsilon = \frac{1}{fg} \tag{1.118}$$

where

$$f = th(a\Delta\tau) \tag{1.119}$$

$$g = \sqrt{\frac{GP}{K_r}}(\Lambda - \Delta\tau) \tag{1.120}$$

Accurate to ε^2, the quantities R_N, R_N^*, and R_∞ can be given by the equations

$$R_N = ar_N \approx \frac{a}{fg^2}\frac{f^2g^2N+1}{fgN+1} \tag{1.121}$$

$$R_N^* = ar_N^* \approx \frac{a\left[\left(f^2g^2+1\right)N+1\right]}{f\left[fg^2\left(f+g\right)N+1+g^2\right]} \tag{1.122}$$

$$R_\infty \approx a\frac{1+fg}{f+g} \tag{1.123}$$

Dependence of the time-averaged radical concentration on irradiation time can be calculated using the equation

$$\langle R_N \rangle = \frac{1}{t}\int_0^t R(t)\,dt = \Sigma_1 + \Sigma_2 \tag{1.124}$$

$$\Sigma_1 = \frac{1}{K_r l}\left[\ln(chaK_r\Delta t) + \frac{th(aK_r\Delta t)}{N+1}\sum_{j=0}^{N} r_j\right] \tag{1.125}$$

$$\Sigma_2 = \frac{1}{K_r l(N+1)}\sum_{j=0}^{N}\ln\left[ar_j^*K_r(l-\Delta t)+1\right] \tag{1.126}$$

The results of calculations given in Figures 1.45 through 1.47 relate to the pulse characteristics of the linear electron accelerator ELU-4 (l=0.005 s, Δt=5 · 10^{-6} s, maximum dose per pulse 200 Gy).

Calculations have shown that the main contribution to radical accumulation is associated with the second term in Equation 1.120. Figure 1.45 shows that the

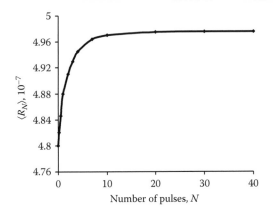

FIGURE 1.45   Dependence of $\langle R_N \rangle$ on $N=t/l$. $P=2 \cdot 10^6$ Gy/s, l=0.005 s, Δt=5 · 10^{-6} s. (From Zaikin, Y.A. et al., *Radiat. Phys. Chem.*, 76, 1404, 2007a; Zaikin, Y.A. et al., Effect of electron beam characteristics on the rate of radiation-thermal cracking of petroleum feedstock, *Proceedings of the 8th International Topical Meeting on Nuclear Applications and Utilization of Accelerators*, Pocatello, ID, American Nuclear Society, pp. 701–707, 2007b.)

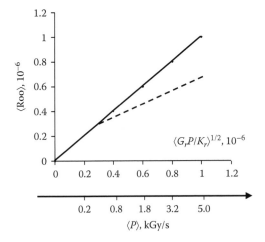

FIGURE 1.46   Dependence of the time-averaged radical concentration in quasi-stationary mode on the square root of the time-averaged dose rate for pulse radiation (5 μs, 200 s⁻¹) and continuous radiation at the same dose rate. $P=3 \cdot 10^5$–5 · 10^6 Gy/s. (From Zaikin, Y.A. et al., *Radiat. Phys. Chem.*, 76, 1404, 2007a; Zaikin, Y.A. et al., Effect of electron beam characteristics on the rate of radiation-thermal cracking of petroleum feedstock, *Proceedings of the 8th International Topical Meeting on Nuclear Applications and Utilization of Accelerators*, Pocatello, ID, American Nuclear Society, pp. 701–707, 2007b.)

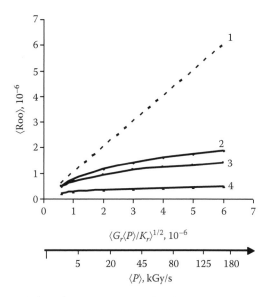

FIGURE 1.47 Dependence of the time-averaged radical concentration in quasi-stationary mode on the square root of the time-averaged dose rate for pulse radiation (3 μs, 300 s^{-1}— the upper curve; 5 μs, 200 s^{-1}—the middle curve; 3 μs, 60 s^{-1}—the lower curve) and for the continuous irradiation at the same dose rate. (From Zaikin, Y.A. et al., *Radiat. Phys. Chem.*, 76, 1404, 2007a; Zaikin, Y.A. et al., Effect of electron beam characteristics on the rate of radiation-thermal cracking of petroleum feedstock, *Proceedings of the 8th International Topical Meeting on Nuclear Applications and Utilization of Accelerators*, Pocatello, ID, American Nuclear Society, pp. 701–707, 2007b.)

quasi-stationary mode sets in very quickly during a time much less than that characteristic of cracking reaction. At higher dose rates, the quasi-stationary mode sets in still faster.

For the same electron beam pulse characteristics ($\Delta t = 5 \cdot 10^{-6}$ s, $l = 5 \cdot 10^{-3}$ s), the steady-state radical concentration increases as the dose rate in a pulse becomes higher. R_∞ has the values of $5.0 \cdot 10^{-7}$, $9.2 \cdot 10^{-7}$, and $1.4 \cdot 10^{-6}$ at the dose rates of $2 \cdot 10^6$, $2 \cdot 10^7$, and $2 \cdot 10^8$ Gy/s, respectively.

Figure 1.46 shows that in the case of continuous irradiation, the steady radical concentration is equal to $(G_r \langle P \rangle / K_r)^{1/2}$. In the case of pulse irradiation, dependence of the steady time-averaged radical concentration on $P^{1/2}$ is close to logarithmic.

The difference between these two quantities increases as the dose rate becomes higher. In the range of low dose rates, the dependence of R on $(G_r \langle P \rangle / K_r)^{1/2}$ can be approximated by a straight line with the slope smaller than unity. As the average dose rate decreases, the slope coefficient tends to unity, and dependences of R on $(G_r \langle P \rangle / K_r)^{1/2}$ become close for continuous and pulse irradiation.

The same dependence is shown in Figure 1.47 for the other pulse characteristics of an electron accelerator ($l = 1/300$ c, $\Delta t = 3 \cdot 10^{-6}$ c, maximum dose rate in a pulse—25 Gy). Calculations show that at the same averaged dose rate, RTC product yields differ insignificantly for the two sets of pulse characteristics considered above.

However, the first set of pulse parameters provides five times higher maximum dose rate that facilitates observation of the dose-rate-sensitive phenomena.

The lower curve in Figure 1.47 illustrates radical accumulation in the other mode used in our experiments ($l = 1/60$ s, $\Delta t = 3 \cdot 10^{-6}$ s). In this case, increase in the time between two pulses causes considerable decrease in the steady average radical concentration compared with the two pulse modes considered earlier.

The three pulse irradiation modes considered earlier are characterized by the following steady-state values of the radical concentrations: $5.2 \cdot 10^{-7}$ (3 μs, 300 s^{-1}), $5.0 \cdot 10^{-7}$ (5 μs, 200 s^{-1}), and $1.9 \cdot 10^{-7}$ (3 μs, 60 s^{-1}). It is obvious that the last mode (3 μs, 60 s^{-1}) is not favorable for the observation of the reactions sensitive to the dose rate, such as radiation cracking at lowered temperatures.

1.8.1.2   Radical Recombination in Arbitrary-Order Reactions

Description of the radical accumulation can be complicated by the effects of track and spur overlapping. Generally, it can be described by the introduction of an effective arbitrary order of reaction into the equation of radical kinetics:

$$\frac{dR}{dt} = \begin{cases} G_r P - K_r R^\alpha, & t_N \le t < t_r^* \\ -K_r R^\alpha, & t_N^* \le t < t_{N+1} \end{cases}, \quad 1 \le \alpha \le 2 \tag{1.127}$$

This case is complicated the fact that $\int \left(1 - (R^2/a^2)\right) dR$ does not allow the expression in quadrature for the irrational α values. For the rational values of parameter α, it can be reduced to a very tangled combination of elementary functions of logarithm and arctangent type.

However, the analytical approach allows obtaining recurrence relations between R_N and R_N^*, convenient for the analysis of radical accumulation and dose rate dependence of the cracking rate:

$$R_{N+1} = \frac{R_N^*}{\left[1 + (\alpha - 1)(\Lambda - \Delta\tau)\left(R_N^{*(\alpha-1)}/a^2\right)\right]^{1/(\alpha-1)}}$$

$$= \frac{\left[R_N^{(\alpha-1)}\left(1 + \Delta\tau/R_N\right)^{(\alpha-1)} - \cdots\right]^{1/(\alpha-1)}}{1 + (1/a^2)(\alpha - 1)(\Lambda - \Delta\tau)\left[R_N^{(\alpha-1)}\left(1 + \Delta\tau/R_N\right)^{(\alpha-1)} - \cdots\right]^{1/(\alpha-1)}} \tag{1.128}$$

Equation 1.128 is valid for any R_N values beginning from $N = 1$.

The values of R_0^* and R_1 can be calculated from the approximate equations

$$R_0^* = \Delta\tau\left(1 - \frac{1}{(\alpha + 1)}\frac{(\Delta\tau)^\alpha}{a^2}\right) \tag{1.129}$$

$$R_1 = \frac{R_0^*}{\left[1 + \left(1/a^2\right)\left(\alpha - 1\right)\left(\Lambda - \Delta\tau\right)R_0^{*(\alpha-1)}\right]^{1/(\alpha-1)}} \tag{1.130}$$

A simple iteration process using Equations 1.128 through 1.130 allows calculation of R_N and R_N^* for any N values and subsequent determination of the time-averaged concentration of radiation-generated radicals.

In liquid hydrocarbons, effects of spur overlapping on the rate of radical generation are weakly defined; therefore, $\alpha \approx 2$. As shown in the previous sections, the part of radicals recombining in the first-order reactions ($\alpha = 1$) becomes greater in the vicinity of the cracking start temperature.

Analysis of the radical accumulation with an arbitrary order of recombination is more important for polymer systems where molecular segments are low—mobile and spur effects are very significant. In this case, the analytical approach of this chapter allows the calculation of radical concentration as a function of irradiation dose and dose rate with the experimentally determined fractional order of reaction. Suppression of radical reactions at high dose rates often attributed to spur overlapping can also be caused by specific characteristics of pulse irradiation.

In polymer systems, these two quite different contributions to the observed weaker dependence of the cracking rate on the dose rate can be separated on the basis of simplified equations for radical accumulation.

For the very low values of radical recombination rates ($K_r \sim 10^{-4}$–10^{-3} s^{-1}) characteristic for solid polymers, Equation 1.126 can be reduced to the simple expression (Zaikin et al. 2007a)

$$\frac{d\langle R\rangle}{dt} = G_r \cdot \langle P\rangle - K_r \cdot \langle R\rangle^\alpha \tag{1.131}$$

with the initial condition

$$R(0) = 0$$

$\langle R\rangle$ and $\langle P\rangle$ are the time-averaged values of the radical concentration and the dose rate, respectively.

1.8.1.3 Radiation Generation of Excited Molecular States

In the case of LTRC (PetroBeam process), light hydrocarbon radicals generated by ionizing irradiation play the role of chain carriers similar to RTC. However, in contrast with the thermal molecular excitation in RTC, chain propagation in the cold cracking is associated with interaction of radicals and radiation-excited molecules. Therefore, the reaction rate, W, is proportional to the production of concentrations of light radicals, R, and radiation-excited molecules, C^*. In the steady-state mode of continuous electron irradiation, dependence of the cracking rate on the dose rate can be described as

$$W \sim RC^* \sim P^{3/2} \tag{1.132}$$

This dependence is weaker in any pulse irradiation mode. Its determination additionally requires description of accumulation of excited molecules during irradiation of a liquid hydrocarbon system. Termination of the radiation-generated radical pairs obeys the first-order kinetics:

$$\frac{dC^*}{dt} = \begin{cases} G^*P - K^*C^*, & t_N \leq t < t_r^* \\ -K^*C^*, & t_N^* \leq t < t_{N+1} \end{cases}$$ (1.133)

The time-averaged concentration of excited molecules can be calculated from Equation 1.128 where R_N and R_N^* should be replaced by C_N and C_N^*, and α set equal to unity:

$$C_N^* = a^2 \left[1 - \left(1 - \frac{C_N}{a^2} \right) e^{-(\Delta\tau/a^2)} \right]$$ (1.134)

$$C_N = a^2 e^{-\frac{\Lambda - \Delta\tau}{a^2}} \left(1 - e^{-\frac{\Delta\tau}{a^2}} \right) \left(1 - e^{-(N\Lambda/a^2)} \right)$$ (1.135)

$$C_\infty = a^2 e^{-\frac{\Lambda - \Delta\tau}{a^2}} \frac{1 - e^{-(\Delta\tau/a^2)}}{1 - e^{-(\Lambda/a^2)}}$$ (1.136)

1.8.1.4 Comparison with Experimental Data

In Figure 1.48, theoretical calculations are compared with the data of low-temperature radiation processing of high-viscous crude oil.

The feedstock was irradiated at the electron accelerator ELV-4 (2 MeV, continuous irradiation) and linear electron accelerator ELU-4 (4 MeV, pulse irradiation), in a wide range of dose rates at a temperature of 50°C.

Equations 1.121 through 1.126 and 1.134 through 1.136 were used for the calculation of the reaction rates in the continuous and pulse irradiation modes. The cracking rate was conventionally defined as the rate of decomposition of the heavy oil residue boiling out above 450°C. The same set of constants characteristic of hydrocarbons was used in the calculations as that used in Sections 1.1 through 1.4.

Figure 1.48 shows a good agreement of calculation with the experimental data.

Generally, the equations given earlier are valid for both radical and ion mechanisms of cracking reactions. Although the radical mechanism of RTC is commonly accepted, we have conducted special experiments with irradiation of crude oil with added methanol.

Methanol is known as a rapid proton donor to the anion radical converting the latter to a neutral propagating free radical (Mishra and Yaga 1998). It is also known for the quenching effect on propagating free ionic species (Odian et al. 1961). It could be expected that in the case of the noticeable contribution of the ion mechanism

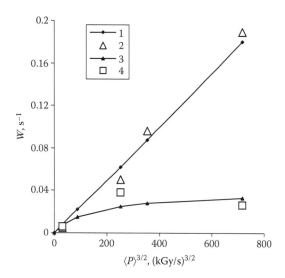

FIGURE 1.48 Dependence of the cracking reaction rate of the high-viscous crude oil on the dose rate of electron irradiation: 1—calculation using Equations 1.21 through 1.26 and 1.33 through 1.136; 2—experimental data (continuous electron irradiation); 3—experimental data (pulse electron irradiation; $l = 0.005$ s; $\Delta t = 5 \cdot 10^{-6}$ s). (From Zaikin, Y.A. et al., *Radiat. Phys. Chem.*, 76, 1404, 2007a; Zaikin, Y.A. et al., Effect of electron beam characteristics on the rate of radiation-thermal cracking of petroleum feedstock, *Proceedings of the 8th International Topical Meeting on Nuclear Applications and Utilization of Accelerators*, Pocatello, ID, American Nuclear Society, pp. 701–707, 2007b.)

to the total yields of cracking, methanol addition would suppress ion initiation and accelerate radical reactions.

High-viscous crude oil was irradiated at the electron accelerator ELU-4 with 4 MeV electrons at a temperature of 30°C. Methanol concentration was varied in the range of 0.1–1.5 mass%. Effect of methanol addition was qualitatively the same and monotonously increased with the methanol concentration.

Figure 1.49 shows the fractional contents of the feedstock and the products of its radiation processing. In the case of methanol addition, the conversion is somewhat lower for the heavy residue boiling out at temperatures higher than 450°C. However, the total yield of light fractions boiling out below 350°C increases almost twice when 1.5 mass% methanol is added.

Figure 1.49 demonstrates that the degree of feedstock conversion and the hydrocarbon contents of the liquid product can be purposefully changed by using special additives. Probably, the observed changes are caused by methanol-induced suppression of ion reactions in the heavier fraction of the feedstock and accelerated radical reactions in its lighter part. However, there is no evidence that ion reactions increase the total radiation-induced conversion of heavy oil.

The same dependence of the cracking rate on dose rate in a very wide range of dose rates indicates the same radical mechanism of the chain reaction. The high mobility of light radicals in liquid hydrocarbons even at the moderate temperatures

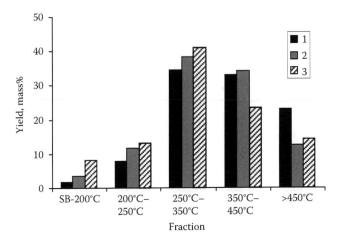

FIGURE 1.49 Effect of methanol addition on PetroBeam processing of high viscous crude oil at 30°C: 1—feedstock; 2—liquid product after processing of the neat feedstock; 3— liquid product after processing of the feedstock with added 1.5% methanol. $P = 14$ kGy/s, $D = 450$ kGy. (From Zaikin, Y.A. et al., *Radiat. Phys. Chem.*, 76, 1404, 2007a; Zaikin, Y.A. et al., Effect of electron beam characteristics on the rate of radiation-thermal cracking of petroleum feedstock, *Proceedings of the 8th International Topical Meeting on Nuclear Applications and Utilization of Accelerators*, Pocatello, ID, American Nuclear Society, pp. 701–707, 2007b.)

enfeebles track and spur phenomena. As a result, no suppression of radical chain reactions is observed at the heightened dose rates of ionizing irradiation.

This analysis shows that

1. The continuous irradiation mode is most favorable for the observation of low-temperature and cold cracking of hydrocarbons with the exception of the case of extremely heavy or strongly polymerized oil.
2. The dependences $\langle R \rangle \sim (\langle P \rangle)^{1/2}$ and $C^* \sim P$ are valid for continuous irradiation and can be approximately applied to relatively low dose rates of a pulse mode. The proportionality coefficients tend to $(G_r/K_r)^{1/2}$ and G^*/K^*, respectively, at low dose rates and decrease as the dose rate grows. At high time-averaged dose rates of pulse irradiation, the dependence of $\langle R \rangle$ on $(\langle P \rangle)^{1/2}$ becomes nearly logarithmic.
3. In liquid hydrocarbons, increase in the dose rate of pulse irradiation always causes noticeable decrease in the time-averaged quasi-stationary concentrations of radicals and excited molecules and, therefore, decrease in radical reaction rates compared with those observed under continuous irradiation at the same overage close rate.
4. In polymer systems with pronounced spur effects at the heightened dose rates, the simplified equation (1.131) can be used for the analysis of radical accumulation with an arbitrary reaction order of radical recombination.

2 Experimental Studies of Radiation- Thermal Cracking in Hydrocarbons

2.1 EARLY STUDIES OF HYDROCARBON RADIOLYSIS

Since the early 1950s, researchers have studied the effects of various kinds of radiation on hydrocarbons (HCs), in general, and petroleum products, specifically. This included studies of kerosene, aviation fuel, and crude oils from West Texas, Salt Lake (Rangeley), Los Angeles Basin Crude, and other fields around the world.

Review of the U.S. Patent literature from the mid-1940s through to 1963 (Wentworth and Canfield 1964, Berejka 2003) indicate that a great number of irradiation process patents were issued covering various applications of this technology in the manufacturing of chemicals and in the petrochemical industry. Around 190 U.S. patents in this timeframe were found to be related to irradiation polymerization, graft polymerization, chemical processes and refining, or petroleum processes. Over one-fourth of these (around 56 patents) were issued to Esso Research and Engineering (now Exxon) alone.

Other petroleum or petrochemical companies were also active during this period: Standard Oil of Indiana (Amoco), Shell Oil, Phillips Petroleum, Gulf Oil, Cities Service, and Union Oil (names that are now blurred through mergers and acquisitions).

The methods for oil radiation processing developed in this period of time were predominantly based on low-temperature radiolysis. As a rule, the radiation sources used were isotope gamma. In some of these studies, mixed reactor radiation was used for oil conversion. For example, Fellows (1966) described the use of nuclear fission products to transfer energy for conducting chemical reactions, either in the presence or in the absence of porous or catalytic materials in crude oil dispersion. However, application of neutron reactor irradiation for oil processing is very undesirable because of the possible residual radioactivity of oil samples.

Burrous and Bolt (1963) published their work on the use of refinery cuts as nuclear reactor coolants to see if they would "plate out" on heat transfer surfaces. The objective was to find a less expensive but stable heat transfer fluid. Their work analyzed the use of gamma-ray irradiation of refinery cuts at 315°C using a dosage of $5 \cdot 10^9$ rad. They compared the solubility and properties of the irradiated streams with radiolyzed terphenyls. The best of the refinery stream residues were less soluble and more intractable than the radiolytic polymer from terphenyls. Irradiation of refinery streams produced residue at faster rates than those from terphenyls. Thus, refinery stocks were at least as undesirable as those of polyphenyls. Since the lower cost of using a refinery stream as coolant does not overcome the effects of their decomposition, their use was not justified.

A method for HC conversion using neutron irradiation was developed by Long et al. (1959a). In this process, HCs were converted in a reaction zone in the presence of a significant amount of a hydrogenation catalyst at a temperature below 370°C, which is below the temperature range where the catalyst would have any appreciable effect on the reaction in the absence of radiation. It was found that while the reaction in the nature of hydrogen cleavage occurred, auto hydrogenation took place and the reaction products were largely saturated. As a result, the effects of conventional hydrogenation or hydrocracking of HCs could be obtained using no or minimal amounts (less than 100 standard cubic feet per barrel) of extraneous hydrogen. More gasoline boiling range product, having a high octane number, was obtained, and less polymer was formed when reforming reactions were carried out. The tendency of the HCs to polymerize or decompose into low-molecular-weight gases after being formed into radicals was inhibited by the catalyst.

It was also found that during the irradiation of HCs in the presence of hydrogenation catalysts, the catalyst exerted an appreciable influence on the selectivity of the reaction, even when the reaction temperature was relatively moderate, that is, the temperature was substantially below incipient thermal cracking (TC) or conversion temperatures.

The hydrogenation components of the catalyst used in the process were metals, oxides, sulfides, or salts of a group VI or VIII element. Preferred catalysts comprised an essentially pure alumina base and 1–15 wt% of molybdenum. The catalyst and carrier materials used were normally those which, upon neutron bombardment, produced radioisotopes having short half-lives or small neutron capture cross section, so that very little material could become radioactive. The process was preferably carried out with the liquid-phase reactions and the pressure used had to be sufficient to maintain liquid-phase conditions.

The air-cooled, natural uranium, graphite moderate research reactor of the Brookhaven National Laboratory was used in tests. Paraffinic gas oils boiling in the range of 200°C–370°C were irradiated with the mixed neutron–gamma irradiation from a pile operating at a total power of 24 MW for 10 days each run. The conversion expressed as material converted out of the feed boiling range was above 10%. Generally, the process described is rather similar to hydrocracking except that hydrocracking does not take place at such lowered temperatures without a catalyst or without irradiation.

Schultze and Suttle (1959) proposed a method for HC conversion under ionizing electromagnetic irradiation in the presence of deuterium. Following irradiation, the deuterium was separated from the converted product, and the product was recovered. In the practice of this invention, the deuterium was employed in a moderately high concentration from about 1 to about 10 mol% of the material processed. The reactants were exposed to an intense electromagnetic irradiation in which energy exceeded 2.26 MeV for a time sufficient to take an absorbed dose of $10–10^6$ Gy.

To directly use electromagnetic irradiation with a higher absorption rate and to produce more intensely ionizing particles, about 0.01% of the gamma flux from a nuclear reactor was converted into heavy particles, protons, and neutrons. This might be accomplished by placing a reactant loaded with deuterium between the reflector and biological shield of a nuclear reactor. This particular mode permitted the chemistry of the reaction to define the temperature. The temperature was varied from 38°C to 1650°C, or even 2760°C, depending on the type of reaction and the type of reactants processed. The feedstocks might include mono-olefins, diolefins, naphthenes, aromatics, paraffins, and mixtures thereof.

According to this procedure, styrene and butadiene were polymerized to form a rubber. The polymerization was affected by emulsifying 1 mol styrene and 1 mol butadiene with 5–10 mol of water, the emulsion being stabilized by the addition of 0.01–0.1 mol% soap. In this operation, about one-fifth of the water was D_2O. The emulsion was then exposed to an intense field of gamma irradiation with the dose in the range of 0.1–1000 kGy. After completion of the polymerization under irradiation, the reactants were recovered by breaking the emulsion by the addition of sodium chloride, the heavy water being recovered by evaporation and recycled for reuse.

Long et al. (1959b) also proposed another process wherein HCs were converted by neutron irradiation in the presence of an acid center HC conversion catalyst or cracking catalyst at lowered temperatures. The products obtained had a decreased molecular weight and an increased amount of branched HCs.

Four types of oil samples (light paraffinic gas oil, light naphthenic gas oil, highly aromatic gas oil, and n-hexadecane) were irradiated in the nuclear reactor channel with the mixed neutron–gamma irradiation for 10 days. Al and Si/Al were used as the cracking catalysts. The Si/Al catalyst had demonstrated a higher selectivity to gasoline and diesel fuel and produced a more aromatic and less olefinic product.

The yields of the fractions boiling out below 320°C were 3.9–27.6 wt%; the conversion to gas was 1.2–7.6 wt% on feed. The relatively high iso- to normal pentane ratio indicated highly branched products in the gasoline range, which contributed to high octane numbers (calculated research octane numbers were 92–96). The high yields of gasoline, diesel fuel, and lubricants were obtained at temperatures where the catalysts used have no activity in the absence of radiation. It should be noted that the application of neutron irradiation did not exclude residual radioactivity of oil products.

The method developed by Noddings et al. (1959) relates to treating catalyst materials with high-energy ionizing radiation. The method was used for processing calcium nickel phosphate catalysts to improve the process of dehydrogenation of aliphatic olefins containing from four to six carbon atoms and having at least four carbon atoms in the unsaturated chain of the molecule.

The calcium nickel phosphate catalysts were subjected to ionizing irradiation with the dose sufficient for maintaining their high selectivity for the dehydrogenation of normal butylenes to form conjugated diolefins, for example, butadiene-1,3, in an amount of at least 75% based on the butylenes consumed in the reaction. It was found that spent or used calcium nickel phosphate catalysts having their selectivity lowered to a point where the catalyst can no longer be profitably used, for example, having selectivity of 70% or lower, could readily be revived under Co-60 gamma irradiation or converted to a catalyst of a higher selectivity.

Sutherland and Allen (1961) had patented a method for the activation of a solid substrate by ionizing radiation to use it in converting straight-chain paraffin HCs to lower or same-molecular-weight straight-chain and branched-chain alkanes. It was found that when a solid zeolite substrate is admixed with a cobalt ion and then formed in-exchange zeolite is irradiated, after the HC is absorbed therein, increase in the yields of branched and low-molecular-weight straight-chain compounds was observed. The yields of the products increased with the increase in the amount of HC distilled onto the solid substrate.

After a known quantity of pentane was distilled into an ion-exchange substrate, NaX or silica gel, and sealed off under liquid nitrogen, irradiations were performed in tubes with gamma rays from a Co-60 source at 25°C. The exposition dose rates were 0.25–0.35 MR/h. Total doses were 17.0 MR for the runs using silica gels and 21.6 MR for all others. Untreated substrate $Na_2O \cdot 3CaO \cdot 4Al_2O_3 \cdot 8SiO_2$ (CaA) was also irradiated.

After irradiation, the tubes were open under vacuum, and the gaseous products pumped through a liquid nitrogen trap. The temperature of the sample, which was initially that of the room, was gradually increased to 400°C, which was maintained for 2 h to ensure thorough removal of the radiolysis products. At least 95% of the permanent gas came over below 200°C. The gas, which contained only methane and hydrogen, was analyzed over copper oxide. In the runs with CaA structures, only a slight variation was observed from the results obtained upon the irradiation of bulk liquid pentane (Figures 2.1 and 2.2).

It was found from the studies of radiolysis on bulk pentane that if there were no interaction between pentane and solid, and the specific radiolysis yield were the same in the adsorbed state as in liquid pentane, then a plot of G_T for each product against the electron fraction of pentane would be a straight line running from the origin to the liquid G-value at 100% pentane. This line is shown in Figures 2.1 through 2.4 as the "liquid line." When a yield is above that line, it means that the amount of product formed by exposure of adsorbed pentane to a given dose of gamma irradiation is greater than what would be obtained from the exposure of the same quantity of pentane in the ordinary liquid form to the same irradiation dose.

The hydrogen and methane yields on the solid NaCoX are shown in Figures 2.1 and 2.2, respectively. The high yields obtained on NaCoX probably indicated that a great deal of energy transfer occurred from the solid to the absorbed pentane.

Figure 2.3 shows the radiation-chemical yields of lower HCs obtained from NaCoX. The relative amount of isopentane was some what greater than that in the radiolysis of bulk liquid pentane but chain branching was not very considerable. From the results obtained where silica gel is used as a substrate, it was noted that a much higher yield of branched chain product (isopentane and isobutane) was obtainable (Figure 2.4). However, in these experiments, there was a very definite fall in the total amount of products as the percentage of pentane on the gel substrate increased. This meant that the first portion of the pentane to go on the silica gel showed a high yield of isomerization of the carbon chain under irradiation. When more pentane was put on, additional pentane did not undergo isomerization. Moreover, its presence prevented the occurrence of this reaction in the more tightly bound molecules. It was suggested that isomerization reaction occurs only with more or less isolated

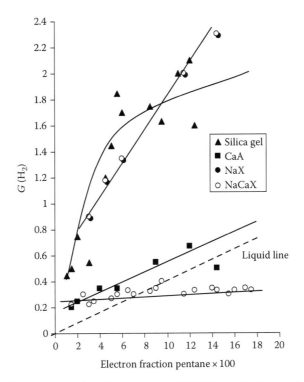

FIGURE 2.1   Hydrogen *G*-values (molecules per 100 eV absorbed energy) obtained with various substrates. (From Sutherland, J.W. and Allen, A.O., Radiolysis of organic compounds in the adsorbed state, US Patent 3002911, 1961.)

molecules of pentane on the silica surface and is quenched by the presence of neighboring pentane molecules.

A mechanism for energy transfer discussed by Sutherland and Allen (1961) consists of the motion through the solid of the so-called sub-excitation electrons, which have energies far above thermal but still below the lowest electronic energy level present in the material. Such electrons can lose their energy only relatively slowly in the material. They may presumably travel considerable distances, if they can escape capture by the positive charges or "holes" formed by the irradiation along with the electrons. If such electrons reach the surface, they might still have energy enough to excite the pentane molecules to upper electronic states, if such states exist in pentane at lower energy levels than the lowest excited energy level present in the solid. In this theory, the existence of low-lying levels can be expected in the solid to decrease the yield of pentane decomposition. However, this certainty did not agree with the finding of very much more decomposition in NaCoX.

Another possible mechanism of energy transfer discussed in the work (Sutherland and Allen 1961) is migration of positive charges or holes in the electron distribution through the solid to the pentane, which would then become ionized. This mechanism principally allows facilitation of positive charge transfer to the pentane by cobalt ions.

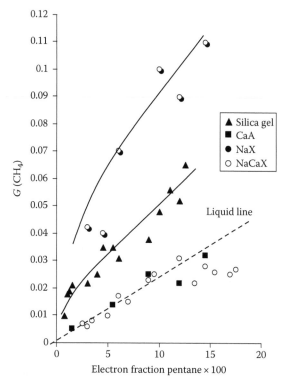

FIGURE 2.2 Methane G-values obtained with various substrates. (From Sutherland, J.W. and Allen, A.O., Radiolysis of organic compounds in the adsorbed state, US Patent 3002911, 1961.)

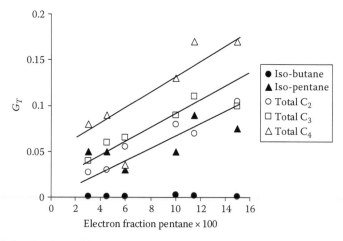

FIGURE 2.3 G-values of lower HCs obtained with NaCoX. (From Sutherland, J.W. and Allen, A.O., Radiolysis of organic compounds in the adsorbed state, US Patent 3002911, 1961.)

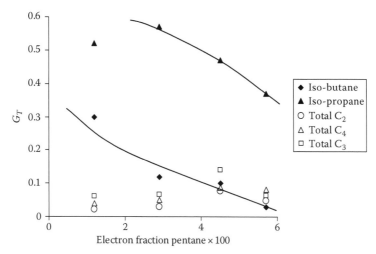

FIGURE 2.4 G-values of lower HCs obtained with silica gel substrate. (From Sutherland, J.W. and Allen, A.O., Radiolysis of organic compounds in the adsorbed state, US Patent 3002911, 1961.)

Tarmy and Long (1959) patented a method for heavy oil radiation processing to prevent their propensity to form stable emulsions when subsequently treated with an aqueous medium. Heavy oil residue was subjected to gamma irradiation from a Co-60 source with doses in the range 16.7–120 MR at the temperatures 40°C–350°C. The samples were then dissolved in benzene and treated with water, aqueous pyridine, and dilute HCl. The irradiated samples did not emulsify while the nonirradiated samples did. Obviously, this method could have a very narrow domain of applicability because of the strong trend of heavy oil to polymerization under long-time irradiation even at moderate temperatures (see Section 1.7).

Ruskin (1959) proposed a process for removing petroleum from petroleum-bearing sands by irradiating them with 100 million to 400 million R and recovering the irradiated petroleum released from the sands.

It was noted that petroleum globules in wells are molecularly bound by long-range weak forces to the crystal lattice structure of the silicate carbonate framework and still further held by polar attraction of connate water. By gamma irradiation of 100–400 million R, these long-range forces can be disturbed, and the very lattice structure of the silicates and carbonates can be altered freeing the bound petroleum. A still further factor is the change in viscosity of the petroleum creating a readier flow out of the rock or shale. The disintegration of the silicate–carbonate structure also contributes to the secondary oil recovery from wells. It was expected that the softening of the rocky composition will promote the release of reservoir gases, further facilitating the promotion of gas pressure and increasing the flow of petroleum. In the case of oil wells that have long been unproductive, irradiation gamma sources were suggested to be kept in the wells for a long period of time.

To carry out this procedure, a Co-60 source of gamma irradiation was encased in a specially constructed canister so arranged that the radiation source was completely shielded. The dimensions were such as not to exceed the dimensions of the average oil well pipe. The radiation source was permanently fixed in the canister bottom while the shielding cylinder around the canister was readily separable from the canister. Thus the radiation source could be lowered into the well for the desired period and then reshielded before removal. The source and the cylinder were housed in within a lead cylinder so that the movements of the canister and source were at no time exposed to possible contact.

In one of the examples of the method application, a 3 in. limestone core was irradiated with 100 million R in the presence of an acetylene atmosphere. A marked increase in darkening and odor of petroleum was detected. The core could be readily crushed by the hand at the edges.

In another example, 500 g of Athabasca oil-bearing sand were suspended in water and irradiated with 200 million R. As a result, a layer of oil separated measuring 42 cm^3.

Other methods developed by Ruskin (1961) relate to HC reforming and cracking processes. It was found that gamma irradiation of HCs in the vapor phase at low temperatures such as that of liquid nitrogen, that is, $-175°C$, and under, and moderate pressure such as 5–20 atm, results in the formation of higher-molecular-weight HCs. The HCs produced were liquids utilizable for high-compression engines and for the other uses of liquid HCs.

For example, propane gas was treated with gamma radiations from a Co-60 source in a gas bomb at liquid nitrogen temperature and initial pressure of 175 psi until the pressure sharply fell. As a result, a series of HC was produced ranging from C_4 to C_{12}. The resultant liquid contained at least eight different components.

To produce cracking of HCs, Ruskin (1961) followed the same procedure with the exception that the liquid nitrogen was introduced into the bomb and then the hydrogen was added in the vapor phase under lower pressure. The whole sample was subjected to gamma irradiation from the Co-60 source. It was recommended to conduct this process in a continuous circular stream where the liquid nitrogen vaporizes and mixes with the HC vapor so that the cracked HC continues to accumulate. The HC cracking was attributed to cojoint action of the gamma irradiation and the active nitrogen species induced by gamma irradiation. The separation of HCN from the other reaction products could be accomplished by fractionation or other commonly known means.

For example, 15 cm^3 liquid nitrogen and 100 cm^3 n-pentane were introduced into a 500 cm^3 gas bomb. The bomb was then subjected to gamma irradiation; the pressure in the bomb was 5 psi. The resultant products were methane, ethane, and butane along with other fractions not determined.

In the other example, a cobalt bomb was used to reform propane gas. The propane gas was introduced into the bomb under 175 psi gas pressure. A bath of liquid nitrogen around the bomb maintained the temperature at $-175°C$. The Co-60 source within the bomb irradiated the propane gas. As the irradiation level attained 75 million R, the pressure dropped down to almost atmospheric. As a result of irradiation, a series of HCs having from 4 to 12 carbon atoms were obtained. The primary compound was 2-methyl pentane.

A method for converting low-octane HCs to high-octane HCs at moderate temperatures was proposed by Folkins (1962). In this method, HCs boiling within boiling range, such as hexane, heptane, octane, nonane, methylcyclopentane, cyclo-hexane, and their mixtures, are heated to about the threshold thermal decomposition temperature of the HC material. The threshold temperature was defined as that temperature at which 1% by weight of the HC decomposes when held at this temperature for 10 min. The HC mixed with hydrogen is contacted with a dehydrocyclization catalyst, such as platinum, palladium, or other noble metal deposited on alumina or silica–alumina. While the HC–hydrogen mixture is in contact with a catalyst, it is exposed to high-energy radiation.

As an example of the process, n-heptane was mixed with hydrogen in a mol ratio of 1 and passed through a fixed-bed catalyst consisting of molybdenum on activated alumina at 150 psi pressure and at 480°C. A 25% yield of toluene was obtained and the gas made 6%. Under the same reaction conditions, except that the reactor was exposed to radiation of Co-60 with intensity of about 235,000 R/h, a toluene yield of 35% was obtained with the same gas production of 6%.

A method for the conversion of normally gaseous, or light-boiling, HCs to normally liquid, higher-boiling HCs was developed by Wilson and McCauley (1965). In this process, light HCs, such as propane, n-butane, and their admixtures, were converted directly to liquid products suitable as motor fuel blending components.

A feedstock comprising substantially olefin-free paraffin having from three to four carbon atoms, inclusive per molecule, was sealed in a pressure-containing bomb under sufficient pressure to maintain a liquid phase at room temperature and charged to a reaction zone where it was exposed in the liquid phase to radiation from spent nuclear reactor fuel elements. The preferred dosage was 10^8–10^9 rep (roentgen equivalent physical). After radiation processing, liquid products were separated from the reaction mixture. In one of the examples, radiation processing n-butane and propane, mixed in the proportion 1:1, resulted in 2% production of the following liquid composition (vol.%): C_5—25, C_6—25, C_7—25, C_8—25, C_9—Nil.

A similar approach was later used in the method for gaseous alkane processing proposed by Gafiatullin et al. (1997a). In this method, alkanes were irradiated in the gaseous phase with 3–6 MeV electron beam at the beam current density of 5–70 μA/MeV m^2.

A cyclic process was carried out in a closed loop at a pressure of 1–4.6 atm/MeV depending on the electron energy. The flow rate of the reacted mixture was higher than 4 g/kW s in the circuit and depended on the electron beam power. After irradiation of the circulated mixture, the reaction products were separated: the condensed products were removed and hydrogen was extracted. The reacted mixture free from the surplus hydrogen and condensed products was mixed with fresh feedstock to provide constant pressure and then returned to the reaction zone under the electron beam. The hydrogen content in the reacting mixture did not exceed 5 mass%. The condensed products contained isomers of saturated HCs from hexane to hexadecane.

Radiation processing of the oil-well gas resulted in the yields of light products C_6–C_{15} of 1.2 kg per 1 kW of consumed energy. The full conversion of the feedstock yielded 97.4 mass% target product and 2.6 mass% hydrogen.

However, in the 1960s–1970s, the process interests of the chemical companies and of the petroleum and petrochemical companies in radiation technologies did not evolve into viable businesses. The petrochemical companies were producing products on a continuous basis with their various catalysts and thermal reactors. At that time, the reliability of industrial electron beams was not as acclaimed as today, so that the risk of using electron beam processing in a 24 h, 7 days a week operation was considerably high. Much research work was conducted using low-dose-rate Cobalt-60 irradiators and in batch-type operations. Generally, none of these studies showed any possibility of a commercially viable process based on any form of ionizing radiation.

In these early studies, the concept of low-temperature radiolysis covered any structural or chemical changes that occurred in HCs under the action of ionizing irradiation. However, further studies have shown that mechanisms of radiation-chemical reactions in HCs undergo considerable changes at heightened temperatures and high dose rates of irradiation.

The studies of the high-temperature radiolysis reported at the 5th World Petroleum Congress, New York (Hoare et al. 1959) were the turning point in the development of the new concepts of HC radiation cracking.

Hoare et al. (1959) reported the results of their studies on the thermal decomposition of normal and iso-pentanes. The experiments were conducted with and without radiation from 0% to 100% conversion in a static system. The rate of pressure build-up and the changing composition of the gases were studied over a range of initial pressures from 30 to 80 mmHg and at temperatures ranging from 400°C to 550°C. Radiation dose rates of 10 R/h were employed. In the range studied, neither temperature and pressure, nor radiation had any appreciable effect on product pattern.

However, the degree of conversion had a great influence on product distribution. As the reaction proceeded, olefins initially formed were used up to form light paraffins, aromatics, and tars. In the absence of radiation, the decomposition rates for both normal and iso-pentanes increased exponentially with temperature, but the pressure dependence of the decompositions was more complex, especially for iso-pentane. At the very low dose rate used in this work, radiation had no apparent effect on the decomposition rates of normal pentane. With iso-pentane on the other hand, at the lower pressures and temperatures, radiation decreased the rate of decomposition, while at the upper end of the temperature and pressure range studied, the effect was reversed and the rate was increased. Note that contribution of the radiation component of radiation-thermal cracking (RTC) to the overall process was negligibly small at the extremely low dose rates used in this study (see Section 1.2).

The observed data were explained in terms of a molecular decomposition accompanied by a radical chain reaction terminated in the gaseous phase. It was supposed that the radical chain reaction may be initiated both thermally and by radiation and involved the cracking of pentyl radicals. Competing with this cracking reaction, however, is a wall reaction of pentyl radicals with olefins, the magnitude of the wall effect being a function of the ratio of the lifetime of the radicals to their mean diffusion time to the walls. At degrees of conversion above about 60–70, secondary reactions based on unsaturated radicals, such as allyl, could replace the role of the pentyl radicals in the propagation of the reaction chain. These lead to the decrease in olefin concentration and to the production of aromatics found experimentally.

One of the first studies of the high-temperature radiolysis was represented by Lucchesi and coworkers (1958, 1959). It was noted that experimental studies of the cracking of HCs initiated by nuclear irradiation emphasized the effect of such process variables as temperature, feed composition, and phase. More detailed later studies (Topchiev and Polak 1962, Lavrovskiy 1976, Zaikin and Zaikina 2008a) have shown that dose and dose rate of ionizing irradiation should be added to this list of the most important variables. Depending upon conditions, each variable can have important effects in radiation chemistry.

At the end of the 1950s, a rather wide range of radiation type and intensities had been used for HC irradiation, including x-rays, Co-60 gamma rays, mixed pile radiation, and alpha particles. Based on these studies, three broad areas of HC radiation chemistry were discussed in the papers (Lucchese et al. 1958, 1959, Topchiev and Polak 1962, Lavrovskiy 1976)—the low-temperature non-chain reactions of paraffins, the high-temperature chain reactions of paraffins and paraffin–olefin mixtures, and such low-temperature catalytic reactions as the acid-catalyzed isomerization of paraffins. At low temperatures, irradiated paraffins undergo dehydrogenation and, at reasonable conversions, such complex side reactions as decomposition, polymerization, and rearrangement. At high temperatures, paraffins such as n-hexadecane undergo a long chain cracking reaction.

Lucchesi et al. suggested that the principal role of radiation is to accelerate the thermal chain reaction. For example, the Al–Co-catalyzed isomerization of n-hexane and methylcyclopentane was accelerated by nuclear radiation. The products obtained were typical of those made under nonradiation conditions, and the effect of radiation was equivalent to the addition of a chemical promoter (olefins) to the reactants. The data available at the end of the 1950s supported the view that, in the chain radiation chemistry of HCs, radiation acts as an accelerator of typically thermal reactions. There was no evidence of radically new chain reactions propagated by species peculiar to the radiation initiation technique.

However, later studies (Topchiev and Polak 1962, Zaikin and Zaikina 2008a, Zaikin 2008, 2013a) have shown that RTC cannot be reduced to the intensification of thermal reactions. It is sufficient to note that fractional and HC contents of the cracking products, including HC types, considerably differ in the cases of thermal and radiation-thermal (RT) processing. The same reactions can be initiated or propagated by thermal or radiation action, but the probability of these reactions and their rates are different in thermal and radiation conditions. The specific conditions of the basic RTC reactions give rise to side effects and reactions characteristic just of this process. Moreover, low-temperature radiation cracking (Zaikin and Zaikina 2008a, Zaikin 2008, 2013a) proceeds with the different mechanism of chain initiation in the temperature range where thermally activated chain cracking reactions are impossible because of the thermodynamic limitations.

Gamma irradiation of a high-paraffin wax was carried out by Henley and Repetti (1959). The wax samples were melted and placed into Pyrex test tubes. All runs were conducted in a static system using a Co-60 isotope as a source of gamma radiation. The dose rate was as low as 46,000 R/h.

To make a run, the system was inserted in its heating jacket and the power was turned on. The duration of the initial heating period varied between 30 and 50 min; the latter duration was necessary to reach 480°C. The automatic temperature controller then maintained the temperature level of the setting. At the conclusion of a

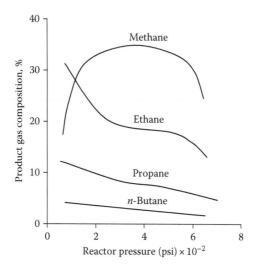

FIGURE 2.5 Variation of product composition with reactor sampling procedure. (From Henley, J.B. and Repetti, R.V., Effect of gamma radiation upon the hydrocracking of a heavy paraffin, Presented before the Division of Gas and Fuel Chemistry ACS, Atlantic City, NJ, pp. 161–168, 1959.)

run, the final temperature and pressure were recorded, and the reactor was withdrawn from the heater and allowed to cool to ambient temperature. Figure 2.5 shows the variation of product composition with reactor sampling procedure.

Figure 2.6 shows a plot of reactor temperature versus pressure. The graph shows that until about 860°F (460°C) is reached, nothing happens other than heating of the gas in the reactor. At 460°C, cracking begins, and because of the additional gas formed, the pressure begins to rise rapidly. The maximum duration of a cracking run was only 5 h, which was not sufficient to attain equilibrium between liquid and gas phases. Figure 2.6 shows a typical time dependence of pressure for a run in progress. Because of the nearly constant rate for run times exceeding 3 h, no attempt was made to obtain data for very long durations. Instead, runs of 2–5 h were made with duplicates.

A typical product spectrum for the cracking runs is shown in Figure 2.7. Under conditions of this experiment, the products were almost identical for radiation and nonradiation runs.

Figure 2.8 shows the dependence of the number of gas moles per gram of wax initially present, S, on the run time. The initial pressure in each run was about 1000 psi.

In Figure 2.9, G-values defined as the number of gas molecules formed per 100 eV of incident radiation energy (not absorbed as commonly used) is plotted versus irradiation time. The G-values were in the range 60,000–80,000 tending to decrease with the increase in run duration.

A method for high-temperature pile irradiation of the n-heptane–hydrogen system was developed by Miley and Martin (1961). A pilot unit for the high-temperature irradiation of flowing reactants was designed for the beam port of a nuclear reactor. RTC of the n-heptane–hydrogen system was studied with molal H_2/C_7 ratios from 0 to 5

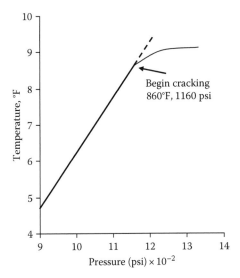

FIGURE 2.6   Dependence of temperature on pressure in the reactor. (From Henley, J.B. and Repetti, R.V., Effect of gamma radiation upon the hydrocracking of a heavy paraffin, Presented before the Division of Gas and Fuel Chemistry ACS, Atlantic City, NJ, pp. 161–168, 1959.)

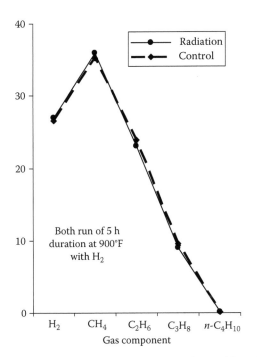

FIGURE 2.7   Typical spectra of products formed by hydrocracking. (From Henley, J.B. and Repetti, R.V., Effect of gamma radiation upon the hydrocracking of a heavy paraffin, Presented before the Division of Gas and Fuel Chemistry ACS, Atlantic City, NJ, pp. 161–168, 1959.)

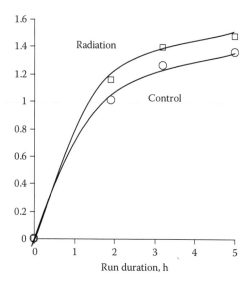

FIGURE 2.8   Variation of S with the duration of 400°C hydrocracking runs. (From Henley, J.B. and Repetti, R.V., Effect of gamma radiation upon the hydrocracking of a heavy paraffin, Presented before the Division of Gas and Fuel Chemistry ACS, Atlantic City, NJ, pp. 161–168, 1959.)

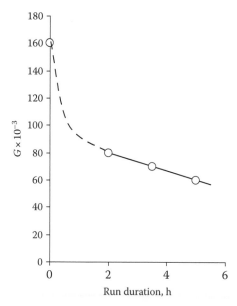

FIGURE 2.9   Variation of G with the duration of 480°C hydrocracking runs. (From Henley, J.B. and Repetti, R.V., Effect of gamma radiation upon the hydrocracking of a heavy paraffin, Presented before the Division of Gas and Fuel Chemistry ACS, Atlantic City, NJ, pp. 161–168, 1959.)

at 17.6 kg/cm^2, 320° to 400°C, 2 to 8 min residence time, and up to 3600 rep/min. Although conversions were low, the decomposition rate was significantly increased by radiation giving G-values > 10^3. The high G-values indicated a radiation-induced chain cracking reaction. Product distributions were not significantly altered by radiation. Radiation yields were found to be reasonably linear with total dose from 0 to 14 krep, but a twofold increase was observed in passing from a molal H_2/C_7 ratio of 0 to 0.5.

In the early 1960s, the phenomenon of RTC of HCs was described and analyzed in the pioneer works by Polak, Topchiev, Lavrovsky, and other researchers (Topchiev et al. 1959, Topchiev and Polak 1962, Lavrovskiy 1976), who revealed the basic mechanisms of radiation-induced chemical conversion in HCs and regularities of the chain reactions of RTC in particular. It was the first evidence of highly efficient self-sustaining chain reactions of HC decomposition initiated by combined radiation and thermal action. The discovery was practically important because only chain reactions can provide the high rates of deep oil processing necessary in industrial conditions.

Since then, different aspects of ionizing radiation application to deep radiation processing of oil feedstock are the subject of research and technological investigations. Up to the beginning of the 1990s, these studies (Brodskiy et al. 1961, Gabsatararova and Kabakchi 1969, Panchenkov et al. 1981), with a few exceptions (Mustafaev 1990, Zhuravlev et al. 1991), were limited by consideration of radiation-induced decomposition of model HCs and light oil fractions.

2.2 EXPERIMENTAL STUDIES OF RADIATION-THERMAL CRACKING OF HYDROCARBONS

2.2.1 GENERAL CHARACTERISTICS

Desirable conditions of any cracking process assume rejection of using high pressure and temperature that considerably raise production costs and make its safety lower. The TC of normal alkanes usually takes place at high temperatures (e.g., 500°C–600°C for heptane). According to the generally accepted theory (Topchiev and Polak 1962), this process is accomplished in two steps: (1) initiation of the reaction by radicals produced by dissociation of a molecule of the starting compound; and (2) propagation of the chain. The latter consists of dissociation of the large initial free radicals into an olefin and shorter radicals and the reaction of the latter with starting molecules.

The chain propagation process results in the formation of a reaction product molecule and of a new large radical. Because of its low activity, this radical cannot interact with another starting molecule, that is, it cannot propagate the chain. However, it can dissociate into an olefin molecule and a shorter and correspondingly more active radical that will be able to propagate the chain. The initiation step requires the activation energy of about 250 kJ/mol, that is, the reaction can proceed at an appreciable rate only at the temperatures of 500°C–600°C. The chain propagation step, controlled by dissociation of the radical, requires the activation energy of about 80 kJ/kg, that is, it requires a much lower temperature. Therefore, there are two main conditions and two stages necessary for chain cracking reaction:

1. Formation and maintenance of relatively low concentration of chain carriers (light radicals, such as H*, CH_3*, C_2H_5*, necessary for cracking initiation)
2. Formation and maintenance of sufficient concentrations of excited molecules necessary for chain propagation caused by interaction of radicals with excited molecules and their disintegration

In the case of TC, both stages of the process are thermally activated. In the case of RTC, the first stage is radiation-initiated: chain carriers are created by irradiation. The absorbed electron energy, necessary for the formation of radical concentration sufficient for cracking initiation, is less than 0.4 kJ/mol. The propagation stage of RTC is still thermally activated.

The radical mechanism of RTC assumes that radiation-induced chain initiation does not depend on temperature, and the rate of radical generation depends only on irradiation dose rate. Conventional sources of ionizing radiation, such as electron accelerators and isotope sources, provide generation of radicals in concentrations sufficient for the initiation of chain reactions in HCs.

Application of RTC releases the most energy-consuming stage of cracking associated with the initiation of chain carriers, thereby lowering the cracking temperature by 200°C–250°C compared with the thermal process (Figure 2.10).

The total energy expended in the RT process is significantly less (at least by ~40%) than that required by the standard heating thermal process alone because initiation energy E for the radiation-induced cracking is much lower than that for TC.

Fractional contents of the overall products of RTC and TC are compared in Table 2.1.

In the case of RTC, contents of valuable commodity products are essentially higher. Therefore, RTC application provides additional energy savings for the production of the designed products, proportional to the ratio of their yields for the two kinds of processing. Subject to this factor, energy consumption of RTC for the production of gasoline (boiling temperature below 200°C) and diesel fuel (boiling temperature in the range 200°C–360°C) is at least 60% lower than that for the thermal process.

There are five main types of radical reactions that considerably contribute to the RTC of liquid HCs (Bugaenko et al. 1988):

$$R + R \rightarrow H \qquad \qquad \text{—Recombination}$$

$$R + R \rightarrow OL + RH \qquad \text{—Disproportionation}$$

$$R \rightarrow R_1 + OL \qquad \qquad \text{—Dissociation} \qquad \qquad (2.1)$$

$$R + RH \rightarrow R_2 + RH \qquad \text{—Molecule break}$$

$$R + OL \rightarrow \text{Polymerization}$$

where
  R is a radical
  OL is an olefin
  RH is a HC molecule

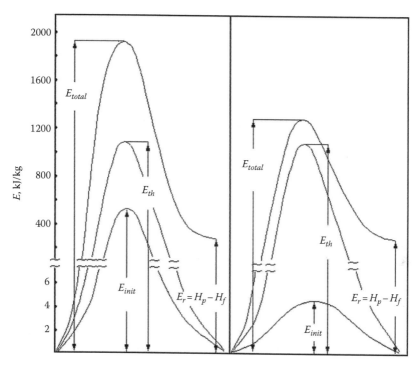

FIGURE 2.10 Heat balance diagram for conventional thermal cracking of fuel oil at 650°C and radiation-thermal cracking at 400°C. (Zaikin et al. 2003.) $p^{20} = 0.95$ g/cm³, $M = 390$ g/mol; E_{total} is total consumed energy; E_t is energy consumed for feedstock heating up to reaction temperature (400°C); E_{init} is energy consumed for cracking initiation; $E_r = H_p - H_f$ is the increase in the product energy after processing.

TABLE 2.1
Hydrocarbon Fractions of Refined Products from Fuel Oil

Boiling Temperature, °C	Feed Fractions Mass%	TC Mass%	RTC Mass%
<200	None	10	15
200–300	2	15	27
300–360	8	15	18
360–450	38	25	20
>450	52	30	10
Gases	None	5	10

Source: Zaikin, Y.A. et al., *Radiat. Phys. Chem.*, 67(3–4), 305, 2003.

RTC is characterised by the high reaction rate (see Section 1.2). The total rate of RTC can be found from

$$W_{RTC} = \frac{k_p}{k_{t2}^2} \sqrt{I_T + I_R}$$

(2.2)

where I_T and I_p are rates of thermal and radiation initiation of cracking, respectively:

$$I_T = k_i\, e^{-E_i/kT}$$

$$I_R = GP$$

(2.3)

The ratio of the rates of the overall RT process and mere thermal process is

$$\frac{W_{RTC}}{W_T} = \sqrt{1 + \frac{I_R}{I_T}}$$

(2.4)

The ratio W_{RTC}/W_T was calculated in the book (Topchiev and Polak 1962) using the following characteristic values of the constants: $k_i \approx 4 \cdot 10^{-10}$ cm³/molecule; $s\ (\sim 10^{12}\ \text{s}^{-1})$; $E_i \approx 250$ kJ/mol; $E_p \approx 80$ kJ/mol; $G=7$ radicals per 100 eV. Table 2.2 shows that W_{RTC}/W_T decreases as the temperature goes up and increases at higher dose rates.

Combination of a high rate of chain propagation and a high ratio of the total rates of radiation and thermal processes is provided in the temperature range of 350°C–550°C for the most part of petroleum feedstock.

Dependence of the product G-values on temperature and irradiation dose rate is given by

$$G, \text{molecules/100 eV} = \frac{100\,e\,N_A}{PM} W = \frac{100\,e\,N_A}{PM} \frac{k_p}{\sqrt{k_{t2}}} \left(k_i e^{-E_i/kT} + GP \right)^{1/2}$$

(2.5)

TABLE 2.2

Ratios of the Total Rates of the Radiation and Thermal Processes for Heptane

Temperature, °C	Ratios of the Rates of Radiation and Thermal Cracking Initiation	Ratios of the Total Rates of Radiation and Thermal Processes
550	820	28.6
600	41	6.5
650	3.7	2.2
700	0.41	1.2
800	0.008	1.00

Source: Topchiev, A.V. and Polak, L.S., *Radiolysis of Hydrocarbons*, The Academy of Sciences of the U.S.S.R., Moscow, Russia, 1962.

---

where

e is the electron charge

M is the product molecular mass

In the temperature range of the prevailing RT process, characteristic G-values of RTC liquid products lie in the range of 1,000–20,000 molecules per 100 eV of absorbed radiation energy.

Different mechanisms of cracking initiation and propagation in thermal and RT processes result not only in the different reaction rates but also in different HC contents of the cracking products. For example, olefins are the natural products that limit the yields of light liquid fractions for any cracking process, including TC, thermocatalytic cracking (TCC), and RTC. However, very high olefin concentrations in liquid fractions are usually undesirable, except special cases.

In different types of initiated cracking, olefin concentrations in products are very different. The highest olefin content is observed after the conventional TC. The olefin concentration in the TCC liquid product is noticeably lower. Due to the effect of the catalyst, the processes of hydrogen redistribution and skeleton isomerization in HCs are much more pronounced in the case of TCC, and the probability of the olefin decomposition is much higher. RTC results in still lower olefin concentrations compared with TCC. In this case, the role of the catalyst is played by ionizing radiation.

RTC allows processing of practically any kind of oil feedstock including heavy oil residues boiling above 600°C. However, the feedstock origin is of critical importance for the selection of the process conditions and properties of the target products. The most important are chemical composition of the feedstock, temperature limits of the distillate fraction boiling, and contents of pitches, asphaltenes, and high-molecular paraffins in the high-boiling vacuum distillates.

Electron accelerators are used as the sources of radiation for basic initiation of RTC reactions in HCs. Isotope gamma sources can also be used for oil radiation processing; however, they are characterised by much lower production rates.

2.2.2   RT Cracking of Light Oil at Low and Moderate Dose Rates of Ionizing Irradiation

The first experiments on RTC of liquid alkanes in static and flow conditions were carried out by Polak, Topchiev, Brodskiy (Brodskiy et al. 1961, Topchiev and Polak 1962), and other researchers in the early 1960s.

In the work (Topchiev and Polak 1962), alkane samples were irradiated with electrons having an energy of 900 keV in static conditions in the walls of molybdenum glass with the wall thickness of about 1 mm in a volume of about 15 mL. The dose rate of electron irradiation was $2 \cdot 10^{19}$ eV/s per 1 mL of liquid heptane.

It was shown that the dose dependence of hydrogen, methane, and unsaturated compound accumulation deviates from linearity only at the high irradiation doses (Figure 2.11). Comparison of the data on RTC conducted with gamma rays from a Co-60 source and electrons having an energy of 900 keV did not show any noticeable difference.

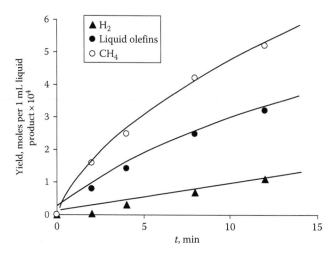

FIGURE 2.11 Kinetics of accumulation of *n*-heptane RTC products at 200°C: (From Topchiev, A.V. and Polak, L.S., *Radiolysis of Hydrocarbons*, The Academy of Sciences of the U.S.S.R., Moscow, Russia, 1962.)

The dependences of radiation-chemical yields of methane and HCs C_2–C_5 on the process temperature are shown in Figure 2.12.

It can be seen from Figure 2.12 that relatively slow increase in the yield of fraction C_2–C_5 changes to a steep ascent at the temperature of 300°C. The temperature of methane yield has a similar character. Simultaneously, the yields of unsaturated HCs in fraction C_2–C_5 increased and reached 62% at 400°C.

RTC of heptane proceeds at temperatures much lower than those characteristic of pure TC of heptane, which has noticeable rates at temperatures above 500°C. The yields of liquid unsaturated compounds increased from 2 molecules/100 eV at room temperature to 350 at 450°C. These facts together with the constancy of the RTC rate allowed assumption that RTC proceeded with the participation of the unbranched chains according to the commonly accepted mechanism of TC.

It was noted that secondary reactions that accompany high-temperature irradiation of *n*-heptane in static conditions distort the real nature of the process. In this connection, RTC of *n*-heptane was studied in the work (Topchiev and Polak 1962) in both static and flow conditions. The reactor of the flow facility was a pipe of stainless steel having an inner diameter of 50 mm and a length of 250 mm. Heating was carried out from the outer side of the reactor. The electron beam was introduced into the reactor through 200–250 μm thick aluminum or beryllium foil.

Figure 2.13 shows temperature dependence of the rate of gaseous product formation for the cases of pure TC and RTC. A component responsible for the initiating action of irradiation at the corresponding temperatures is separately shown.

Although the RTC yields were much higher than those of TC products, it was noted (Topchiev and Polak 1962) that the yields obtained were considerably lower than those expected in flow conditions because of the strongly nonuniform distributions of radiation dose and temperature associated with the reactor design.

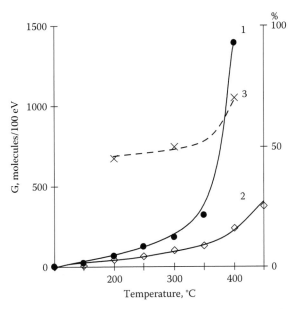

FIGURE 2.12　Dependence of radiation-chemical yields of HCs: fraction C_2–C_5 (1), methane (2), and percentage of unsaturated HCs in fraction C_2–C_5 (3) on RTC temperature. (From Topchiev, A.V. and Polak, L.S., *Radiolysis of Hydrocarbons*, The Academy of Sciences of the U.S.S.R., Moscow, Russia, 1962.)

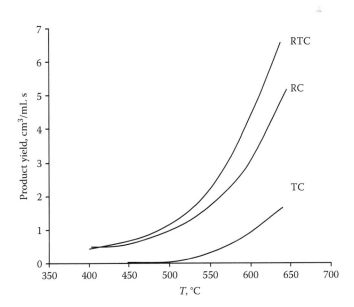

FIGURE 2.13　Dependence of gaseous products yields on temperature for RTC and TC of *n*-heptane. RC is radiation component of RTC. (From Topchiev, A.V. and Polak, L.S., *Radiolysis of Hydrocarbons*, The Academy of Sciences of the U.S.S.R., Moscow, Russia, 1962.)

To clear up the prospects of RTC application to different types of feedstock, RTC of gasoline with the end of boiling at 140°C, gasoline boiling below 200°C, crude oil from Tatarstan field, Russia, and middle and heavy gas oils was studied in the work (Topchiev and Polak 1962). All the experiments were conducted at nearly atmospheric pressure. The dependences obtained for the yields of RTC products for these types of oil feedstock were similar to those observed for n-heptane (Figure 2.13).

Figure 2.14 shows the dependence of the original feedstock conversion in the work (Topchiev and Polak 1962) on process temperature at the constant dose rate and irradiation dose. In the case of Krasnodar natural-gas gasoline processed at 550°C, RTC conversion was an order of magnitude higher than that observed in TC; however, the absolute conversion value was small. At 600°C, the RTC conversion reached about 34%, the RTC to TC ratio of about 3:1, and the conversion due to the radiation component about twice higher than that caused by the thermal component. This ratio can be made still higher by an increase in the dose rate.

An important statement made in the work (Topchiev and Polak 1962) is that a linear dependence of the conversion on irradiation dose and dose rate can be characteristic of a narrow range of these parameters. Their variation in a wider range may reveal the conditions where contribution of the radiation component of RTC can be considerably increased. Further studies (Zaikin and Zaikin 2008a, Zakin 2013a) confirmed this statement.

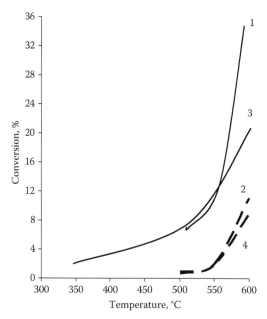

FIGURE 2.14 Dependence of feedstock conversion under RTC and TC on temperature for Krasnodar natural gas gasoline and gasoline fraction with the end of boiling at 140°C. 1—RTC and 2—TC of Krasnodar natural gas gasoline. 3—RTC and 2—TC of gasoline with end of boiling at 140°C. (From Topchiev, A.V. and Polak, L.S., *Radiolysis of Hydrocarbons*, The Academy of Sciences of the U.S.S.R., Moscow, Russia, 1962.)

In the work by Brodskiy et al. (1961), the experiments on decomposition of oil fractions under combined thermal and radiation action were conducted in the channel of a water–water–type nuclear reactor. Oil was irradiated in static conditions (quartz ampoules) and in flow conditions using a circulation facility. The static experiments were intended to the study of the basic regularities in RT decomposition of oil. To verify these results in dynamic conditions and to study the processes of pitch formation and accumulation of the radiolysis products, additional experiments were conducted using a circulation loop placed in a channel of the nuclear reactor. The tests were conducted at temperatures up to 360°C and at pressures up to 6 atm.

Two types of straight-run kerosene–gas oil fractions (200°C–350°C and 190°C–325°C) were used as feedstocks. Both of the fractions contained relatively high concentrations of naphthene and aromatic HCs. For desulfurization before radiation processing, the fractions were subjected to hydrostabilization using a cobalt–alumina–molybdenum catalyst at a pressure of 60 atm and a volume rate of 1 h⁻¹.

At relatively low temperatures below 360°C, a noticeable degree of HC conversion was reached at very high irradiation doses (up to 6 MGy). Because of the low dose rate of gamma radiation in the reactor (about 2.8 Gy/s) and, therefore, low cracking rate, a longtime radiation processing was required to take a high irradiation dose (each experimental run took 10 days).

The results of gas oil radiation processing are shown in Figures 2.15 and 2.16.

The experiments have shown that, in the temperature range of 310°C–330°C, decomposition processes similar to those in TC take on an important role. Below 310°C–330°C, recombination of the radicals forming under ionizing radiation is predominant. At temperatures above this range, the reactions of radical decomposition with C–C bond breaks proceed at a much higher rate. Intensification of the cracking processes becomes apparent in the increase in methane concentration in the gaseous products of oil radiation processing and decrease in the hydrogen content. It is accompanied by increase in the absolute yields of other HC gases (Figure 2.16).

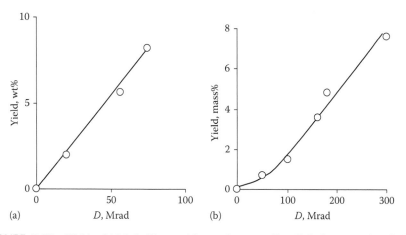

FIGURE 2.15 Yield of high-boiling residue under gas oil radiolysis versus irradiation dose: (a) fraction 200°C–350°C; (b) fraction 190°C–350°C. (From Brodskiy, A.M. et al., *Neftekhimiya (Oil Chem.)*, 3(3), 370, 1961.)

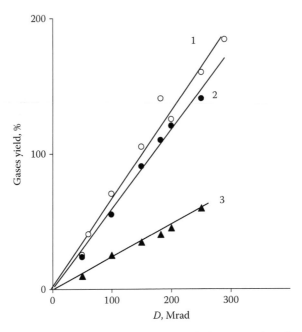

FIGURE 2.16 Gaseous product yield obtained by radiation processing of gas oil (fraction 190°C–325°C) versus irradiation dose: 1—total gas yield; 2—hydrogen yield; 3—methane yield × 10. (From Brodskiy, A.M. et al., *Neftekhimiya (Oil Chem.)*, 3(3), 370, 1961.)

In the work (Brodskiy et al. 1963), similar conditions of oil radiation processing in ampoules were used in RTC studies of the naphthene–gas oil fraction and solutions of benziprene (10^{-6}–10^{-2} M) and anthracene (10^{-3}–10^{-2} M) in this fraction. The decomposition rate of liquid HCs sharply increased at temperatures of 290°C–310°C corresponding to the transition from mode I (simple radiolysis) to mode II (RTC).

The chemical compositions of the products obtained by RTC of the gas oil fraction and activation energies for different modes of radiation processing are shown in Tables 2.3 and 2.4. The activation energy of the overall decomposition process is equal to 2–4 kcal/mol in mode I and 24±2 kcal/mol in mode II. The value of the activation energy in mode II decreases with the inhibitor addition and makes 17±2 kcal/mol. Addition of these types of inhibitors practically does not shift the cracking start temperature and affects only the yields in mode I. Activation energy in mode II changes insignificantly with the transfer from methane to ethane. The obtained values of activation energy corresponding to the energies of radical decomposition showed that the limiting stage of HC decomposition in mode II was associated with the reactions of radical decomposition in a cage.

In the work (Brodskiy and Lavrovskiy 1963), a condition was stated under which the chemical action of ionizing irradiation becomes negligibly small and decomposition of organic compounds proceeds in the same manner as that just under heating, even if the heating is caused by irradiation.

TABLE 2.3

Chemical Composition of Products Obtained by RTC of Gas Oil (Vol%)

Gas Composition	Temperature, °C					
	250	280	300	315	320	350
H_2	80.00	76.76	70.06	64.64	67.07	52.94
CH_4	5.39	5.88	9.78	6/02	6/27	10.60
$C_2H_6+C_2H_4$	5.40	7.33	5.72	7.69	8.69	15.57
C_2H_2	0.38	0.14	0.07	0.12	0.12	0.11
C_3H_8	2.25	3.61	4.75	6.04	6.01	8.88
C_3H_6	1.92	1.56	2.43	3.87	3/66	4.47
C_3H_4	0.17	0.08	0.10	0.17	0.29	0.25
$i\text{-}C_4H_{10}$	0.27	0.38	0.52	0.75	0.59	0.71
$n\text{-}C_4H_{10}$	1.19	1.41	2.48	3.29	2.20	2.37
$i+\alpha\,C_4H_8$	0.75	0.65	1.41	2.52	1.58	1.49
$\beta\text{-}C_4H_8$	0.23	0.17	0.30	0.47	0.36	0.26
$\beta\text{-}C_4H_8+C_4H_6$	0.18	0.16	0.23	0.41	0.40	0.21
ΣC_5	1.86	1/87	2/15	4.01	2/76	2.14

Source: Brodskiy, A.M. et al., *Kinetika I Kataliz (Kinet. Catal.),* 4, 337, 1963.

TABLE 2.4

Activation Energy in Modes I and II for Different Products of Radiation Processing

Sample	Mode	C_1–C_4	CH_4	C_2	C_3H_6	C_3H_8	C_4H_{10}
Gas oil	I	2	4	4	4	7	4
	II	24	26	24	18	18	15
Benziprene solution	I	2	2	2	2	2	2
$(10^{-3}$ M)	II	17	19	24	12	18	12

Source: Brodskiy, A.M. et al., *Kinetika I Kataliz (Kinet. Catal.),* 4, 337, 1963.

The rate of radical thermal generation is higher than the rate of their generation by ionizing radiation if

$$\frac{N}{M} \cdot e^{-E_i/kT} > GP \qquad (2.6)$$

where

    N is the Avogadro number

    M is the molecular mass

Equation 2.6 can be considered as a condition that radiation action cannot anymore provide a considerable chemical effect on decomposition of chemical compounds. Actually, even at the temperature higher by a small margin that the RTC start temperature and even at low dose rates, the RTC rate is many times higher than the rate of the proper radiolysis. The rate of the thermal radical generation rapidly increases with temperature. As soon as it becomes higher than the rate of their radiation generation, a chemical action of radiation may be neglected. Inequality (2.6) can be rewritten in the form

$$T > \frac{E_i}{k \ln\left(k_i \, N / MGP\right)} \tag{2.7}$$

Condition (2.7) strongly depends on activation energy, E_i; all other quantities can be assigned as accurate within an order of magnitude. Substitution of the constant values obtained in the experiments on RTC of the straight-run gasoline fractions resulted in the estimate of $T > 800$ K.

In this derivation, it was considered that radiation contribution to RTC is limited by the reaction chain initiation. However, calculations given in Chapter 1 show that contribution of radiation to chain propagation at temperatures above 500°C is negligibly small, that is, inequality (2.7) retains its validity.

Radiolysis and RTC of n-butane were studied by Matsuoka et al. (1974) in static and flow conditions. Irradiations were performed with a 1.5 MeV collimated electron beam from an electron accelerator.

The effect of temperature on the radiolysis and RTC of n-butane (density was equal to 1.35 g/cm³) and of n-butane with 20 mol% ammonia (density of n-butane was $1.1 \cdot 10^{-3}$ g/cm³) was studied at temperatures from 17°C to 421°C. The dose rate averaged over the length of the cell was 187 Gy/s. The absorbed dose was varied between 5 and 55 kGy, corresponding to reaction temperature. The conversion was in the range of 0.10%–2.00%.

The main products formed in the RTC region, where chain reactions proceeded, were methane, ethane, ethylene, and propylene, which were in accordance with those formed by the TC. In Table 2.5, the product distributions obtained by TC, RTC, and RTC with ammonia addition at 421°C are compared with each other.

In the static experiments above 350°C, some TC of samples occurred during the time of preheating, and the irradiation yields reported were corrected for the contribution of TC; the maximum correction amounted to 15% at 421°C. No considerable difference in the main product distributions was observed between TC and RTC (Table 2.5).

The addition of 20 mol% ammonia, which is well known to scavenge carbonium ions, alters neither the decomposition rate in the RTC of n-butane (Figure 2.17) nor the distribution of the main products. Figure 2.17 shows that the value of activation energy for propylene formation in the RTC region varies with temperature and is not constant. A similar phenomenon was reported by Hoehlein and Freeman (1970) in the RTC of diethylether, and they explained the phenomenon by the change of the type of termination reaction in the radical chain mechanism due to different

TABLE 2.5

Comparison of the Product Distributions Obtained by TC, RTC, and RTC with 20 mol% Ammonia Addition at 421°C (Mol%)

Product	Radiation-Thermal Cracking	Radiation-Thermal Cracking NH$_3$	Thermal Cracking
H$_2$	1.48	1.72	0.70
CH$_4$	36.1	36.9	37.0
C$_2$H$_6$	12.1	11.7	12.2
C$_2$H$_4$	12.3	12.3	12.8
C$_3$H$_6$	35.3	25.7	36.4
1-C$_4$H$_8$	0.35	0.37	0.22
trans-2-C$_4$H$_8$	0.37	0.31	0.30
cis-2-C$_4$H$_8$	0.23	0.19	0.18
Other products	1.73	0.79	0.20

Source: Matsuoka, S. et al., *Can. J. Chem.,* 5(2), 2579, 1974.

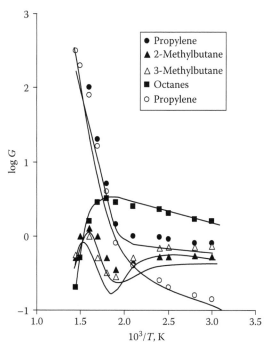

FIGURE 2.17 Arrhenius plots of the yields of (●) propylene, (▲) 2-methylbutane, (Δ) 3-methylbutane, and (■) octanes from the decomposition of *n*-butane with 20 mol% ammonia, and (○) propylene from radiolysis of *n*-butane. (From Matsuoka, S. et al., *Can. J. Chem.,* 5(2), 2579, 1974.)

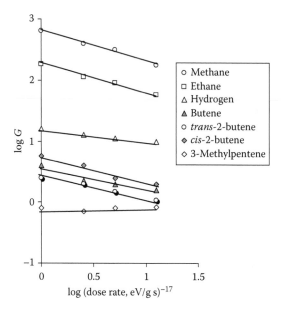

FIGURE 2.18   Dose rate dependencies of the yields at 390°C. (From Matsuoka, S. et al., *Can. J. Chem.*, 5(2), 2579, 1974.)

temperatures. The chain length of RTC obtained from Figure 2.17 was 42 at 421°C. The analysis of RTC kinetics in the paper (Matsuoka et al. 1974) supports this explanation.

Figure 2.18 shows the dose rate dependence of the yields of various products. Dose rate was varied in the range of 15.5–184 Gy/s. The absorbed dose was not strictly fixed and was in the range of 1.9–5.5 kGy.

Figure 2.18 demonstrates decrease in the yields of products with the dose rate at the approximately given irradiation dose. However, any increase in dose in the plot corresponds to an increased dose rate. Being recalculated to the fixed irradiation time, the same data would demonstrate an increase in the product yields approximately proportional to the square root of the dose rate in accordance with RTC theory based on the radical mechanism.

Generally, the analyses of RTC kinetics and dependences of the product yields on irradiation dose and dose rate confirmed that the main products from the RTC of *n*-butane were produced by the radical chain mechanism. It supported the conclusion that the RTC of propane proceeded by radical mechanism, inferred by Pereverzev et al. (1968) from the effect of nitric oxide addition.

Among the products of the RTC of *n*-butane, a series of octane products was found, such as 3,4-dimethylhexane (3,4DiMHx), 3-methylheptane (3-MHp), and *n*-octane (*n*-Oct). The temperature dependence of these octane yields are shown in Figure 2.19. The other octanes were probably formed by an ionic process, since their yields were suppressed by the addition of ammonia.

The effect of temperature on the yield of *n*-butane with ammonia is given in Figure 2.20. In the temperature range between 85°C and 220°C, the methane yield

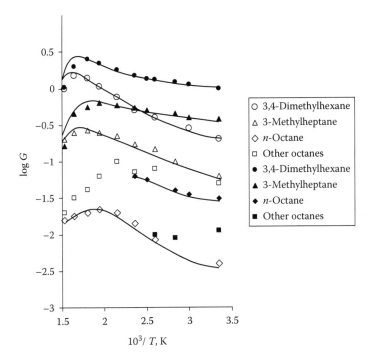

FIGURE 2.19 Temperature dependencies of the yields of individual octanes: (○) 3,4-dimethylhexane, (△) 3-methylheptane, (◇) n-octane, and (□) other octanes. Open and filled marks correspond to the yields of individual octanes in the absence or presence of 20 mol% ammonia, respectively. (From Matsuoka, S. et al., *Can. J. Chem.*, 5(2), 2579, 1974.)

increases with an apparent activation energy of 3.1 kcal/mol. This was attributed not to the thermal decomposition of $C_4H_{10}^+$ ion but to a hydrogen atom abstraction from the n-butane molecule by the methyl radical.

The yields of propylene molecules from TC and RTC per 100 eV consumed energy in a flow experiment are shown in Figure 2.21. The calculated activation energies for propylene formation in the individual temperature ranges of RTC varied in the range from 14 to 28 kcal/mol. A lower value of 12 kcal/mol obtained at high temperatures was probably due to the buildup of the product propylene, which suppresses the decomposition of the rate of n-butane.

Above 410°C, the activation energy is between 14 and 12 kcal/mol. Kinetic analysis had shown that it corresponded to the following reaction of the chain termination:

$$C_2H_5 \cdot + C_2H_5 \cdot \rightarrow \text{Products} \tag{2.8}$$

The analysis of the experimental data on the RTC of n-butane in static and flow conditions had shown that the major part of the butane yields aroused from the propagating radicals in the following reactions:

$$C_4H_9 \cdot \rightarrow H + C_4H_8 \tag{2.9}$$

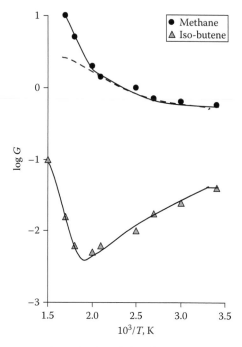

FIGURE 2.20 Temperature dependencies of the yields of (●) methane and (▲) iso-butene with 20 mol% ammonia. Broken line shows the calculated yield in the paper. (Matsuoka et al. 1974.)

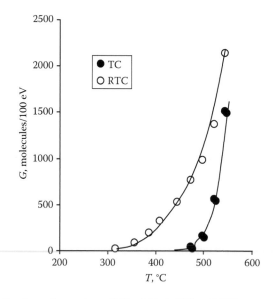

FIGURE 2.21 G-values of propylene, C_3H_8 yields in TC and RTC versus temperature.

$$H + C_4H_{10} \rightarrow H_2 + C_4H_9 \qquad (2.10)$$

Thus, in the RTC of n-butane at the temperature of 390°C and the dose rate of 184 Gy/s, 50%–70% of the product butenes resulted from the decomposition of butyl radicals that were generated by the radical chain propagation process. In the TC of n-butane, the same reactions may also cover the major part of the butane formation when conversion is low.

The earlier-discussed studies of the RTC of HC in the gaseous phase show that this process is to a great extent similar to TC. The essential difference was found in the alkane isomerization. For example, RTC of butane (Matsuoka et al. 1975) resulted in the formation of butane isomers and butylene, which was attributed to a chain process. Neither isobutene nor isobutylene were formed as a result of TC.

A much more considerable difference between TC and RTC is observed in the condensed matter due to the competing polycondensation and isomerization reactions, cage effect, formation of radical pairs and their contribution to the chain propagation, etc. The heavier the oil, the greater is the difference between the TC and RTC mechanisms and the product distributions. This difference becomes most significant in the case of low temperatures and high dose rates.

The effect of dose rate on the formation of unsaturated compounds in the process of n-heptane radiation cracking was studied in the work (Gabsatararova and Kabakchi 1969).

Co-60 facility was used as a source of gamma radiation. Irradiation of heptane samples was carried out in the sealed degasified ampoules of molybdenum glass preheated to the reaction temperature. In the whole temperature range of 250°C–450°C used in these experiments, n-heptane remained in the gaseous state. The pressure in the ampoules at temperatures from 250°C to 450°C was varied from 5.0 to 7.2 atm, respectively.

The isotherms in Figure 2.22 show the dependence of the bromine number for the irradiated sample on the absorbed dose at the dose rate of 6.3 Gy/s. The observed nonlinearity in the initial part of the curve was explained by a low concentration of the unsaturated products at low absorbed doses so that they did not enter secondary reactions. In the high-dose region, unsaturated products enter secondary processes as radical acceptors.

Figure 2.23 shows temperature dependence of radiation-chemical yields of unsaturated products determined from the initial region of the isotherms for different dose rates. At the given dose rate, radiation-chemical yields of unsaturated products increase as the temperature increases with a higher rate at lower dose rates. The same behavior can be seen in Figure 2.24 where accumulation of the unsaturated products is represented as a function of dose at the temperature of 450°C for the different dose rates.

It should be emphasized that decrease in the rate of accumulation of cracking products was observed at the given dose but not at the given time of irradiation. Reprocessing of these data (Zaikin 2008, Zaikin and Zaikina 2008a) has shown that, at the given irradiation time, accumulation of the unsaturated compounds increases with the increase in dose rate. At the low dose rates of gamma irradiation used in the study by Gabsatarova and Kabakchi (1969) (0.63–8.9 Gy/s) and heightened irradiation temperature, radiation contribution to RTC propagation is very small, and the

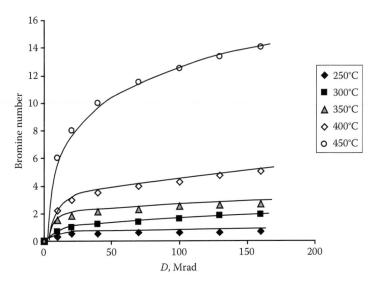

FIGURE 2.22 Accumulation of unsaturated products (in the units of g bromine/100 g irradiated product) at the dose rate of 0.63 Gy/s and different temperatures. (From Gabsatarova, S.A. and Kabakchi, A.M., *Khimiya Vysokikh Energiy (High-Energy. Chem.)*, 3, 126, 1969.)

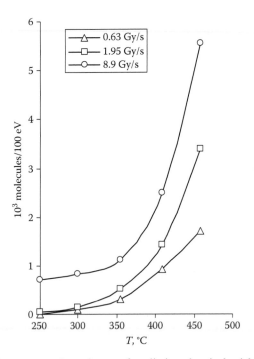

FIGURE 2.23 Temperature dependence of radiation-chemical yields of unsaturated compounds at different dose rates. (From Gabsatarova, S.A. and Kabakchi, A.M., *Khimiya Vysokikh Energiy (High-Energy. Chem.)*, 3, 126, 1969.)

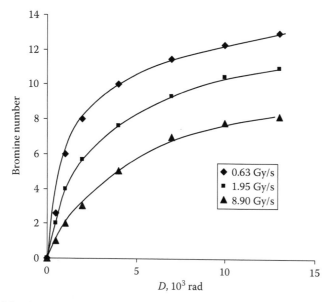

FIGURE 2.24 Accumulation of unsaturated products (in the units of g bromine/100 g irradiated product) at the temperature of 450°C and different dose rates. (From Gabsatarova, S.A. and Kabakchi, A.M., *Khimiya Vysokikh Energiy (High-Energy. Chem.)*, 3, 126, 1969.)

rate of RTC at the given irradiation time is proportional to the square root of dose rate in complete accordance with RTC theory (Section 1.2).

RTC of *n*-hexadecane under gamma irradiation was studied in the work (Panchenkov et al. 1981). The reaction was carried out in the autoclave after its purging with helium for the removal of gases from the reaction volume. Irradiation was conducted using a Co-60 facility. The dose rate was varied in the range from 7.8 to 16.7 Gy/s; the maximal absorbed dose was 20 kGy. The pressure in the autoclave depended on temperature and experimental conditions but did not exceed 10 MPa in all the cases.

The yields of RTC products and the degree of *n*-hexadecane conversion at different process temperatures are shown in Figure 2.25. The trial experiments without irradiation have shown that TC in these conditions can be neglected.

The radiation-chemical yields of *n*-hexadecane decomposition are shown in Figure 2.26 as functions of inverse temperature. The *G*-value at 723 K approximately equal to 1000 molecules/100 eV corresponds to the data earlier obtained for other HCs while activation energy is close to the value characteristic of chain propagation in TC. It indicates the major energy consumption at the stage of chain propagation and the predominating contribution of the radiation component to cracking initiation at this temperature.

The studies of the dose effect on the yields of the main RTC products (Figure 2.27) indicate a linear increase in the total conversion and predominant yields of lighter HCs C_5–C_{10} compared with that of heavier HCs C_{11}–C_{12}. Concentration of the unsaturated compounds reaches the value of 17 mass% at the dose of 20 kGy.

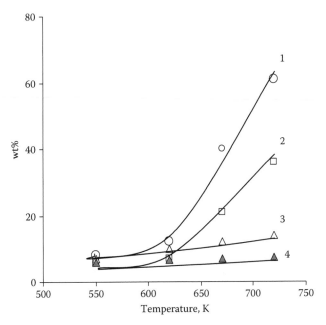

FIGURE 2.25 Dependence of *n*-hexadecane conversion and RTC product yields on temperature at $P = 16.7$ Gy/s and $D = 20$ kGy. 1—conversion; 2—yields of C_5–C_{10}; 3—yields of gaseous products. (From Panchenkov, G.M. et al., *Khimiya Vysokikh Energiy (High Energy Chem.)*, 15, 426, 1981.)

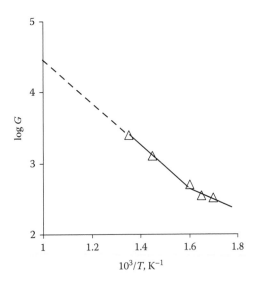

FIGURE 2.26 Dependence of log *G* for products of *n*-hexadecane decomposition on $1/T$ at $P = 16.7$ Gy/s and $D = 20$ kGy. (From Panchenkov, G.M. et al., *Khimiya Vysokikh Energiy (High Energy Chem.)*, 15, 426, 1981.)

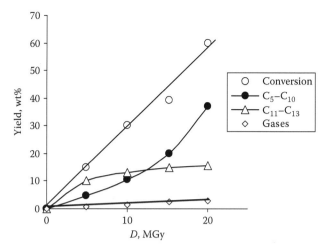

FIGURE 2.27 Dependence of *n*-hexadecane conversion and RTC product yields on absorbed dose at a temperature of 723 K and dose rate of 6.7 Gy/s. (From Panchenkov, G.M. et al., *Khimiya Vysokikh Energiy (High Energy Chem.)*, 15, 426, 1981.)

The results shown in Figure 2.28 demonstrate some decrease in the yields of RTC products as dose rate increases at the given dose of 20 kGy. However, reprocessing of these results in the works (Zaikin 2008, Zaikin and Zaikina 2008a) has shown that, at the given irradiation time, the yields of RTC products increase with the dose rate so that dependence of the cracking rate on dose rate is well described by

$$\frac{W}{\sqrt{P}} = a\,P + b \tag{2.11}$$

It indicates a considerable contribution of radiation to propagation of the RTC chain reaction.

A study of *n*-hexadecane behavior under gamma irradiation was carried out in the later work by Wu et al. (1997a). The study was intended to examine the product compositions of *n*-$C_{16}H_{34}$ after liquid- and gas-phase irradiation and to obtain more specific information on the process kinetics.

n-Hexadecane with a purity of above 99.9% was irradiated in two types of ampoules of Pyrex glass. One sort of ampoules was about 10 mL in volume, into which 5 mL of *n*-$C_{16}H_{34}$ was added. The other sort of ampoules was about 50 mL in volume with 0.5 mL of $C_{16}H_{34}$ added. The cracking in these two types of ampoules was carried out in the liquid and gas phases, respectively. Argon under the pressure of 5 atm was introduced before cracking to balance the pressure inside and outside the glass ampoule to prevent its breaking. A Co-60 facility was used as a source of gamma radiation.

Higher yields of *n*-alkanes were observed in liquid-phase RTC. At low temperature and low conversion, the product distribution after radiation processing was very similar to that for pure TC, that is, there was nearly an equimolar distribution of *n*-alkanes and 1-alkenes, and about 200 mol scission products were formed due to decomposition of 100 mol *n*-hexadecane. However, the yields of scission products

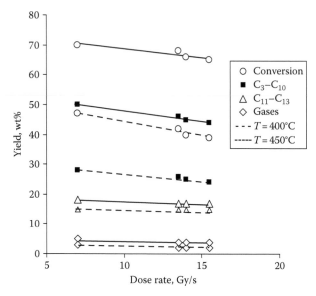

FIGURE 2.28 Dependence of *n*-hexadecane conversion and yields of RTC products on dose rate at the dose of 20 kGy; $T=400$ (- - -) and 450 K (–). (From Panchenkov, G.M. et al., *Khimiya Vysokikh Energiy (High Energy Chem.)*, 15, 426, 1981.)

per 100 mol cracked *n*-hexadecane decreased to about 100 mol at higher temperatures or after longer residence time due to an appreciable yield of addition products.

Contrary to the liquid-phase irradiation, the selectivity of 1-alkenes was higher for the gas-phase radiation processing, even at lower temperature. More than 200 mol of scission products were formed upon decomposition of 100 mol *n*-hexadecane in the gas phase irrespective of the temperature and residence time.

In order to characterize phase and temperature dependences of the product selectivity, the products were classified into three parts: gas (C_1–C_4), light (C_5–C_{15}), and addition products ($C^{16}+$). Plots of weight percentages of the three parts versus residence time are shown for liquid-phase and gas-phase RTC in Figures 2.29 and 2.30, respectively.

The fraction of gas products was very low, for example, about 5 wt% at 330°C. As residence time increased, there was an increase in the addition products and a decrease in C_5–C_{15} products at the given higher temperature. The proposed explanation of this effect was that addition of alkenes to parent radicals was favored at higher alkene concentrations, which could be achieved after longer irradiation. It was noted that the contents of the addition products decreased while the contents of other products increased with increasing temperature.

Figure 2.30 shows that the contents of both gas and C_5–C_{15} products are nearly independent of the residence time at 330°C in the case of gas-phase irradiation. At higher temperatures, longer residence time leads to an increased yield of gas products and a decreased yield of C_5–C_{15} products.

In contrast to the earlier-described experiments by Panchenkov et al. (1981), radiation behavior of *n*-hexadecane studied by Wu et al. (1997a) is not characteristic of

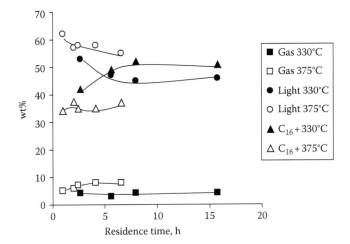

FIGURE 2.29 Correlation between the weight percentage of various group products and residence time for liquid-phase RTC. $P = 0.13$ Gy/s. (From Wu, G. et al., *Ind. Eng. Chem. Res.*, 36, 1973, 1997a.)

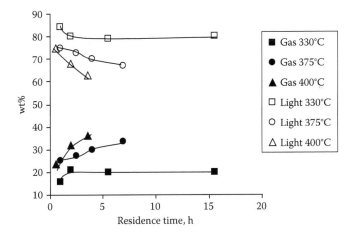

FIGURE 2.30 Correlation between the weight percentage of various group products and residence time for gas-phase RTC. $P = 0.16$ Gy/s. (From Wu, G. et al., *Ind. Eng. Chem. Res.*, 36, 1973, 1997a.)

RTC, especially in the case of liquid-phase radiation processing. The distinguishing characteristics of RTC, noted practically in all RTC studies, are strong gas evolution with the yields proportional to irradiation dose (or residence time) and considerable increase in the contents of light fractions. The same is valid for the conventional TC. As distinct from these observations, Figure 2.29 demonstrates very small gas evolution and decrease in the yields of light fractions with increasing irradiation dose.

The difference in studies by Panchenkov et al. (1981) and Wu et al. (1997a) is that a 100 times lower dose rate is used in the latter case. It means that the dose of

20 kGy was taken after 20-min irradiation in the study by Panchenkov et al. (1981), while 15.5 h were needed in the experiment by Wu et al. (1997a) to take the dose of 8 kGy. The low dose rate and, therefore, long-time irradiation at heightened temperatures create favorable conditions for the competing polycondensation reactions that always accompany RTC (Mustafaev 1990, Zaikin and Zaikina 2008a). The role of these reactions in the study by Wu (1987a) is confirmed by the formation of a great amount of addition products.

This interpretation is supported by the molar distribution of *i*-alkenes and *n*-alkanes in the product of liquid-phase irradiation of *n*-hexadecane represented in the paper [Wu et al. 1997a] (Figure 2.31).

Figure 2.31 shows noticeable decrease in the contents of unsaturated HCs after radiation processing. Olefins are the natural products of both thermal and radiation cracking. Increase in olefin yields is the sign of cracking while decrease in their concentration is the sign of polycondensation. Thus, the study by Wu et al. (1997a) demonstrates a considerable role of polycondensation reactions in low dose rate radiation processing of HCs. However, this type of reactions was not taken into account in the analysis of RTC kinetics represented in the papers (Wu et al. 1997a,b).

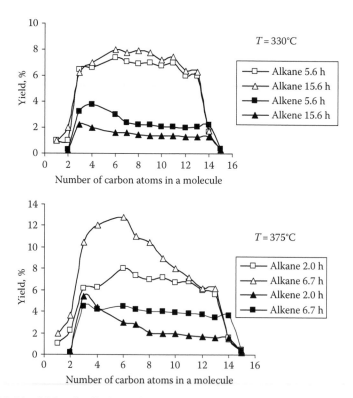

FIGURE 2.31 Molar distributions of 1-alkene and *n*-alkane products as a function of carbon numbers for liquid-phase RTC. The yield is expressed in mole per cracked 100 mol *n*-$C_{16}H_{34}$. Dose rate: 0.13 Gy/s. (From Wu, G. et al., *Ind. Eng. Chem. Res.*, 36, 1973, 1997a.)

2.3 RADIATION-THERMAL CRACKING OF AROMATIC AND NAPHTHENE HYDROCARBONS

The RTC regularities discussed in the previous sections mainly relate to alkanes whose behavior, to a great extent, determines the results of radiation cracking of crude oils and oil products. RTC of aromatic and naphthene compounds was explored less sufficiently. RTC of terphenyls, aromatic compounds consisting of a central benzene ring substituted with two phenyl groups, was studied in more detail.

Ortho-Terphenyl   Meta-Terphenyl   Para-Terphenyl

Tomlinson et al. (Tomlinson 1966, Tomlinson et al. 1966a) studied radiation and RT decomposition of the terphenyls and the partly hydrogenated terphenyl mixture. A focal point of these studies was the evaluation of a terphenyl mixture as a low-melting coolant for a nuclear reactor.

The initial decomposition rates of *ortho*- and *meta*-terphenyl under the mixed (fast neutron and gamma) irradiation are shown in Table 2.6. The data indicated that

1. The rate of radiolytic decomposition of *ortho*- and *meta*-terphenyl increased approximately 10-fold as the temperature increased from 100°C to 450°C. Over the same temperature range, the decomposition rate for *meta*-terphenyl increased approximately threefold.
2. Below 300°C, the rate of *ortho*-terphenyl decomposition by the mixed fast neutron and gamma irradiation was twice that for gamma rays alone. The corresponding ratio for *meta*-terphenyl was 2.5. These data indicate that the reactor fast neutron radiation was about 3 times as damaging to *ortho*-terphenyl and 4.5 times as damaging to *meta*-terphenyl as was gamma irradiation.
3. At temperatures above 350°C, the decomposition rates appeared to be the same for fast neutrons and gamma radiation but decreased with increasing dose rate.

Irradiation of *meta*-terphenyl with 1.5 MeV electrons was performed using a Van de Graaf electron accelerator. Electron irradiation allowed covering a wider range of dose rates than was available for the reactor irradiations. The temperature and dose rate dependences of radiation-chemical yield of *meta*-terphenyl decomposition at temperatures above 350°C are given (Figure 2.32).

The variation of the *G*-values with electron dose rate is illustrated in Figure 2.33.

TABLE 2.6

Initial Radiolytic Decomposition Rates of
***Ortho-* and *Meta*-Terphenyl**

Temperature, °C	Gamma Rays	Fast Neutron and Gamma Rays	
	0.2 W/g	0.1–0.3 W/g	1 W/g
Ortho 250	0.23	0.5	0.5
350	0.72	0.78	0.75
400	1.5	1.5	0.9
450	5	4	1.3
Meta 182	0.15		
250		0.4	0.4
360	0.3	0.6	
385		0.7	0.44
397	0.9	0.8	
420		0.95	0.51
455	~1.7	~1.7	

G, Molecules Decomposed per 100 eV Energy Absorbed

Source: Tomlinson, M., Radiation and thermal decomposition of terphenyls, Atomic Energy of Canada, Report AECL-2641, 20pp, 1966.

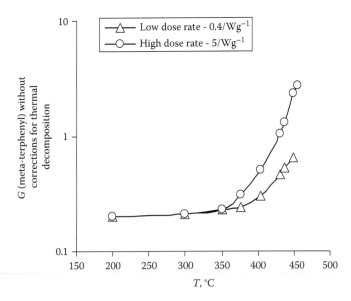

FIGURE 2.32 Temperature dependence of *G*-values for *meta*-terphenyl decomposition under electron irradiation. (From Tomlinson, M., Radiation and thermal decomposition of terphenyls, Atomic Energy of Canada, Report AECL-2641, 20pp, 1966.)

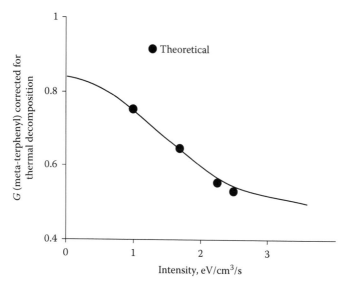

FIGURE 2.33   Radiation intensity dependence of *G*-values for *meta*-terphenyl decomposition. (From Tomlinson, M., Radiation and thermal decomposition of terphenyls, Atomic Energy of Canada, Report AECL-2641, 20pp, 1966.)

The observations of this study indicate that at the temperatures above 350°C, there is a change in the mechanism of radiolysis accompanied with the increase in *G*-values for terphenyl decomposition at these high temperatures. It should be emphasized that this observation does not mean cracking suppression or any decrease in the rate of cracking reactions at high dose rates. Actually, G_d-values for a material decomposition expressed are defined by the equation

$$G_d = -\frac{1}{P}\frac{dC}{dt} = \frac{W}{P}$$

(2.12)

where
   C is the concentration of the original fraction
   P is the dose rate
   W is the initial cracking rate

Figure 2.33 shows that the cracking rate of *meta*-terphenyl, *W*, increases by 4000 as the dose rate increases by four orders of magnitude. For comparison, the theory of alkane radiation cracking, described in Chapter 1, predicts proportionality of *W* to P^α where α may vary between 0.5 and 1.5 depending on the radiation contribution to the chain propagation.

The high boiler and residual coolant concentrations as functions of irradiation dose of 1.5 MeV electrons are shown in Figure 2.34.

The same regularities were observed in the experiments on the irradiation of the partially hydrogenated terphenyl mixture HB-40, which was of interest as a reactor

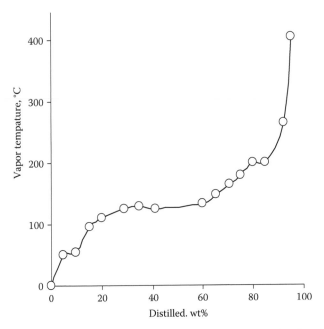

FIGURE 2.34 Distillation curve (lower) and molecular weight distribution (upper) of an HB-40 coolant sample. (From Tomlinson, M., Radiation and thermal decomposition of terphenyls, Atomic Energy of Canada, Report AECL-2641, 20pp, 1966.)

coolant (Tomlinson et al. 1966b, 1967). HB-40 is a complex mixture of aromatic and saturated groups. Its distillation curve is shown in Figure 2.35. The products of HB-40 decomposition were measured in four groups, according to their volatilities: gases (H_2 to C_5 HCs), volatiles (benzene to biphenyl (not included)), residual coolant (biphenyl to *para*-terphenyl), and high boilers (HB; less volatile than *p*-terphenyl). The principal radiation decomposition products were HB, although gases and volatiles were also formed in significant amounts.

The effect of dose rate on HB-40 decomposition under electron irradiation is shown in Table 2.7.

Table 2.7 shows that the average rate of high-boiling fraction accumulation, $W_{HB} = C_{HB\,average}/t$, increases by 52 times while the dose rate increases by 35.2 times. However, more detailed data on cracking kinetics are needed for the determination of the instant cracking rate dependence on dose rate.

The initial decomposition rates for electron irradiation of HB-40 are compared in Table 2.8 with the results of reactor irradiation where 62% of the energy absorbed was due to recoil protons from the fast neutron component of the radiation and the remainder was due to Compton and photoelectrons from gamma rays.

The decomposition rates for reactor irradiation were about twice the rates for electron irradiation. This suggests that HB-40 resembles terphenyl mixtures in being more sensitive to the damage by fast neutrons than to that by gamma rays and electrons.

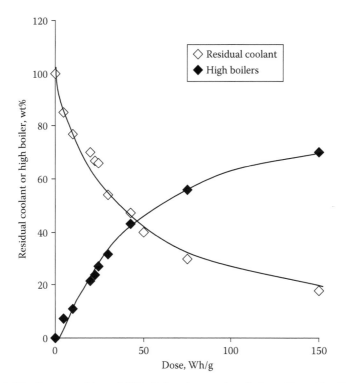

FIGURE 2.35 Decomposition of HB-40 by electron irradiation. (From Tomlinson, M. et al., *Nucl. Sci. Eng.*, 26, 547, 1966b.)

TABLE 2.7

Effect of Dose Rate for HB-40

Electron Irradiation at 350°C

Mean dose rate, W/g	8.8	0.25
Irradiation time, h	0.88	30.9
Target current, μA	100	3.0
High Boilers, wt%	9.6	6.4
Total gas produced, cm^3/g	13.1	13.2
Hydrogen content, vol%	89.2	88.0

Source: Tomlinson, M. et al., *Nucl. Sci. Eng.*, 30, 14, 1967.

Accumulation of the high-boiling fraction during irradiation was accompanied by the increase in the viscosity of the overall product (Figure 2.36).

The high-boiling fraction formed in irradiated HB-40 possessed relatively low thermal stability and was partly decomposed to more volatile materials on further heating. The same behavior was demonstrated by the industrial terphenyl mixture

TABLE 2.8

Comparison of Electron and Reactor Radiation Decomposition of HB-40

	Electron Irradiation	Reactor Irradiation
Energy absorbed from fast neutrons, %	0	62
High boilers, g/(k Wh A)	15.4±1.6	36±7
Coolant decomposed, g/(k Wh A)	33±4	1.6±0.5
Gas produced, g/(k Wh A)	1.8±0.3	4.0

Source: Tomlinson, M. et al., *Nucl. Sci. Eng.*, 30, 14, 1967.

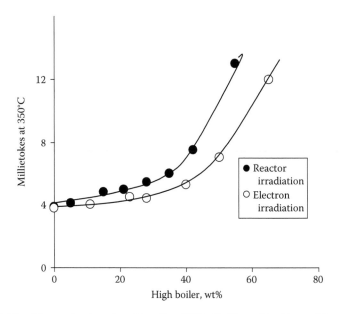

FIGURE 2.36 Kinematic viscosity of irradiated HB-40. (From Tomlinson, M. et al., *Nucl. Sci. Eng.*, 30, 14, 1967.)

Satowax OM, as shown in Figure 2.37. The initial decrease in HB content at 425°C was followed by further polymerization at about the same rate as that for unirradiated material. Similar results were obtained at 350°C and 400°C on a more protracted time scale.

A study of the temperature, intensity, and post-irradiation effects in *meta*-terphenyl decomposition under electron irradiation was conducted by Wushke and Tomlinson (1968). Terphenyl samples were irradiated with 1.35 MeV electrons in static conditions. The dependences of G-values for *meta*-terphenyl decomposition on

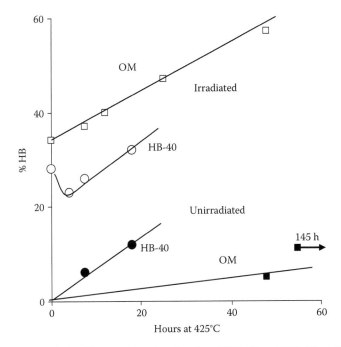

FIGURE 2.37   Pyrolysis of irradiated and nonirradiated HB-40 and OM. (From Tomlinson, M. et al., *Nucl. Sci. Eng.*, 26, 547, 1966b.)

temperature and beam intensity were similar to those reported earlier in the paper (Tomlinson 1966).

It was suggested that radiolytic decomposition of *meta*-terphenyl was a second-order reaction, and an additional thermal process is superimposed at high temperatures. The following system of equations was proposed to describe the resulting RT process:

$$-\frac{dC}{dt} = G_p PC^2 + kC \tag{2.13}$$

The solution of this equation is

$$\ln\left(\frac{C_p P + kC}{C}\right) - \ln\left(\frac{G_0 P + kC_0}{C_0}\right) = k(t - t_0) \tag{2.14}$$

where
   C, C_0 are weight fractions of *meta*-terphenyl at times t, t_o
   G_p is the initial radiation-chemical yield for *meta*-terphenyl decomposition
   P is the dose rate
   k is the rate constant for the first-order thermal decomposition

The constants G_p and k were computed as a best fit for the available data. It appeared that, in the frames of this model, the rate of the thermal process k strongly depended on the dose rate of irradiation. It indicated a self-contradiction of the model.

A more detailed study of *meta*- and *ortho*-terphenyl decomposition under reactor and electron irradiation was carried out by Boyd and Tomlinson (1968). Radiation-chemical yields for *ortho*- and *meta*-biphenyl decomposition under reactor irradiation are shown in Figures 2.38 and 2.39 as functions of temperature.

A marked increase in the total terphenyl decomposition above 350°C was accompanied by increased formation of biphenyl and gases. G-values for biphenyl formation varied from 0.002 at 100°C–200°C to 0.8 for *ortho*- and 0.3 for *meta*-biphenyl radiation decomposition. In the same temperature range, total gas G-values increased from 0.04–0.05 up to 0.2 and 0.15 for *ortho*- and *meta*-biphenyl decomposition, respectively.

The variation of the decomposition yield for *meta*-terphenyl at 440°C is shown over a wide range of electron beam intensity in Figure 2.40.

For electron irradiation, decomposition varied inversely as $\sqrt{P}$. The least mean squares fit yielded $G = (0.383 \pm 0.07) + (0.277 \pm 0.01)/\sqrt{P}$. The reaction rate can be written in the form

$$W = GP = a\sqrt{P} + bP \tag{2.15}$$

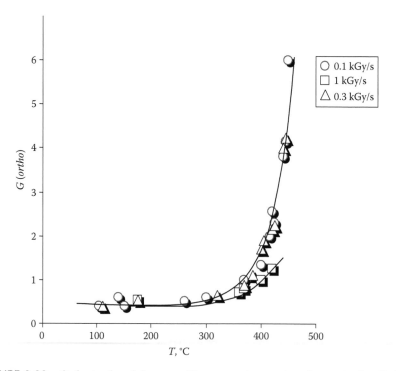

FIGURE 2.38 *Ortho*-terphenyl decomposition versus temperature for reactor irradiation: (○) 0.1 kGy/s, (□) 1 kGy/s, and (△) 0.3 kGy/s. (From Boyd, A.W. and Tomlinson, M., Radiolysis of ortho- and meta-terphenyl, Atomic Energy of Canada, Report AECL-2730, 20pp, 1968.)

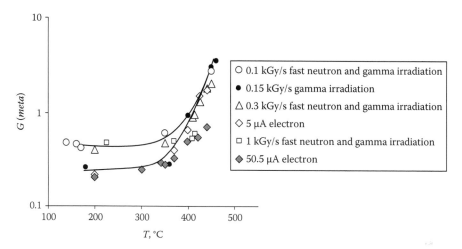

FIGURE 2.39 *Meta*-terphenyl decomposition versus temperature. (From Boyd, A.W. and Tomlinson, M., Radiolysis of ortho- and meta-terphenyl, Atomic Energy of Canada, Report AECL-2730, 20pp, 1968.)

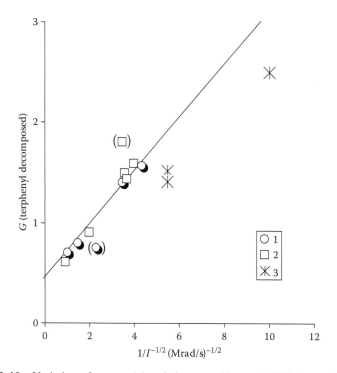

FIGURE 2.40 Variation of *meta*-terphenyl decomposition at 440°C: 1—small electron irradiation cell; 2—large electron irradiation cell; and 3—reactor irradiation. (From Boyd, A.W. and Tomlinson, M., Radiolysis of ortho- and meta-terphenyl, Atomic Energy of Canada, Report AECL-2730, 20pp, 1968.)

Note that the steady-state concentration of radiation-generated radicals (or ions) is proportional to $\sqrt{P}$. Therefore, it can be suggested that the first term in the equation is responsible for the first-order reaction of radicals (or ions) with terphenyl molecules while the second term is responsible for radical (or ion) recombination.

Scarborough and Ingalls (1966) studied the pyrolysis and high-temperature radiolysis (400°C–482°C) of liquid *ortho*-terphenyl. The pyrolytic decomposition was found to follow simple first-order kinetics with an activation energy of 71.1 kcal/mol. The radiolytically induced decomposition occurred with an apparent activation energy of about 22 kcal/mol. The ratio of biphenyl formation to terphenyl disappearance was about 0.35 for pyrolysis but increased as temperature increased during radiolysis with 1 meV electrons (Figure 2.41).

The results were interpreted in terms of a thermal spike model, which attributes the increased terphenyl decomposition at high temperatures to transient excess temperature at the site of the radiation interaction events. However, the estimate had shown that the spikes are not large enough to account for the yields if all the reactions were to take place in the spikes. Therefore, the model required that a chain reaction continued outside the thermal spike.

In view of the theory of radiation-induced chain reactions, involvement of the thermal spikes into a reaction mechanism seems superfluous since radiation energy

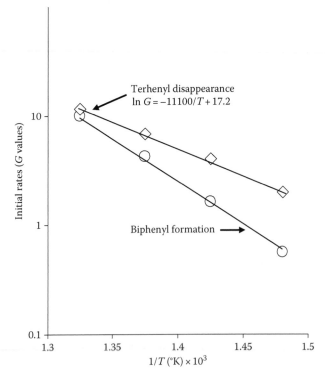

FIGURE 2.41 Initial rates of *ortho*-terphenyl decomposition versus temperature. (From Scarborough, J.M. and Ingalls, R.B., *J. Phys. Chem.*, 71, 486, 1966.)

is quite sufficient for primary decomposition reactions and maintenance of the high concentrations of chain carriers. The activation energy of 22 kcal/mol for the chain propagation calculated in the paper (Scarborough and Ingalls 1966) is characteristic of the radiation cracking by a radical mechanism.

From the point of heavy oil industrial processing, most important is conversion of asphaltenes, the heaviest and hardly processed aromatic compounds. We conducted a series of experiments to study asphaltene conversion in different model solutions. As a result of radiation processing, asphaltene conversion was 30%–50%.

In one of the runs, 23 mass% asphaltenes were dissolved in 63 mass% kerosene and 14 mass% chloroform. This solution was irradiated in the sealed aluminized polyethylene bags at the 4 MeV linear electron accelerator ELU-4 at the time-averaged dose rate of 6 kGy/s. The sample was processed at the temperature of 70°C. Irradiation dose was 780 kGy. The fractional content of the liquid product was analyzed by means of gas–liquid chromatography (Figure 2.42).

The data of Figure 2.42 show 53% conversion of the fraction C_{30}–C_{43} boiling out in the temperature range 450°C–580°C.

The effects of irradiation conditions on mechanisms of chemical conversion in naphthene and aromatic HCs were studied on the examples of two high-viscous crude oils with very different compositions and properties: heavy oil from Karazhanbas field (Zaikina and Mamonova 1999, Zaikin et al. 2001) and high-paraffin oil from Kumkol field, Kazakhstan.

Karazhanbas oil pertains to the sort of crude from some of Kazakhstan fields that is hardly extracted, transported, and processed due to its high viscosity (99.7 mm²/s), density (0.93–0.95 g/cm²), and considerable contents of sulfur (about 2 mass%) and vanadium (100–120 mcg/g). This sort of oil has a low point of solidification (–18°C) as a result of high contents of pitch-aromatic components combined with a small amount of solid paraffins (1.4–1.5 mass%).

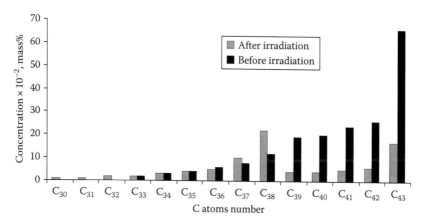

FIGURE 2.42 Dependence of fraction concentrations in asphaltene solution on the number of carbon atoms in a molecule before and after electron irradiation. (From Scarborough, J.M. and Ingalls, R.B., *J. Phys. Chem.*, 71, 486, 1966.)

Concentration of fractions with the boiling temperature up to 200°C is 1.3–6.5 mass%, the content of fractions boiling in the temperature range of 200°C–350°C is 15.1–20.2 mass%, and that in the range of 350°C–490°C is 22.6–25.7 mass%. According to chromatography data, the HC content of the gasoline fraction with boiling temperature up to 100°C is represented by isoparaffin (29.5 mass%), naphthene (30.2 mass%), and aromatic HCs (15.0 mass%); concentration of n-paraffins (mainly HC with C_8–C_9 chain length) is 5.9 mass% (Nadirov 1995).

RT processing of Karazhanbas oil with 2 MeV electrons was performed in two different modes: $T=425$°C, $P=1.5$ kGy/s (mode 1), and $T=375$°C, $P=1.25$ kGy/s (mode 2), where P is the dose rate.

In most part of the works on RTC in the oil feedstock, the theoretical treatment of experimental results is limited by the analysis of radiation-induced reactions for saturated HCs of the aliphatic series subjected to radiation destruction to the greatest extent. However, Karazhanbas oil is characterized by high contents of naphthene and aromatic HCs represented by alkyl-substituted compounds. Therefore, optimization of its RTC conditions required consideration of conversion mechanisms for cyclic and aromatic compounds essentially affecting the contents and characteristic properties of RTC products.

The predominant direction in the reactions proceeding under irradiation of such a multicomponent system as heavy oil is affected by the simultaneous presence of the different types of HC and non-HC components. The resulting direction of the radiation energy transfer depends on the reactivity and on the structure of the molecules in the complex composition that affects both radiation resistance of the separate components and prevailing directions of the reactions. In particular, less stable alkanes with a higher degree of branching can "protect" other components of the system against radiation.

Dependence of cyclic alkane contents in gasoline fractions extracted from liquid RTC products on irradiation dose in modes 1 and 2 is shown in Figure 2.43 according to gas-liquid chromatography data.

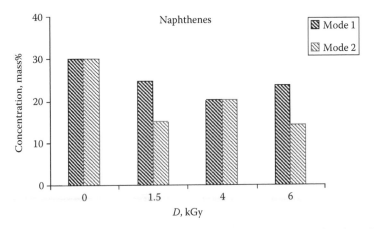

FIGURE 2.43 Dependence of naphthene concentrations in the gasoline fraction of liquid RTC products on irradiation dose in two modes: mode 1: $T=425$°C, $P=1.5$ kGy/s; mode 2: $T=375$°C, $P=1.25$ kGy/s. (From Zaikin, Y.A. et al., Radiat. *Phys. Chem.*, 60, 211, 2001.)

In mode 1, characterised by a heightened rate of HC decomposition, the contents of the cyclic alkanes in the gasoline fractions of RTC products do not change much in comparison with those in the gasoline fractions of the feedstock. The prevailing direction of decomposition processes in this irradiation mode is connected with behavior of alkyl substituents of the ring. The cyclic alkanes with long alkyl chains are subjected to chain disintegration during RTC through the reactions of the same type as those for alkanes. Availability of substituents in the cyclic alkane ring increases yields of C–C bond breaks compared with breaks of C–H bonds. Together with complete dealkylation of substituents (2.16), (2.17), breaks of side chains (2.18), (2.19) are possible for alkyl cyclic alkanes with a more complicated substituent than CH_3:

$$\text{(cyclopentane)}-C_4H_9 \longrightarrow \text{(cyclopentyl radical)} + C_4H_8 \qquad (2.16)$$

$$\text{(cyclopentane)}-C\text{-}C\text{-}C\text{-}C\text{-} \longrightarrow \text{(cyclopentane)}-C^{\bullet} + C_3H_6 \qquad (2.17)$$

$$\text{(cyclopentane)}-C_2H_5 \longrightarrow \text{(cyclopentyl radical)} + C_2H_4 \qquad (2.18)$$

$$\text{(cyclopentane)}-C\text{-}C \longrightarrow \text{(cyclopentane)}-C^{\bullet} + CH_4 \qquad (2.19)$$

Unsaturated HCs C_2, C_3, and C_4 appearing in these reactions participate then in polymerization and copolymerization that lead to isoparaffin formation in light RTP products.

In mode 2, characterized by lower values of the reaction temperature and the dose rate, considerable decrease in the contents of cyclic alkanes is observed in the gasoline fractions of RTC liquid products. Together with the disclosure of the naphthene rings and destructive processes in the alkyl substituents, increase in the lifetime of such radicals as Alk-C_6H_{11} and R-C_6H_{12} can be responsible for dimerization reactions and accumulation of dimers in fractions with boiling temperature higher than that in gasoline fraction.

The cyclic alkanes that appeared in the gasoline fractions after the RTC of Karazhanbas oil in the irradiation modes discussed earlier were represented mainly by cyclopentane derivatives (Table 2.9). Their high yields were obviously caused by isomerization of a six-member ring into the five-member ring similar to reactions described in detail (Lavrovskiy 1976) for HC catalyst processing. Bicyclic alkanes transformed mainly into monocyclic alkanes with a high yield of cyclopentane derivatives.

In all irradiation modes used, the number of the five-member naphthenes in the gasoline fraction of RTC products obtained from Karazhanbas oil did not change much while the amount of the six-member naphthenes considerably decreased.

TABLE 2.9

Concentrations of Naphthene Hydrocarbons (Mass%) in the Gasoline Fraction with the Start of Boiling at 100°C Obtained by RTC of Karazhanbas Oil in Two Irradiation Modes

		Mode 1			Mode 2	
Hydrocarbons	Feedstock $D=0$	$D=1.5\,kGy$	$D=4\,kGy$	$D=6\,kGy$	$D=4\,kGy$	$D=6\,kGy$
Naphthenes	30.80	25.37	20.02	24.48	11.82	13.70
Cyclopentane derivatives	17.26	21.08	16.20	19.59	8.34	12.00
Cyclohexane derivatives	11.90	3.06	2.54	3.83	3.18	1.32

Source: Zaikin, Y.A. et al., *Radiat. Phys. Chem.*, 60, 211, 2001.

Comparison of the gasoline fraction composition in the original oil with that in the products of radiation processing allowed only indirect conclusions on the probable reactions in different types of HCs. It was suggested that a change in the number of six-member naphthene cycles in the gasoline fractions of RTC products was associated with the following competing reactions proceeding under irradiation: (1) reactions of the ring opening that predominate in the six-member naphthenes compared with similar reactions in the five-member cycles, and (2) isomerization of the six-member cycles to five-member cycles.

The mechanisms of the naphthene cycles opening under irradiation were described in detail (Cserep et al. 1985). It is known that there are preferential atomic positions for the ring opening in alkylcycloalkanes: because of the cycle geometry, its opening occurs in the β position with respect to the carbon atom with a neighboring substitution group.

One of the possible mechanisms of the cycle opening (Cserep et al.) consists in the radiation-assisted formation of the hot hydrogen atoms capable of breaking a cycloalkane ring:

$$\text{(cyclohexane)}{-}CH_3 + H \longrightarrow CH_3\text{-}CH_2\text{-}CH_2\text{-}CH_2\text{-}CH_2\text{-}\overset{\cdot}{C}H_3 \overset{|}{CH_3} \tag{2.20}$$

The behavior of an unbranched radical formed in this reaction is probably similar to that of an unbranched radical, namely, it undergoes fragmentation with formation of short isoalkyl radicals, iso-alkane precursors. It corresponds to a high concentration of these HCs in the gasoline fraction of the condensates obtained by RTC of Karazhanbas oil.

$$CH_2\text{-}CH_2\text{-}CH_2\text{-}CH_2\text{-}CH_2\text{-}CH_2 \longrightarrow i\text{-}C_5 + i\text{-}C_6 \tag{2.21}$$
$$\overset{|}{CH_3}$$

Otherwise, an unbranched radical can enter bimolecular reactions without fragmentation:

$$CH_3\text{-}CH_2\text{-}CH_2\text{-}CH_2\text{-}CH_2\text{-}CH_2 + RH \longrightarrow$$
$$\vert$$
$$CH_3$$

(2.22)

$$\longrightarrow CH_3\text{-}CH_2\text{-}CH_2\text{-}CH_2\text{-}CH_2\text{-}CH_3 + RH (\text{-}H)$$
$$\vert$$
$$CH_3$$

A deep destruction of the naphthene cycles accompanied by the formation of unsaturated HCs (such as ethylene and propylene) similar to that observed in the high-temperature thermal processes (such as cracking and pyrolysis) is less probable (Brodskiy et al. 1961).

According to the following schemes, formation of various alkyl-substituted five-member naphthenes in the gasoline fraction of RTC products (Figure 2.44) is associated with isomerization of a six-member ring:

(2.23)

Thus, the main reactions proceeding in alkylcycloalkanes of Karazhanbas crude oil under irradiation resulted from splitting of the side chains, dealkylation of the five-member cycles, degradation of the six-member naphthene ring, and its isomerization.

Unlike other petroleum HCs, aromatic compounds are characterised by very high radiation resistance in the condensed phases. For example, the highest yield of benzene decomposition due to radiolysis is only one-seventh of the decomposition yield for such saturated HCs as cyclohexane. Polynuclear aromatic compounds display a still higher stability. To a more or less extent, it relates to the radiolysis of aromatic compounds with the side alkyl chains (Cserep et al. 1985)

According to GLC data given in Figure 2.44, aromatic HC of the gasoline fraction in the original Karazhanbas oil are represented by various methyl-substituted

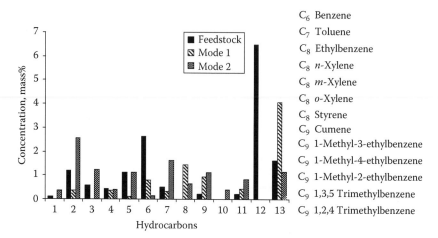

FIGURE 2.44 Concentrations of aromatic compounds in the gasoline fraction of RTC products obtained from Karazhanbas crude oil.

benzenes. About 40 mass% of aromatic HCs fall to the share of 1,3,5-trimethylbenzene (6.5% with respect to the mass of the gasoline fraction) and about 20 mass% is the fraction of *o*-xylene (2.6% with respect to the mass of the gasoline fraction). The remainder includes benzol, toluene, *m,p*-xylenes, 1,2,4-trimethylbenzene, and styrene. HC compounds with the methyl group are represented only by ethylbenzene (0.58 mass%).

Though aromatic rings display rather high stability during irradiation, reactions characteristic for this class of HCs involve alkyl substituents in aromatic rings, namely, these are the reactions of dealkylation and cracking of alkyl groups. Since radiation activates both C–C and C–H bonds in molecules of aromatic HCs, it depends on the molecular structure and irradiation conditions to know which of these types of reactions would prevail. Dealkylation is rather difficult for methyl- and ethylbenzenes where reactions of disproportionation predominate. Alkylbenzenes containing from three to five carbon atoms in the main chain are mostly subjected to dealkylation. Radiation-induced reactions in alkylbenzenes with longer alkyl chains lead to more complicated contents of the reaction products.

The weakest C–C bond (230–260 kJ/mol) conjugated with the aromatic ring is observed in aromatic HC of $C_6H_5CH_2$-R type. The bond energy decreases as the number of carbon atoms in R increases. Alkylbenzenes disintegrate according to the following probable scheme:

$$C_6H_5\text{-}CH_2\text{-}CH_2\text{-}R \rightarrow \underset{\text{Benzyl radical}}{C_6H_5\text{-}CH_2\text{·}} + \text{·}CH_2\text{-}R \qquad (2.24)$$

Benzyl radicals are less reactive and do not propagate the chain. As a result of recombination, dibenzyls appear through the reaction:

$$2\,C_6H_5\text{-}CH_2\text{·} \rightarrow C_6H_5\text{-}CH_2\text{-}CH_2\text{-}C_6H_5 \qquad (2.25)$$

Simultaneous formation of toluene (Figure 2.44) in mode 2 is indicative of the benzyl radical reaction with olefins:

$$+C_6H_5\text{-}CH_2\text{·}+ CH_2= CH\text{--} CH_2\text{-}R \rightarrow C_6H_5\text{-}CH_3 + CH_2= CH\text{--}CH\text{--}R \quad (2.26)$$
$$\underset{\text{Toluene}}{}$$

Separation of a H atom due to the break of the weakest bond conjugated with the ring forms a radical C_6H_5-CH-CH$_2$-R. Disintegration of this radical leads to styrene formation (Figure 2.44):

$$C_6H_5\text{-}CH\text{-}CH_2\text{-}R \rightarrow C_6H_5\text{-}CH=CH_2 + R\text{·} \quad (2.27)$$
$$\underset{\text{Styrene}}{}$$

According to GLC data (Figure 2.44), aromatic HCs of the gasoline fractions are mainly represented by methyl derivatives of benzol. The number of compounds where the ethyl group is present as a substituent is much lower, while an HC compound with the isopropyl group was observed only in one case.

In all irradiation modes, the gasoline fraction of RTC liquid products did not contain 1,3,5-trimethylbenzol (Figure 2.44), while its concentration in the gasoline fraction of the feedstock was maximum. Most probably, it is the result of C–C bond breaks under irradiation accompanied by detachment of methyl substituents. The fact that accumulation of m-xylene was not observed in RTP products shows that several methyl groups were detached from the benzol ring. On the other hand, high contents of a polymethylated benzene and 1,2,4-trimethylbenzene in the products of RTC in mode 1 indicate to intense reactions of methyl group redistribution and alkylation of the benzene ring by the methyl groups appeared during irradiation. Probable separations of CH_3 groups and their attachments are the competing conjugated reactions:

$$(2.28a)$$

The same conclusion is valid for the ethyl groups.

Irradiation in mode 2 leads to the formation of the maximum toluene concentration, some decrease in the contents of xylans, and considerable reduction in the amount of polymethylated aromatic HC (Table 2.12). The basic reactions in this irradiation mode are probably connected both with cracking of the side chains and with the separation of an alkyl group as a whole. Though the first direction is dominating in cracking reactions, detachment of side substituents occur too, in accordance with certain increase in benzol concentration in the gasoline fractions obtained in this mode.

In view of the low contents of heavy oil residua with boiling temperature higher than 350°C in products of Karazhanbas oil irradiation, it is obvious that the rate of polycondensation reactions is low. Conversion of polycyclic aromatic HC during RTP covers splitting of polycyclic structures, disproportionation of alkyl substituents, and dealkylation.

Increase in irradiation dose from 2 to 6 kGy leads to considerable decrease in the share of aromatic HC in the gasoline fractions both in mode 1 and in mode 2 (Table 2.12) due to accumulation of alkyl aromatic radicals and, therefore, intensification of dimerization reactions according to the following scheme:

$$2C_6H_5\text{-}CH_2 \rightarrow \left(C_6H_5\text{-}CH_2\right)_2 \qquad (2.28b)$$
$$\underset{\text{Dibenzyl}}{}$$

In the case of dimethyldiphenyl, these reactions can be written in the following form:

$$C_6H_5\text{-}CH_3 + H \rightarrow C_6H_6\text{-}CH_3 \qquad (2.29)$$

$$C_6H_5\text{-}CH_2 + C_6H_6\text{-}CH_3 \rightarrow CH_3\text{-}C_6H_4\text{-}C_6H_4\text{-}CH_3 + H_2 \qquad (2.30)$$
$$\underset{\text{Dimethyldiphenyl}}{}$$

Products with molecular mass higher than dimer masses can appear due to the interaction of two dimers or in multistage addition reactions.

A by-product of Karazhanbas oil radiation processing was a heavy coking residue. It is known that coke forms as a result of a series of subsequent condensation reactions giving the products with growing molecular mass and growing degree of aromatization according the following scheme: HCs → resins → asphaltenes → coke (Lavrovsky 1976).

In the case of RTC, destructurization reactions in polycyclic structures lead to the formation of intermediate products with aromatization degree less than that in the feedstock: the lower being asphaltene aromatization, the lower is the coke yield. The following chain reactions describe appearance of products with high molecular mass:

$$R^{\bullet} + P(A) \rightarrow RH + P_1^{\bullet}(A^{\bullet}) \qquad (2.31)$$

$$P_1^{\bullet}(A^{\bullet}) + P(A) \rightarrow PP_1^{\bullet}(AA^{\bullet}) \qquad (2.32)$$

where
    P is an intermediate high-molecular product
    A is a molecule of a native asphaltene

The most important reactions of aromatic HCs in the RTC process proceed through the formation of an intermediate phenyl radical C_6H_5. The rate constant of the dimerization reaction is small.

$$C_6H_{5^-} + C_6H_{5^-} \rightarrow C_6H_5 - C_6H_5 \qquad (2.33)$$

Radicals C_6H_5 enter reaction with a preferable addition of H, CH_3, and, to a less extent, C_2H_5. The latter are formed in considerable concentrations in the RTC of heavy Karazhanbas oil.

However, increase in irradiation time (and dose) leads to a considerable decrease in the concentration of aromatic compounds in the gasoline fraction both in mode 1 and in mode 2 (Figure 2.45). Since aromatic rings are radiation resistant, increase in irradiation time leads to accumulation of alkyl aromatic radicals in the system and, therefore, furthers dimerization reactions, for example, according to the following scheme:

$$C_6H_5\text{-}CH_3^* — \text{\textbackslash\textbackslash} - \rightarrow C_6H_5\text{-}CH_2 + H \qquad (2.33a)$$

$$C_6H_5 - CH_3 + H \rightarrow C_6H_5 - CH_2 + H_2 \qquad (2.34)$$

$$2C_6H_5 - CH_2 \rightarrow \underset{\text{Dibenzyl}}{\left(C_6H_5 - CH_2 \right)_2} \qquad (2.35)$$

In the case of dimethyl phenol, these reactions can be written as follows:

$$C_6H_5 - CH_3 + H \rightarrow C_6H_6 - CH_3 \qquad (2.36)$$

$$C_6H_5 - CH_2 + C_6H_6 - CH_3 \rightarrow \underset{\text{Dimethyldiphenyl}}{CH_3 - C_6H_4 - C_6H_4 - CH_3} + H_2 \qquad (2.37)$$

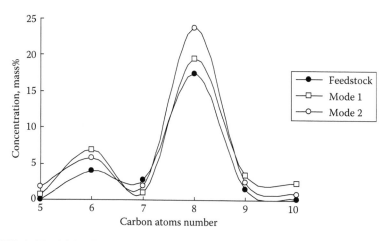

FIGURE 2.45 Molecular mass distribution of naphthenic HCs in fraction SB-180°C. Mode 1: $T = 425°C$, $P = 1.5$ kGy/s; mode 2: $T = 375°C$, $P = 1.25$ kGy/s.

The products with a molecular mass higher than that of dimers can form as a result of interaction of two dimers or in the multistep addition reactions.

In the aromatic HCs having more than one ring, the reactions proceeding in RTC conditions can be accompanied by the breaks of all three types of chemical bonds: breaks of C–H bonds, intramolecular breaks of C–C bonds of a benzene ring, and breaks of C–C bonds between aromatic rings (in the case of polyphenyls). However, the most part of the products is formed through the breaks of C–H bonds that require activation energy of 102 kcal/mol. The breaks of C–C bonds in a benzene ring and between aromatic rings with activation energies of 120–130 and 93–103 kcal/mol, respectively, occur with a lower probability (Gabsatarova and Kabakchi 1969).

Realization of RTC of Karazhanbas crude oil in different irradiation modes allowed obtaining gasoline fractions with maximal concentrations of different aromatic HCs, such as polymethylated structures, toluene, and ethylbenzene.

Reactions of aromatic HC condensation require a number of interactions. They lead to the formation of concentrated products and proceed slowly. Relatively low rates of coke formation is a specific symptom of the rule stating that processes of new solid-phase formation are always difficult (Lavrovsky 1976).

Distributions of aromatic compounds in gasoline fractions produced by RTC of Karazhanbas oil independently of irradiation conditions indicated energetic substitution of hydrogen atoms of the ring by methyl, ethyl, or rarely isopropyl groups. Gasoline fractions with maximum contents of various aromatic HCs of polymethylated structure, and also toluene and benzolene, were obtained from Karazhanbas oil by variation of irradiation dose. Increase in irradiation dose promoted polymerization processes accompanied by the appearance of heavy irradiation products.

Transformations of the naphthene HCs were also studied in the experiments on the RTC of crude oil from Kumkol field, Kazakhstan, remarkable for low contents of heavy aromatic compounds. The HC content of Kumkol oil is mainly represented by paraffin–naphthene structures having a total concentration of 77%–80%, 21%–36% of this amount falling to the share of naphthene HCs. According to the data of gas–liquid chromatography, the gasoline fraction thermally distilled from the original crude oil contained up to 25% naphthenic HCs (cycloalkanes) (Nadirov et al. 1997a).

The data of gas–liquid chromatography on the HC contents of the gasoline fraction extracted from crude Kumkol oil and from the liquid product of its RT processing are shown in Table 2.10. According to these data, RTC leads to increase in the cycloalkane and cycloolefin concentrations, while concentration of the normal alkanes becomes lower than that in the original feedstock.

Obviously, irradiation of Kumkol crude oil was accompanied by very intense breaks of the C–C bonds in the high-molecular paraffinic structures. As a result, accumulation of the light short-chain n-paraffin HCs was observed in the liquid RTC product. Such considerable increase in the n-paraffin concentrations in the liquid fraction was not observed in crude oils with high contents of heavy aromatics (Zaikin et al. 2004). It can be suggested that RTC of this type of crude oil was accompanied by the breaks of several C–C bonds in the excited molecules.

According to the data of gas–liquid chromatography, naphthenic HCs in the light RTC products were represented by different alkyl derivatives of cyclopentane and cyclohexane with the chain length of an alkyl substituent up to C_4.

TABLE 2.10

Hydrocarbon Group Contents of the Gasoline Fraction (SB-180°C) Extracted from Kumkol Crude Oil and Products of Its Radiation Processing

Sample	Hydrocarbon Contents, Mass%		
	n-Paraffins	Naphthenes	Cycloolefins
Feedstock (Kumkol crude oil)	42.5	25.3	—
Sample 1 (*T*=425°C, *P*=1.5 kGy/s, *D*=2 kGy)	31.8	36.2	0.5
Sample 2 (*T*=400°C, *P*=2.5 kGy/s, *D*=2 kGy)	34.4	35.7	2.0
Sample 3 (*T*=375°C, *P*=1.25 kGy/s, *D*=6 kGy)	29.0	32.2	0.8
Sample 4 (*T*=370°C, *P*=3 kGy/s, *D*=2 kGy)	34.0	33.8	0.8

Figure 2.45 shows a typical molecular mass distribution of naphthenic HCs in the gasoline fraction obtained by RTC of Kumkol oil. Two maximums in the curves correspond to the HCs C_6 and C_8. In the original crude oil, the first maximum was attributed to cyclohexane. In the product of radiation processing, this maximum results from contributions of both cyclohexane and methylcyclopentane. Appearance of cyclopentane (C_5 naphthenic HC) and methylcyclopentane (C_6 naphthenic HC) is a result of decomposition reactions with C–C bond breaks leading to the formation of both pentyl and hexyl radicals and probable isomerization of the latter (Zaikin et al. 2001).

The second maximum in Figure 2.45 (naphthenic HCs C_8) results from the accumulation of cyclopentane derivatives (methyl ethyl cyclopentane, trimethyl cyclopentane) in the products of radiation processing. No derivatives of cyclohexane C_8 were detected in RTC products. Naphthenic HC C_8 in the original crude oil also contained mainly cyclopentane derivatives.

The bar graph of cyclopentane derivative distribution in the RTC gasoline fraction (Figure 2.46) shows that the original oil contained a considerable amount of trimethyl-substituted cyclopentane (14.35%). Cyclopentane and methyl cyclopentane were absent in the original feedstock. Appearance of these compounds in the irradiation products testifies not only to the energetic breaks of C–C bonds in the long-chain structures but also to the dissociation of the C–C bonds of the alkyl substituents in the rings. The formation of such compounds as propyl-, *i*-propyl, methyl propyl-, butyl-, and methyl butyl cyclopentanes is also a result of radiation-induced cracking reactions. Formation of these compounds due to interactions of radicals,

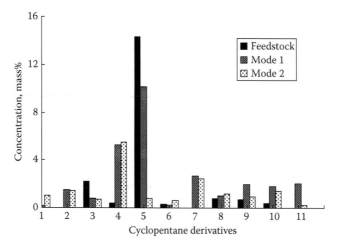

FIGURE 2.46 Distribution of cyclopentane derivatives in the gasoline fraction obtained by RTC of Kumkol oil.

such as cyclopentyl ˙C_5H_9 and butyl ˙C_4H_9 radicals, is low probable because of the steric difficulties.

Generally, the ionizing irradiation of oils gives rise to the competition between benzene ring opening and splitting of the side chains of HC molecules (Panchenkov et al. 1981, Saraeva 1986). In the experiment described earlier, it was difficult to evaluate the contribution of the benzene ring opening to the observed RTC reactions. However, it was supposed that contribution of this reaction was small because of the presence of branched cycloalkanes in the system under irradiation. The breaks of weaker C–C bonds of the alkyl substituents serving as benzene ring "protectors" were considered as a more probable reaction.

Thus naphthenic HCs in the RTC gasoline of Kumkol oil were mainly represented by cyclopentane derivatives. The alkyl groups formed as a result of dissociation of C–C bonds in the paraffin structure of the original crude oil contained more complicated substituents than CH_3.

Cycloolefin HCs, which were absent in the original composition, appear in the RTC products. They are represented by 1,2-dimethyl cyclopentene, 3-ethyl cyclopentene, 3-methyl cyclohexene, 1-methyl cyclooctene, and 4-methyl cyclooctene. According to these results and the data of the works (Saraeva 1986, Zaikin et al. 2001), a mechanism of these compound formations can be interpreted as follows.

In the case of HCs with a ring containing five or greater number of carbon atoms, hydrogen appears in the system as a result of detachment of hydrogen atoms H or molecules H_2 from the excited cyclic molecules forming either under direct excitation or in the charge recombination process. Detachment of hydrogen atoms leads to the formation of cycloalkyl radicals. For example, in the case of alkyl-substituted cyclohexane, this reaction can be written in the form

$$\text{Alk-cyclo-}C_6H_{11}^* \rightarrow \text{Alk} - \text{cyclo} - C_6H_{10} + H \tag{2.38}$$

Further stabilization of the alkyl cyclohexyl radical Alk-cyclo-C_6H_{10} leads to the formation of a cycloolefin in the disproportionation reaction (2.39) or dialkyl cyclohexane in the recombination reaction (2.40):

$$Alk\text{-}cyclo\text{-}C_6H_{10} + R \rightarrow Alk\text{-}cyclo\text{-}C_6H_9 + R\text{-}H \tag{2.39}$$

$$Alk\text{-}cyclo\text{-}C_6H_{10} + R \rightarrow Alk\text{-}cyclo\text{-}C_6H_9\text{-}R \tag{2.40}$$

The rate of disproportionation for these radicals is higher than the rate of their recombination (Cserep et al. 1985). As a result, cycloalkene HCs with branched chains appear in the products of oil radiation processing indicating a secondary importance of the reaction of naphthene ring dehydrogenation.

2.4   RADIATION-THERMAL CRACKING OF HIGH-VISCOUS OIL AND BITUMEN

2.4.1   HEAVY OIL RADIATION PROCESSING

The first studies of RTC in heavy oil and bitumen were carried out in the end of the 1980s and beginning of 1990s.

The effect of gamma irradiation on heavy oil fractions, hydrogenation products, and their mixtures with brown coal was studied in the work (Skripchenko et al. 1986). The test subjects were a heavy fraction of crude oil boiling above 260°C, products of coal hydrogenation (a mixture of 70% heavy distillate fraction having $T_b > 400°C$ and 30% hydrofined fraction boiling in the temperature range of 300°C–400°C), and a mixture of the oil product with coal in the ratio 1:1. The samples were subjected to gamma irradiation at nominal room temperature (heating of the samples under irradiation was not controlled).

The data of IR spectroscopy have shown that the major changes in the oil product structure were observed in its aliphatic part; the changes in the aromatic structure were inconsiderable. A characteristic increase was observed in the ratio CH_2/CH_3 (from 1.3 in the original feedstock to 2.1 in the irradiated product). At the same time, the number of the olefin bonds C=C increased in the irradiated sample.

The HC group content of the oil product did not considerably change directly after irradiation. However, considerable changes were observed after longtime storage of the irradiated product. The asphaltene concentration increased from 7.2% to 16.7% and the lube oil content decreased from 92.8% to 83.3% after 2 month exposure. In the course of storage, the product was strongly oxidized.

To study the effect of gamma irradiation on the fractional contents of oil products, the original oil fraction and the same fraction irradiated during 12.5 h were thermally separated to two distillates boiling above and below 300°C. The overall yield of the light fraction having $T_b < 300°C$ increased from 8.5% to 32.8%.

It should be noted that this method for increase in the light fraction contents is not universal because heavy oils usually have a strong tendency for polymerization during a longtime irradiation with low dose rates (Mustafaev 1990, Zaikin and Zaikina 2008a).

In the studies of RT transformations of the oil product (Skripchenko et al. 1986), oil samples were heated in the autoclave to the given temperature (425°C and 450°C). Then the sample was subjected to gamma irradiation in the isothermal conditions during 2.5 h. A considerable gas evolution indicated thermo-destructive processes under irradiation.

Another type of samples subjected to gamma irradiation was products of coal hydrogenation. The samples were irradiated during 12.5 h. Radiation processing resulted in decrease in the light fraction yield from 23.1% to 14.5%, which was attributed to the availability of olefin bonds and reactive oxygen-containing groups in the irradiated material.

Brown coal and a fraction having boiling temperature $T_b > 260°C$ mixed in the proportion 1:1 were used as feedstock in the studies of the effect of gamma irradiation on coal–oil mixtures. An increase in the yields of the lube fractions from 51% to 58% was observed as a result of irradiation of the coal–oil mixture. The yield of the benzene-soluble (BS) material increased from 51% to 58%. Irradiation substantially affected the process of the oil mixture thermal decomposition. The thermogravimetric analysis has shown that coal irradiated during 45 min was characterised by a total yield of volatile components of 44% when heated to the temperature of 1000°C. The maximum of its basic decomposition was observed at 415°C. The irradiated oil product in the similar heating conditions was decomposed to 95% with the maximum of the mass loss at 420°C.

In the case of the irradiated mixture of coal with the oil product, the maximum of mixture basic decomposition was observed at a temperature of 395°C, which is lower than that characteristic both of irradiated coal and of irradiated oil fraction (Figure 2.47). This effect was attributed to deeper destructive processes in the case of coal–oil mixture irradiation.

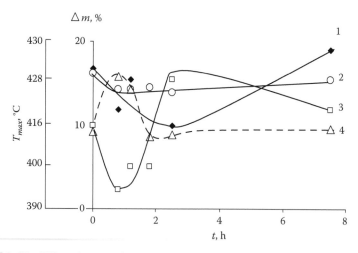

FIGURE 2.47 Effect of gamma irradiation on thermal decomposition of brown coal mixed with the oil product. 1—T_{max} of coal basic decomposition; 2—T_{max} of oil product basic decomposition; 3—T_{max} of coal–oil mixture basic decomposition; 4—deviation from mass additivity. (From Skripchenko, G.B. et al., *Khimiya Tverdogo Tela (Chem. Solids)*, 4, 55, 1986.)

Radiation-chemical conversion of the oil fraction and its mixture with brown coal described in the paper (Skripchenko et al. 1986) is mainly caused by slow processes of the low-temperature radiolysis.

A series of RTC studies in heavy oil fractions was carried out by Mustafaev and coworkers (Mustafaev et al. 1985, 1989, 1990, Mustafaev 1990).

The regularities in the formation of H_2, CH_4, C_2H_4, C_2H_6, C_3H_8, and liquid olefins in the process of RT conversion of pentadecane were studied in the work (Mustafaev et al. 1990). Radiation processing of pentadecane was conducted in the temperature range of 370°C–550°C, dose rate of gamma irradiation, accelerated electrons of 28–9740 kGy/h, and absorbed dose up to 2088 kGy.

Under gamma irradiation at the dose rate of 28 kGy/h and at the dose of 42 kGy, pentadecane conversion to gases varied from 2% to 17% as temperature increased from 370°C to 435°C. The maximal yield of liquid olefins reached 75.35 with respect to the original feedstock mass at the temperature of 435°C. At higher temperatures, the yield of liquid olefins decreased because of the considerable gas formation.

As a result of irradiation with accelerated electrons ($P = 2088$–8352 kGy/h), the maximal yield of liquid olefins was 83.5 mass% at the temperature of 375°C and doses of 700–1000 kGy/h. At higher dose rates, concentration of double bonds decreased due to secondary reactions. Increase in temperature to 400°C–425°C did not cause any increase in the yields of liquid olefins.

In the range of the dose rates of accelerated electrons (2088–8352 kGy/h) and temperatures of 370°C–435°C, the total radiation-chemical yield of gaseous products varied from 23.1 to 50.6 molecules/100 eV, which corresponded to G (-pentadecane) = 2.9–2.6.

In flow conditions of pentadecane radiation processing at the dose rates of 2435–9470 kGy/h, temperatures of 450°C–550°C, and residence time of 1.6–31.2 s, chain cracking reactions of pentadecane decomposition were observed with the radiation-chemical yield G (-pentadecane) = 500–600.

The ratio of the rates of radiation-thermal (RT) and thermal (T) processes was determined as a function of dose rate and temperature. At the temperature of 425°C, dose rate variation from 28 to 8352 kGy/h leads to an increase in the ratio W_{RT}/W_T from 1.9 to 15.3. At the same time, increase in temperature from 450°C to 550°C at the dose rate of 9740 kGy/h caused a decrease in the ratio W_{RT}/W_T from 120 to 87.

In these works (Mustafaev et al. 1985, 1989), it was shown that gamma irradiation of fuel oil in the temperature range of 20°C–500°C gives rise to evolution of H_2, CO, and CH_4 with the following radiation-chemical yields per 100 eV of absorbed radiation energy: $G(H_2) = 0.46$–28.3, $G(CO) = 0.01$–7.5, and $G(CH_4) = 0.03$–286.5. The following activation energies were determined in the temperature range of 300°C–500°C: $E(H_2) = 47.7$, $E(CO) = 24.8$, and $E(CH_4) = 55.7$ kJ/mol.

Radiation-thermal conversion of the oil fraction having boiling temperature $T_b \geq 285°C$ was studied in the work (Mustafaev 1990). More than 40% of this fraction boiled at a temperature above 300°C. Other types of feedstock used in this study were liquid products extracted from a bitumen rock by its heating to 500°C.

Liquid products extracted from bitumen by vacuum distillation were divided into three fractions. Radiation-chemical conversion was studied in each of these fractions.

TABLE 2.11

Characteristics of the Heavy Fraction and Liquid Products of Bitumen Thermal Distillation

Feedstock	Temperature Range of Boiling, °C	Yield, Mass%	Molecular Mass, g/mol	Density, g/cm³	Iodine Number
Petroleum fraction	$T \geq 285$	100	245	0.89	82
I	$T \leq 150$	8.5	90.9	0.84	120.7
II	$150 \leq T \leq 300$	20.2	129.2	0.89	90.2
III	$300 \leq T \leq 350$	18.5	162.8	0.91	84.5
IV	$350 \leq T \leq 400$	23.3	223.8	0.94	82.5

Source: Mustafaev, I.I., *Khimiya Vysokikh Energiy (High Energy Chem.)*, 24, 22, 1990.

Formation of H_2, CO_2, and HCs C_1–C_3 and changes in the number of double bonds in the RTC process were determined.

The main feedstock characteristics are given in Table 2.11. High values of the iodine numbers in the oil fractions studied were attributed to their specific structure and to the effect of dehydrogenation in the process of their distillation and extraction from bitumen.

Radiation-thermal processing was carried out in static conditions using gamma radiation and accelerated electrons in the temperature range of 200°C–500°C at the dose rates of 3–472 Gy/s and irradiation dose up to 850 kGy. Figures 2.48 and 2.49 show kinetics of gas formation from the oil fraction and fraction II (a liquid product extracted from bitumen) after their radiation-thermal processing at 400°C.

The formation rates of the gases formed from the oil fraction are given in Table 2.12 for the case of radiation processing at the dose rate of 472 Gy/s in the temperature range of 200°C–500°C.

The identified gaseous products had the following composition: H_2, 4.4%; CO, 1.0%; CH_4, 46.3%; C_2H_2, 9.4%; C_2H_6, 19.7%; and C_3H_8, 19.2%. The ethylene concentration in the gaseous products reached 13.3%. Radiation-chemical yields of the gaseous products are shown in Figure 2.50.

The highest degree of conversion was observed at the temperature of 500°C and the dose rate of 72 Gy/s. In the linear region of the kinetic curves, conversion of the oil fraction to gases reached 31%. In these conditions, about 54% of all the hydrogen content in the oil fraction was evolved in the form of H_2 and HC gases. At higher doses, the cracking rate decreased and the oil conversion was lower.

The contents of the gaseous products formed due to radiation-chemical conversion of fractions I–IV extracted from the bitumen organic part varied in the following ranges: H_2 5–17%; CO, 1.5–2.5%; CH_4, 8–10%; C_2H_6, 17–21%; C_3H_8, 21–30%. This set of values was close to the composition of gaseous products characteristic of the decomposition of heavy oil fractions.

At the temperatures of 200°C–400°C, irradiation caused substantial increase in the iodine numbers of the heavy oil fraction and fractions I–IV (Figure 2.51).

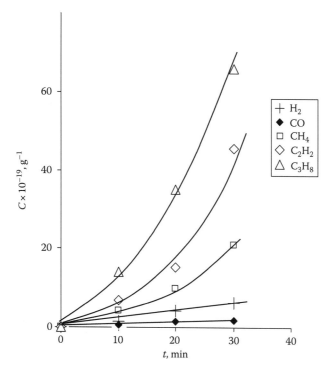

FIGURE 2.48 Accumulation of gases in the course of radiation-thermal conversion of heavy oil fraction $T = 400°C$, $P = 472$ Gy/s. (From Mustafaev, I.I., *Khimiya Vysokikh Energiy (High Energy Chem.)*, 24, 22, 1990.)

A similar decrease in the number of olefin bonds caused by radiation-thermal processing was observed in the work by Skripchenko et al. (1986) cited earlier. At the temperatures of 400°C–500°C, irradiation first leads to an increase in iodine numbers, but increase in the dose above 600 kGy provokes decrease in the iodine numbers.

It was noted that changes in the concentration of olefin bonds in heavy oil are, most probably, a result of competition of the destructive and polycondensation processes under radiation-thermal reaction. In the temperature range of 200°C–400°C and absorbed doses above 10 kGy, the polymerization-type processes predominate and the amount of double bonds decreases. At the temperatures of 400°C–500°C and low doses, the predominating processes are decomposition and dehydrogenation giving rise to an increase in the amount of double bonds. However, even at such heightened temperatures, polycondensation processes become more intense as irradiation dose increases; this is accompanied by a decrease in the iodine numbers.

Radiation-thermal processes in heavy oil were studied in the works (Zhussupov 2006, Yang 2009, Yang et al. 2009, 2010, Alfi et al. 2012a,b).

Hamaca heavy oil was irradiated with 1.35 MeV electrons from a Van de Graaf accelerator in the mode of under beam distillation (Zhussupov 2006). Oil was poured into a can reactor mounted on a heater and placed under an electron beam. The heater

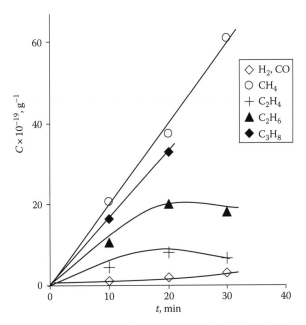

FIGURE 2.49 Accumulation of gases in the course of radiation-thermal conversion of fraction II extracted from bitumen organic part $P=284$ Gy/s, $T=400°C$. (From Mustafaev, I.I., *Khimiya Vysokikh Energiy (High Energy Chem.)*, 24, 22, 1990.)

TABLE 2.12

Effect of Temperature on Formation Rates of the Gaseous Products from Radiation-Thermal Decomposition of Heavy Oil Fraction at $P=472$ Gy/s

	$W \cdot 10^{-15}$, g$^{-1 \cdot}$ s^{-1}					
T, °C	H_2	CO	CH_4	C_2H_4	C_2H_6	C_3H_8
200	14.2	—	9.2	7.6	5.2	—
300	19.2	—	44.3	1.2	18.6	24.4
400	72.3	13.2	729.5	166.8	314.8	335.8
500	234.5	110.5	9000.0	1950.0	5010.0	6315.0

Source: Mustafaev, I.I., *Khimiya Vysokikh Energiy (High Energy Chem.)*, 24, 22, 1990.

---

plate temperature was fixed at a level of about 500°C. Irradiation and heating were started simultaneously. The gas and vapor products were passed through a condenser where liquid and gas products were separated. Irradiation continued until a sufficient amount of condensate was collected. Irradiation resulted in a slight increase in oil viscosity and molecular weight: no changes in oil density were detected.

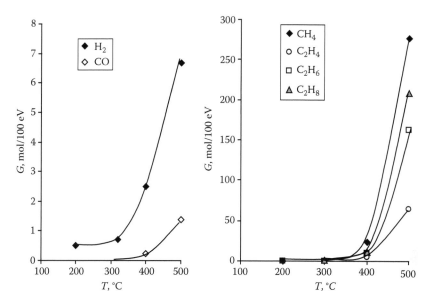

FIGURE 2.50 Temperature dependence of *G*-values for gaseous product formation in the RTC of heavy oil fraction at *P* = 472 Gy/s. (From Mustafaev, I.I., *Khimiya Vysokikh Energiy (High Energy Chem.)*, 24, 22, 1990.)

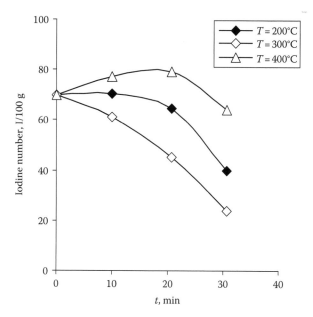

FIGURE 2.51 Changes in the iodine number of heavy oil fraction under radiation-thermal processing. (From Mustafaev, I.I., *Khimiya Vysokikh Energiy (High Energy Chem.)*, 24, 22, 1990.)

A similar experimental technique was used for radiation-thermal processing of *n*-hexadecane (Yang 2009, Yang et al. 2009, 2010, Alfi et al. 2012a,b). However, low process temperature (about 300°C) together with low dose rate (about 0.2 Gy/s) did not allow observation of pronounced RTC effects and obtaining considerable yields of light fractions.

An estimate of the ratio of radiation-thermal and pure thermal components in the RTC of a heavy oil fraction ($T_b \geq 285°C$) was made in the work by Mustafaev (1990) in supposition that this ratio depends only on the relationship of the thermal and radiation-thermal generation of the reaction chain carriers (Figure 2.52).

At the constant dose rate, the ratio W_{RT}/W_T decreases as temperature increases while, at the constant temperature, it increases with the dose rate. This behavior qualitatively corresponds to the available experimental data. Since the values of activation energies for the formation of separate gases (51.8–159.3 kJ/mol) may considerably differ, the ratio W_{RT}/W_T also changes when passing from one gas to another. It should be noted that radiation contribution to propagation of the chain cracking reaction (Chapter 1) leads to increase in the ratio W_{RT}/W_T.

An important conclusion made in the work (Mustafaev 1990) is that irradiation of heavy oil fractions and the organic part of bitumen gives rise to simultaneously proceeding processes of oil decomposition and polycondensation. The cross sections of these processes depend on temperature and irradiation dose rate. The contributions

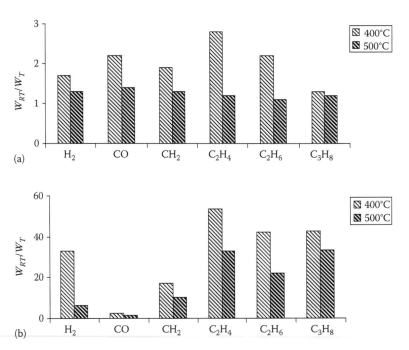

FIGURE 2.52 Effect of dose rate and temperature on the rate ratio, W_{RT}/W_T, of radiation-thermal and pure thermal conversion for the heavy oil fraction ($T_b \geq 285°C$) at the dose rates of (a) 3.1 Gy/s and (b) 472 Gy/s. (Plotted using the data by Mustafaev, I.I., *Khimiya Vysokikh Energiy (High Energy Chem.)*, 24, 22, 1990.)

of thermal and radiation-thermal processes to conversion of the original fractions depend on the combination of dose rate and temperature.

RTC of fuel oil produced from crude oil of Balakhany field (Azerbaijan) was discussed in the paper (Mustafaev and Gulieva 1995). The feedstock had the following characteristics: temperature of boiling start $T_b = 300°C$, average molecular mass $M = 280$, refractive index $n = 1.4747$, density $\rho = 0.927$ g/cm^3, viscosity $\eta = 10.3$ cSt. The elemental composition of fuel oil was represented (mass%) by C, 87.6; H, 11.5; O, 0.06; S, 0.8; and N, 0.09.

Fuel oil was irradiated with electrons at the temperature of 430°C and the dose rate of 556 Gy/s. As a result of RTC, 71% fuel oil was turned to motor fuels including 16% gasoline and 55% diesel fuel. Conversion of fuel oil into gases did not exceed 1.1–4.0 mass%. The maximal yields of light fractions were observed at the dose of 3.5 kGy. The characteristics of diesel fuel produced from fuel oil are represented in Table 2.13.

The material balance of RTC products is shown in Figure 2.53.

Comparison of RTC and thermal catalytic process in the paper (Mustafaev) indicated higher RTC efficiency and expediency of its commercial application.

RTC of heavy oil residues was studied in the work (Zhuravlev et al. 1991). A gas oil fraction of West Siberian crude oil (fraction 350°C–450°C) was subjected to RTC at the temperatures of 300°C–400°C. Parallel to radiation processing, TC was carried out.

RTC was studied in the dose range of $0.5–2 \cdot 10^5$ Gy at the dose rate of 5.1 Gy/s. TC was conducted in similar conditions; the process duration was the same as that

---

TABLE 2.13

Characteristics of Diesel Fuel Produced by RTC of Fuel Oil

Characteristic	Value
Fractional composition	
Start of boiling, °C	200
50% distilled at °C	288
90% distilled at °C	320
End of boiling, °C	335
Cetane number	45
Kinematic viscosity at 20°C, cSt	3.6
Flash point	42
Actual gum concentration, mg per 100 mL fuel	35
Acidity	4.0
Iodine number	76.2
Ash, %	4.1
Coking capacity	0.5
Density at 20°C, g/cm^3	0.854
Water-soluble acids and alkalies	Not detected

Source: Mustafaev, I. and Gulieva, N., *Radiat. Phys. Chem.*, 46(4–6), 1313, 1995.)

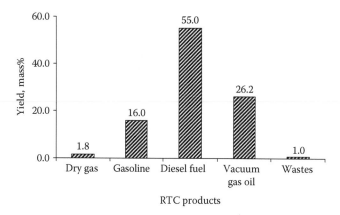

FIGURE 2.53 Mass balance of products obtained by RTC of fuel oil. (Plotted using the data by Mustafaev, I. and Gulieva, N., *Radiat. Phys. Chem.*, 46(4–6), 1313, 1995.)

used in the case of RTC. RTC products were subjected to atmospheric and vacuum distillation. Fraction 350°C–400°C was considered conventionally unconverted. Each experimental run was twice repeated; the relative divergence of the results did not exceed 5%.

A high conversion level was observed at 400°C; the rate of RTC was 1.5 times higher than the decomposition rate in TC. RTC was remarkable for the increased yields of gasoline and diesel fractions. Their total concentration at the process temperature of 400°C and the dose of $1.5 \cdot 10^5$ Gy reached 49.2 mass% compared with 24.9 mass% in the case of TC. The yield of the densification products in RTC was considerably lower than that in TC (18 mass% compared with 31.8%). Dose rate variation from 5.1 to 8.2 Gy/s did not lead to considerable changes in the product yields (Figure 2.54).

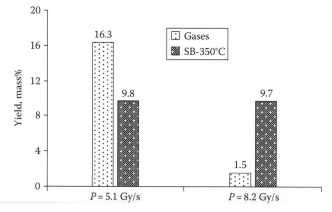

FIGURE 2.54 Yields of RTC products (mass%) at the temperature of 350°C and irradiation dose of 10^5 Gy. The total yield of light fractions (T_b, 350°C) was 88.8 mass%. (Plotted using the data by Zhuravlev, G.I. et al., *Khimiya Vysokikh Energiy (High Energy. Chem.)*, 25(1), 27, 1991.)

Table 2.14 shows physical and chemical characteristics of TC and RTC products. The data related to the original gas oil are given for comparison. It can be seen that fraction 350°C–450°C could only conventionally be considered as "unconverted." Actually, its characteristics differ from those of the original gas oil. Gasoline and kerosene fractions obtained by RTC were lighter compared with the similar fractions produced by TC.

Gasoline fractions obtained in RTC and TC processes at 400°C during 8.25 h were tested for their HC group contents (Table 2.15). RTC gasolines contained higher concentrations of saturated HCs and lower concentrations of aromatic compounds. The olefin contents were practically the same in TC and RTC products.

A more detailed specification of HC groups was obtained in the mass-spectrometric tests of the RTC products ($T = 350°C$, $D = 1.5 \cdot 10^5$ Gy, $P = 5.1$ Gy/s).

The total sulfur concentration in the light RTC and TC fractions was measured by the methods of x-ray fluorescence analysis. Figure 2.55 shows that a higher level of desulphurization was observed in the RTC light fractions.

Based on RTC material balance at 400°C, radiation-chemical yields of the product molecules per 100 eV of absorbed energy were calculated for different irradiation doses. The values of radiation-chemical yields shown in Figure 2.56 indicated a chain cracking reaction.

Activation energy of RTC was determined from the overall conversion of the original gas oil. The calculated activation energy of 84 ± 17 kJ/mol was in good agreement with the known values of activation energy for propagation of the chain cracking reaction. The data of the work (Zhuravlev et al. 1991) confirmed that, in RTC conditions, radiation releases the most energy-consuming stage of cracking initiation ($E_i = 335$ kJ/mol) while the thermal action controls decomposition of high-molecular radicals, that is, chain propagation with activation energy $E_p = 84 \pm 17$ kJ/mol.

RTC of heavy fuel oils (residues of crude oil primary distillation) was carried out in the studies (Nadirov et al. 1994a, 1995, Zaikin et al. 2004b, Zaikin and Zaikina 2008a). Fuel oils were irradiated with 2 MeV electrons in static conditions at the temperatures of 400°C–410°C and the dose rates of 3–4 kGy/s. The fractional contents of the feedstock and the products of its radiation processing are shown in Figure 2.57 for two types of fuel oil (API 7° and 17°). Dependence of the liquid product yield on irradiation dose is shown in Figure 2.58 for a lighter fuel oil (API 22°).

The maximal yields of liquid products were observed at the doses of about 20 kGy (Figure 2.58). Further increase in irradiation dose leads to a decrease in the light fraction yields, indicating considerable radiation-induced polymerization of the product. The optimum conditions of radiation processing allowed production up to 80 mass% motor fuel including up to 20% gasoline and 60% diesel fractions. The by-products of RTC were heavy coking residue (up to 10%) and the gas mixture (up to 10%), containing hydrogen, methane, ethylene, and other HC gases.

RTC of heavy crude oil from Bugulma field (Tatarstan, Russia) having API gravity 19° was studied in a series of our experiments. This type of oil is characterised by the high contents of asphaltenes (>8%), sulfur (>4%), and vanadium (>330 g/ton). The average molecular mass of the crude oil is 410–420 g/mol. Sulfur concentration exceeds 6% in the residues and 32% in the assist gases.

TABLE 2.14

Physical and Chemical Characteristics of Original Gas Oil and Reaction Products

350°C

Characteristic	Original Gas Oil 350°C–450°C	TC		RTC	
		SB-350°C	350°C–450°C	SB-350°C	350°C–450°C
Refractive index	1.5118	1.5068	1.5113	1.5059	1.5155
Density	0.9351	0.9078	0.9134	0.9038	0.9043
Average molecular mass	245	202	241	215	240

400°C

Characteristic	Original Gas Oil 350°C–450°C	TC			RTC		
		SB-180°C	180°C–350°C	350°C–450°C	SB-180°C	180°C–350°C	350°C–450°C
Refractive index	1.5118	1.4908	1.5031	1.5111	1.4901	1.5026	1.5107
Density	0.9351	0.8112	0.8298	0.8913	0.8074	0.8122	0.8847
Average molecular mass	245	135	219	280	133	213	270

Source: Zhuravlev, G.I. et al., *Khimiya Vysokikh Energiy (High Energy. Chem.)*, 25(1), 27, 1991.

Note: Cracking duration, 8.25 h; dose, $1.5 \cdot 10^5$ Gy; dose rate, 5.1 Gy/s.

TABLE 2.15

Hydrocarbon Group Contents of Gasoline Fractions (Mass%)

Hydrocarbons	TC	RTC
Saturated + naphthenic	52.8	57.9
Unsaturated	34.9	33.4
Aromatic	12.3	8.7

Source: (From Zhuravlev, G.I. et al., *Khimiya Vysokikh Energiy (High Energy. Chem.),* 25(1), 27, 1991.)

Note: Fraction, SB-180°C; dose rate, 6.1 Gy/s.

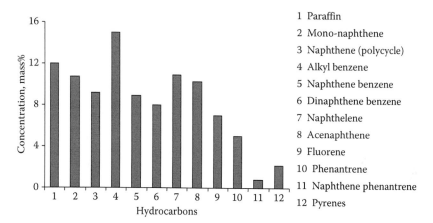

1 Paraffin
2 Mono-naphthene
3 Naphthene (polycycle)
4 Alkyl benzene
5 Naphthene benzene
6 Dinaphthene benzene
7 Naphthelene
8 Acenaphthene
9 Fluorene
10 Phenantrene
11 Naphthene phenantrene
12 Pyrenes

FIGURE 2.55 Data of mass spectrometric analysis for RTC fraction SB-350°C. (Plotted using the data by Zhuravlev, G.I. et al., *Khimiya Vysokikh Energiy (High Energy. Chem.),* 25(1), 27, 1991.)

RTC was conducted at the temperature of 390°C, relatively low for this type of feedstock, and at the dose rate of 3 kGy/s. Oil was irradiated with 2 MeV electrons; relatively high absorbed doses were 102.7 and 130 kGy. Radiation processing was carried out in the mode of under beam distillation. In these conditions, the yields of the liquid condensates did not exceed 60%, but their fractional contents satisfied the requirements of the synthetic oil quality.

Changes in the fractional contents of crude oil after radiation processing are shown in Figure 2.59.

Oil irradiation was accompanied by strong evolution of sulfur-containing gases. Concentration of V_4^+ in RTC residues was up to 200 g/ton due to the accompanying process of vanadium oxidation and its transition to the form of V_2O_5.

The dose dependence of the liquid condensate yield under radiation processing of this sort of oil (Figure 2.60) shows that higher yields of light fractions (and higher product stability) can be provided by using lower irradiation doses.

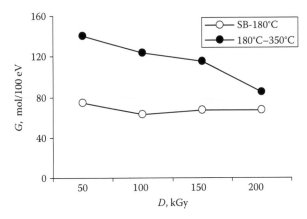

FIGURE 2.56 Radiation-chemical yields of RTC products. (Plotted using the data by Zhuravlev, G.I. et al., *Khimiya Vysokikh Energiy (High Energy. Chem.)*, 25(1), 27, 1991.)

Figure 2.61 shows considerable increase in the yield of fractions C_4–C_{21} at a lower absorbed dose of 44 kGy.

RTC of a heavier crude oil (API 17°) from the same region was conducted at the electron irradiation dose rate of 3.5 kGy/s and at the temperature of 380°C. The yield of the liquid condensate having API 27.5° was 81.2% at the dose of 25 kGy.

The dose dependence of the condensate yield plotted using the data of several experimental runs is shown in Figure 2.62. This dependence corresponds to the cracking kinetics discussed in Section 1.2.

The fractional content of the liquid RTC product (Figure 2.63) shows considerable conversion of the heavy residue and high yields of light fractions.

To increase the level of oil dispersion and to raise the yields of light fractions, heavy Tatarstan oil (API 22°) was bubbled with ozone-containing ionized air in the presence of a bresshtraung x-ray background in the electron accelerator vault. The x-ray dose rate was 16.5 Gy/min and the absorbed dose was 0.5 kGy. The ionized air consumption was 20 mg/s/1 kg feedstock. The pretreated oil was irradiated with 2 MeV electrons at the temperature of 380°C and the dose rate of 3 kGy/s. The absorbed dose of electron irradiation was 98 kGy.

Similar oil pretreatment with ionized air allowed considerable improvement in the fractional contents of the same crude oil irradiated with electrons at a lower temperature but at a higher dose rate (Figure 2.64). The yields of fraction $T_b < 350°C$ were 56.6 mass% from the feedstock or 76.3% from the product irradiated without ionized air bubbling and 83.6% from the product pretreated with ionized air.

RTC of the straight-run fuel oil distilled from Tatarstan crude oil (API 19°–22°) was performed by the same scheme. The fuel oil studied had the following characteristics: $\rho_{20} = 1.003$ g/cm³ (API 7°), $\mu_{80} = 71.1$ cSt, $S_{tot} > 5.0$ mass%, pour point—27°C, coking ability—12.4%. Fuel oil samples were irradiated with 2 MeV electrons at the temperatures of 400°C–420°C and the beam current density of 1.5–1.7 µA/cm².

In this experiment, the liquid condensate yield was 73–75 mass%, and the yields of coking residue and gases were 12%–15% and 10%–15%, respectively.

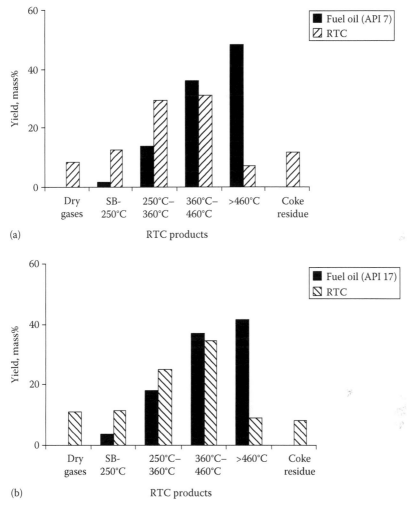

FIGURE 2.57 RTC mass balance for two types of fuel oil having API gravity (a) 7° and (b) 17° for similar irradiation conditions: $T=400°C$, $P=4$ kGy/s, $D=2$ kGy.

The condensates of liquid fractions had a density of 0.9 g/cm², which corresponds to API gravity of 24°–27°, more than three times higher than that of the original fuel oil. Thus, despite a considerable weighting of the feedstock compared with crude oil, the condensates obtained from crude oil and fuel oil had similar fractional contents.

The dose dependence of the condensate yields obtained by RTC of fuel oil distilled from Tatarstan heavy crude oil (Figure 2.65) has the same form as the dose dependence of RTC liquid product yields from the same crude oil (Figure 2.61).

Comparison of Figures 2.65 and 2.66 shows that decrease in irradiation dose from 42 to 25 kGy leads to an increase in the condensate yield while its fractional contents change inconsiderably.

C/H ratio in original fuel oil and RTC products

	H	C	C/H
Feedstock	11.8	84.4	7.2
Condensate (77%)	12.8	81.0	6.3
Coking residue	9.0	83.8	9.3

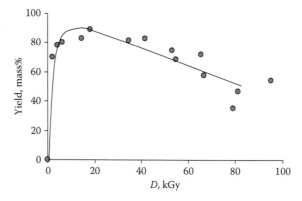

FIGURE 2.58 Dose dependence of RTC product yield from fuel oil M-40 (API 22°) and changes in C/H ratio due to radiation processing.

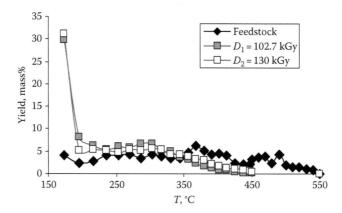

FIGURE 2.59 Fractional contents of high-sulfuric Tatarstan oil (API 19°) before and after radiation processing at 390°C.

Changes in the yields of the cracking liquid product and feedstock conversion with removed light fractions were studied on the example of the two types of fuel oil: heavy oil residue with API 13° and heavier fuel oil having API 7°. Both sorts of residues were distilled from the same crude oil and then subjected to radiation processing in similar conditions ($P=3.5$ kGy/s, $T=410°C$, $D=45$ kGy).

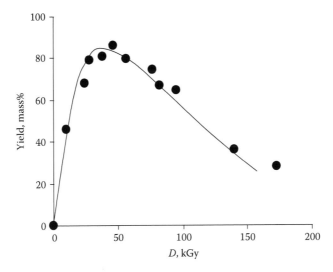

FIGURE 2.60 Dose dependence of the yield of RTC liquid product from Tatarstan crude oil (API 19°).

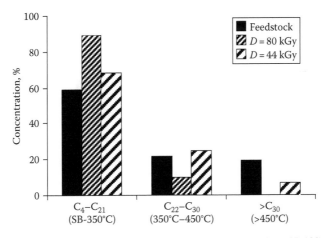

FIGURE 2.61 Changes in the fractional contents of Tatarstan oil (API 19°) at different irradiation doses.

Figure 2.67 shows that RTC of the heavy residue leads to "restoration" of the light fractions removed by distillation.

The yield of the liquid condensate was 71 mass% in the case of lighter fuel oil and 60 mass% in the case of heavier fuel oil. However, comparison of the changes in the fractional contents of RTC liquid products for the two types of fuel oil (Figure 2.68) shows that RTC of heavier fuel oil leads to the formation of a lighter product. A higher conversion level in the case of radiation processing of heavier fuel oil indicates

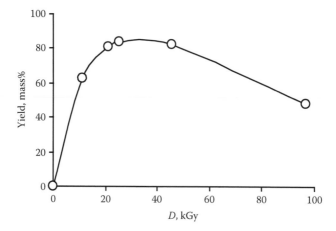

FIGURE 2.62 Dose dependence of the yield of RTC liquid product from heavy Tatarstan crude oil (API 17°).

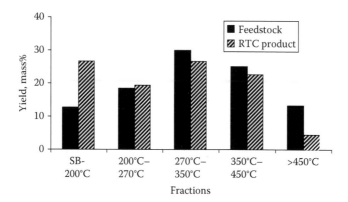

FIGURE 2.63 Fractional contents of heavy Tatarstan crude oil (API 17°) and liquid product of its radiation processing at 390°C.

that the main contribution to the increase in light fraction yields in the RTC of heavy oils comes from the detachment of molecular fragments playing the role of bridges and links in the branched heavy aromatic structures.

To study the opportunities of using water as an additional source of hydrogen in radiation-thermal oil conversion, special experiments were conducted on RTC of fuel oil with water addition to the original feedstock. Before radiation processing in the mode of under beam distillation, a mixture of fuel oil and water was heated to a temperature of 70°C–80°C and thoroughly stirred. As a result of the addition of 8 mass% water, a yield of the liquid RTC product increased by 20% as the irradiation dose decreased by four times. Together with some decrease in the concentration of

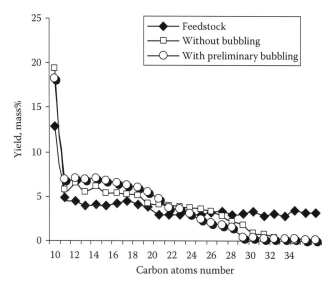

FIGURE 2.64 Molecular mass distribution of fractions in Tatarstan oil (22° API) and the products of its radiation processing ($D = 98$ kGy, $P = 3$ kGy/s, $T = 400$°C).

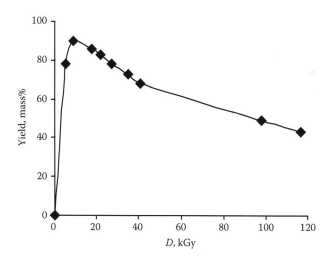

FIGURE 2.65 Dose dependence of the yield of liquid product obtained by RTC of fuel from Tatarstan crude oil.

the heavy residue, a considerable decrease in the naphtha concentration (SB-250°C) was observed in the liquid condensate (Figure 2.69).

The experiments on radiation processing of heavy Karazhanbas oil (Kazakhstan) with water addition to the feedstock were conducted in flow conditions at the temperature of 90°C and the dose rate of electron irradiation of 7.7 kGy/s. Oil was processed in a 0.15 cm thick layer at the flow rate of 12 cm/s. The absorbed dose was

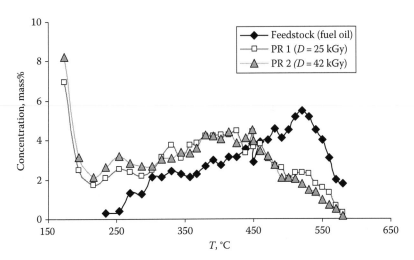

FIGURE 2.66 Fractional contents of fuel oil distilled from Tatarstan crude oil and liquid RTC products: dose rate, 3.2 kGy/s; temperature, 400°C.

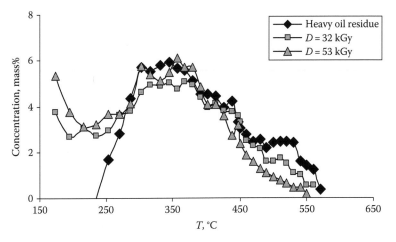

FIGURE 2.67 Fractional contents of fuel oil (API 13°) and products of its radiation processing at 410°C.

18.7 kGy. Before irradiation, 5 and 10 mass% of water were added to the crude oil samples with continuous stirring. The yield of the gaseous products evolved during irradiation did not exceed 4 mass%.

Figure 2.70 shows that addition of 5 mass% water leads to considerable increase in heavy residue conversion and yields of light fractions. As the water concentration increased to 10 mass%, the effect of water addition on product yields disappeared, and fractional contents of the products of oil radiation processing with and without water addition were nearly the same.

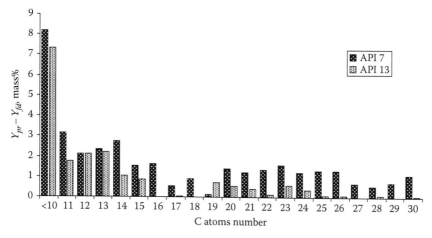

FIGURE 2.68   Changes in the fractional contents (SB-450°C) of the two types heavy oil. $Y_{pr} - Y_{fd}$ is difference of fraction concentrations in the liquid product and original feedstock.

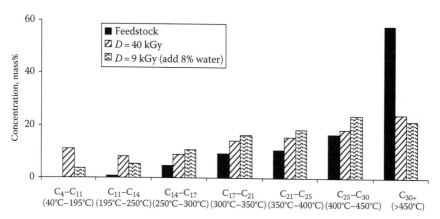

FIGURE 2.69   Fractional contents of fuel oil distilled from Tatarstan crude oil and liquid RTC product with and without water addition to the original feedstock.

2.4.2   EFFECT OF REACTOR MATERIAL AND PROCESS TEMPERATURE ON THE YIELDS OF LIGHT FRACTIONS IN RTC OF OIL FEEDSTOCK

The experiments conducted have shown that a material used for the reactor manufacture affect RTC efficiency.

Figure 2.71 shows fractional contents of the liquid product obtained by RTC of heavy Tatarstan oil (API 19°, sulfur concentration—3.9 mass%) in the reactors made of alloys based on iron and aluminum. RTC was carried out at the dose rate of accelerated electrons of 3 kGy/s and the temperature of 400°C. The cracking product in the reactor made of iron alloy was notable for higher concentration of light fractions.

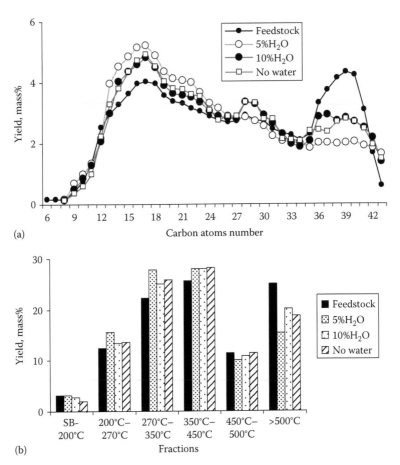

(a)

(b)

FIGURE 2.70 (a) Effect of water addition on the distribution of liquid RTC products from heavy Karazhanbas oil by boiling temperature and (b) concentration of narrow fractions in the liquid product obtained by oil processing in flow conditions.

Figure 2.72 demonstrates a higher concentration of a wide fraction C_4–C_{20} in the condensates obtained by RTC in the reactor made of iron alloy, even at a lower cracking temperature.

It can be seen from Figure 2.73 that at the RTC temperature as low as 360°C, a considerable improvement in the fractional contents of the product is still observed although concentration of light fractions in the condensate becomes lower. In this experiment, feedstock pretreated with ionized air bubbling was irradiated by electrons in a reactor made of an iron alloy. Oil radiation processing in iron reactors was accompanied by more intense oxidation of sulfur compounds and reduction in sulfur concentration in the liquid RTC product. It testified to a catalytic action of the reactor material.

The same behavior was observed in a lighter Tatarstan oil (API 21°) processed in a reactor made of an iron alloy (Figure 2.74).

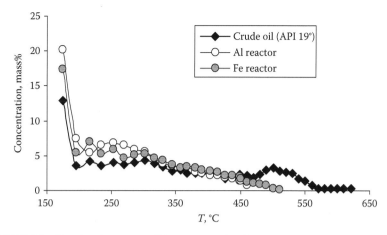

FIGURE 2.71 Fractional contents of Tatarstan crude oil (API 19°) and RTC liquid product from the reactors made of different materials: ——◆—— crude oil, ——○—— reactor made of aluminum alloy ($D = 39.5$ kGy), and ——◉—— reactor made of iron alloy ($D = 40$ kGy).

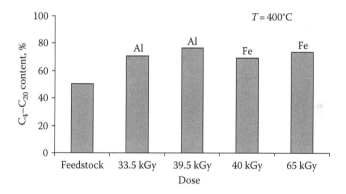

FIGURE 2.72 Yields of liquid RTC product in the reactors made of different materials.

The dose dependencies of the yields of liquid products obtained by RTC of Tatarstan crude oil at temperatures of 390°C–400°C and the dose rate of 3 kGy/s in the reactors made of different materials are shown in Figure 2.75. It can be seen that the yields of liquid condensates were higher by about 10% in the reactor made of aluminum alloy.

The experiments conducted lead to a conclusion that aluminum alloys further higher yields of the total RTC liquid product but, at the same time, bring down concentrations of light fractions in the liquid product compared with the effect of iron alloys.

These observations were confirmed in the experiments with the incorporation of finely ground iron bore chips into the RTC reactor. Tatarstan crude oil (API 19°) was irradiated with electrons at the temperature of 370°C and the dose rate of 3 kGy/s and the dose of 65 kGy. Figure 2.76 shows that incorporation of the iron bore chips leads to considerable improvement in the fractional contents of the condensate.

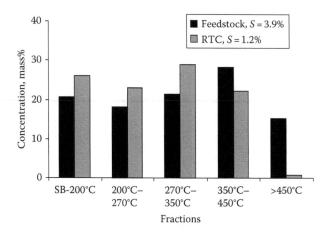

FIGURE 2.73 Fractional contents of the liquid product obtained by RTC of Tatarstan crude oil (API 19°): $T = 360°C$, $P = 2.5$ kGy/s, $D = 98$ kGy.

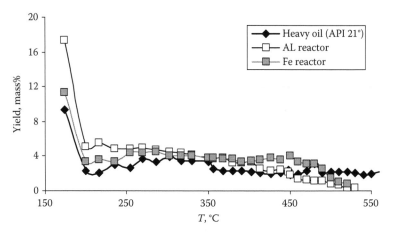

FIGURE 2.74 Fractional contents of the liquid product obtained by RTC of Tatarstan crude oil (API 21°). Al reactor: $T = 390°C$, $P = 2.8$ kGy/s, $D = 44$ kGy; Fe reactor: $T = 400°C$, $P = 2.8$ kGy/s, $D = 46$ kGy.

However, as distinct from the previous experiments with the reactor made of an iron alloy, more considerable increase was observed in the concentration of the fraction boiling in the temperature range of 270°C–350°C.

A strong effect of oxidation processes on RTC intensity in heavy oils noticed in a series of works (Nadirov et al. 1997b, Zaikin and Zaikina 2004a,d) was also observed in our experiments when radiation processing of heavy oil with a high content of heavy aromatic compounds was carried out in the presence of oxidation catalyst MnO_2. MnO_2 powder in concentrations of 1.8–7.7 mass% was contained in the reactor in the suspension state. Oil irradiation with 2 MeV electrons was conducted in the mode of under beam distillation at the temperatures of 400°C–420°C and the

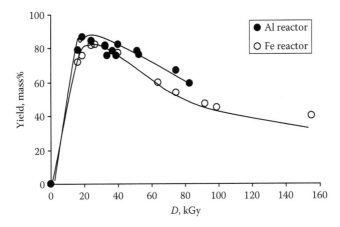

FIGURE 2.75 Dose dependence of the liquid product yields obtained by RTC of Tatarstan crude oil (API 19°) in the reactors made of different materials.

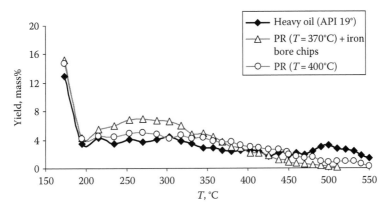

FIGURE 2.76 Changes in the contents of the liquid RTC product obtained from Tatarstan crude oil (API 19°) with iron bore chip incorporation into reactor.

dose rates of 3–4 kGy/s. A dose dependence of the yield of liquid RTC product without catalyst addition is shown in Figure 2.77.

At the given irradiation dose (15 kGy), the liquid condensate yield increased approximately proportionally to the catalyst concentration (Figure 2.78). Increase in the liquid product yield was 14.8% with the addition of 0.17 mass% MnO_2 and 26.6% when 7.7% MnO_2 was added.

Figure 2.79 demonstrates considerable changes in the fractional contents of the liquid RTC product as catalyst concentration is varied.

Figure 2.80 shows that the maximal degree of heavy residue conversion and the maximal concentration of light fractions in the liquid condensate were observed with the MnO_2 concentration of about 2%.

The main contribution to the observed oil conversion comes from the increase in the concentration of a fraction boiling in the temperature range of 270°C–350°C

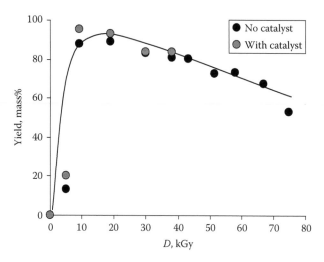

FIGURE 2.77   Dependence of the yield of liquid RTC product obtained from heavy oil.

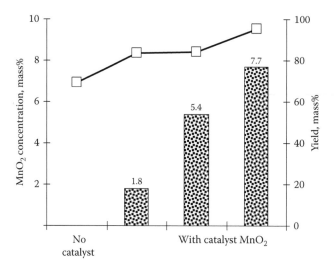

FIGURE 2.78   Dependence of the RTC liquid product yield on MnO_2 concentration.

(Figure 2.81). It indicates the connection of the increase in the yields of light fractions in the presence of MnO_2 with a more intense oxidation of the heavy aromatic part of the feedstock by atmospheric oxygen and products of its ionization present in the reactor. In turn, it facilitates detachment of molecular fragments from the skeleton of the polynuclear aromatic system under ionizing irradiation. This is also confirmed by the changes in the HC contents of the gasoline fraction. A very characteristic is considerable increase in the concentration of relatively heavy aromatic HCs C_8–C_9 (Figure 2.81).

Catalyst addition leads to an increase in the RTC rate and, therefore, to an increase in the concentration of unsaturated HCs (Figure 2.82).

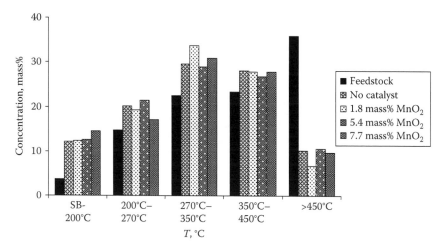

FIGURE 2.79 Fractional contents of the liquid product obtained by RTC of heavy oil with different concentrations of MnO_2.

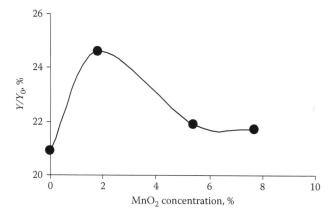

FIGURE 2.80 Dependence of the yield of light fractions ($T_b < 350°C$) on MnO_2 concentration. $Y - Y_0$ is difference of the light fraction concentrations in the liquid RTC product and in the original feedstock.

Dienes and olefins appearing among unsaturated compounds may be formed as a result of detachment of aliphatic fragments, dehydration reactions, and cycloolefin radiolysis (Figure 2.83).

Accumulation of unsaturated compounds at the high catalyst concentrations leads to intensification of the polycondensation reactions and decrease in the cracking rate.

2.4.3 BITUMEN RADIATION PROCESSING

Approaches developed and tested on heavy crude oil and oil residua were applied to processing of still heavier and more viscous feedstock, such as bitumen from different fields of Kazakhstan, Russia, and Canada.

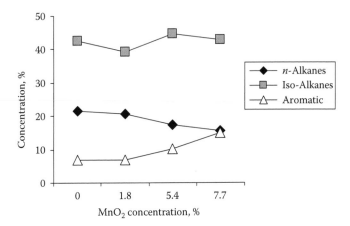

FIGURE 2.81 Dependence of HC group contents in the gasoline fraction obtained by RTC of heavy oil on MnO_2 concentration.

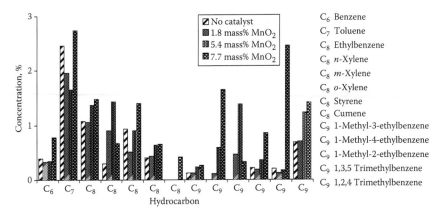

FIGURE 2.82 HC contents of the aromatic component of RTC gasoline fraction for different MnO_2 concentrations.

According to conventional classification, the density of heavy crude oil is about 0.95 g/cm^2 while bitumen density is nearly 1.00 g/cm^2. Oil residua of atmospheric or vacuum fractionation fall into the same class of heavy petroleum feedstock. Traditional methods of deep oil processing are mainly purposed for the processing of oil feedstock with a density up to 0.90 g/cm^3. A general characteristic of bitumen is an enormous molecular mass of its components that can be higher than 500 g/mol, for example, in the case of high concentrations of asphaltenes.

Oil bitumen rocks are important alternative sources of HCs widespread all over the world. Depending on bedding depth, they contain from 5% to 45% bitumen organic part that can be used as valuable feedstock for the production of engine fuels, lubricants, coke, etc. However, economic bitumen utilization comes across technological difficulties and high cost of their processing.

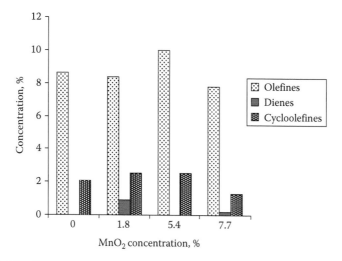

FIGURE 2.83 Concentration of unsaturated HCs in the gasoline fraction obtained by RTC of heavy oil: dose rate, 3 kGy/s; irradiation dose, 15 kGy.

Mustafaev et al. (2004) applied RTC to oil–bituminous rock (OBR) refining. It was noted that use of OBRs as energy source and raw material for chemical synthesis requires beforehand extraction of the organic part from the rock. The approach used in the study by Mustafaev et al. (2004) was to turn the organic part of bitumen to fuel gases from the rock by means of RTC without extraction of the organic part from the rock.

OBR samples from Kirmaku field (Azerbaijan) were used as feedstock. Their specific weight was 1.50–1.56 g/cm³, oil content in the rock was 8.2–9.4 mass%, water saturation was 3.14, and the clay content was on average 25.7. The H/C ratio in the organic mass was 0.14.

The OBR samples were subjected to vacuum drying at 80°C for 0.5 h before irradiation. After injection of water steam into the ampoule with OBR, it was soldered. Irradiation of the ampoule was carried out in the gamma radiation field of Co-60 with the dose rate of 11 Gy/s. Kinetics of H_2, CO, and CH_4 accumulation, as a result of OBR thermoradiation conversion, is shown in Figure 2.84. Radiation-chemical yields of the gases (number of molecules produced per 100 eV absorbed energy) were $G_1 = 2.7$, $G_2 = 5.3$, $G_3 = 1.1$, $G_4 = 1.6$, $G_5 = 4.1$, and $G_6 = 12.5$.

Temperature dependence of the gas formation rate is shown in Figure 2.85. The temperature increase from 20°C to 500°C leads to increase in the G-values of gases from $G(H_2) = 0.054$, $G(CO) = 0.003$, and $G(CH_4) = 0.004$ to $G(H_2) = 52.1$, $G(CO) = 3.9$, and $G(CH_4) = 62.5$. The activation energies of the gases produced were $E(H_2) = 118.2$, $E(CO) = 32.5$, and $E(CH_4) = 76.4$ kJ/mol.

Addition of steam at the pressure of 0.1 MPa did not change the yields of gases in the temperature range of 20°C–300°C. However, the effect of steam injection was observed when the temperature was raised up to 500°C. In this case, the yields of hydrogen and methane increased up to $G(H_2) = 95.1$ and $G(CH_4) = 142.6$ while the yield of CO did not significantly change.

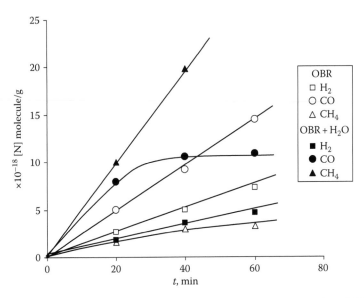

FIGURE 2.84 Kinetics of H_2, CO, and CH_4 accumulation due to conversion of OBR and OBR + H_2O systems. $P = 11$ Gy/s; $T = 400°C$; H_2O pressure, 0.1 MPa. (Plotted using the data by Mustafaev, I. et al., *J. Radioanalyt. Nucl. Chem.*, 262(2), 509, 2004.)

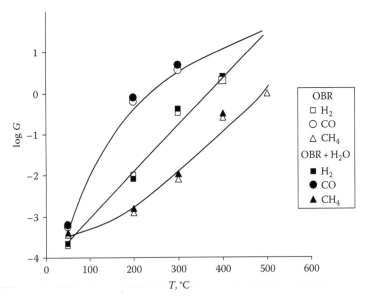

FIGURE 2.85 Temperature dependence of G-values for H_2, CO, and CH_4 produced due to thermoradiation conversion of OBR ($\triangle$, $\circ$, $\square$) and OBR + H_2O system ($\blacktriangle$, $\bullet$, $\blacksquare$). $P = 11$ Gy/s, H_2O pressure—0.1 MPa. (Plotted using the data by Mustafaev, I. et al., *J. Radioanalyt. Nucl. Chem.*, 262(2), 509, 2004.)

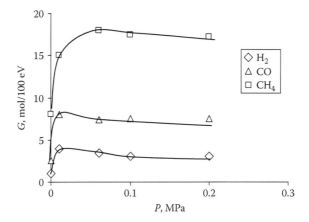

FIGURE 2.86 Dependence of G-values for H_2, CO, and CH_4 produced due to thermoradiation conversion of $OBR + H_2O$ system on water steam pressure. $P = 11$ Gy/s, $T = 400°C$. (Plotted using the data by Mustafaev, I. et al., *J. Radioanalyt. Nucl. Chem.*, 262(2), 509, 2004.)

A monotonous increase in the yields of H_2 and CH_4 up to $G(H_2) = 7.1$ and $G(CH_4) = 16.2$ was observed in the dependence of G-values of fuel gases from $OBR + H_2O$ at the temperature of 400°C on water steam pressure in the pressure range 0–0.025 mPa (Figure 2.86). Further increase in steam pressure up to 0.2 MPa did not change $G(H_2)$ and $G(CH_4)$.

Formation of H_2, CO, and CH_4 due to radiation-induced OBR decomposition was attributed to the radiation decay of the organic part of OBR (2.41) and recombination of primary radicals (2.42):

$$OBR \rightarrow H, CH_3, CO \tag{2.41}$$

$$H + H + M \rightarrow H_2 + M \tag{2.42}$$

$$CH_3 + H \rightarrow CH_4$$

G-values of gases at room temperature do not exceed the value of

$$\sum G_g = G(H_2) + G(CO) + G(CH_4)$$

because of the high radiation stability of the aromatic part of the OBR organic mass. Similar yields of gases were observed in the earlier-discussed study by Skripchenko et al. (1986). Yields of gases additionally increase due to the secondary processes (2.43) accompanied by decomposition of weakly bound oxygen-containing groups:

$$H + OBR \rightarrow H_2 + R'$$
$$CH_3 + OBR \rightarrow CH_4 + R' \tag{2.43}$$

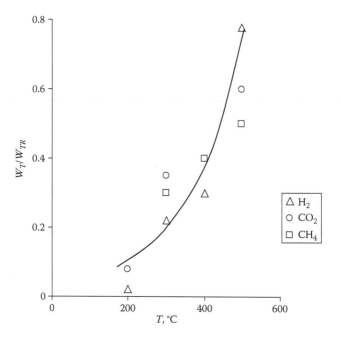

FIGURE 2.87 Temperature dependence of W_T/W_{TR} resulting from thermoradiation conversion of OBR system: $\triangle$, H_2; $\circ$, CO; $\square$, CH_4. $P = 11$ Gy/s. (Plotted using the data by Mustafaev, I. et al., *J. Radioanalyt. Nucl. Chem.*, 262(2), 509, 2004.)

Significant increase in $G(H_2)$ and $G(CH_4)$ was attributed to the formation of the active particle (H, OH, H_2O^*) as a result of water radiolysis. Chain processes of fuel gas formation, due to thermoradiation conversion of OBR and OBR + H_2O systems, proceed in the temperature range 400°C–500°C where the chain length reaches the values $\nu \sim G_g/G_R \sim 100$. Fuel gases are also formed by direct thermal decomposition of OBR at the temperatures above 250°C. However, Figure 2.87 shows that the rate of the thermoradiation (W_{TR}) process is higher than the rate of the thermal process (W_T), that is, $W_T/W_{TR} < 1$, in the overall temperature range studied (up to 500°C).

The cracking of bitumen from the Mordovo-Karmalskoe field (Tatarstan, Russia) was described in the paper (Bludenko et al. 2007). The density of bitumen samples was 0.967 ± 0.014 kg·dm^{-3}, the total sulfur content was 3.7 ± 0.1 wt%, and the asphaltene content was about 15.5 wt%. The initial boiling point was about 180°C. Crude bitumen contained components (about 20 wt%) with boiling points below 370°C (initial point of thermal decomposition). The authors (Bludenko et al. 2007) suggested that there was no necessity to apply cracking to this light fraction. Therefore, the low-boiling fraction was first distilled off at 360°C–370°C, and the heavy residue (about 80 wt%) was used in the radiation experiments.

An ELU-6E linear accelerator served as a source of electron radiation (energy of 8 MeV, pulse duration of 6 μs, pulse generation frequency of 300 Hz, average beam current of ≤800 μA, initial width of scanned beam of 245 mm, and scanning frequency of 1 Hz). The average dose rate was 8 kGy/s.

Samples of bitumen were irradiated in temperature-controlled vessels in two modes with bubbling by various gases. In the first (bubbling) mode (A), the samples were heated to 395°C±6°C only by the electron beam at a high dose rate. During irradiation, the bitumen was continuously stirred by a gas flow, and volatile radiolytic products were removed together with the gas.

The second mode (B) was based on a gas lift procedure. Bitumen at 395°C±6°C was moved from the receiver into the radiation area and then into the separation vessel by means of discrete gas bubbles. Vapor–gas mixture was removed to cooling and condensation, while the liquid was returned to receiver 1 along tube 7. Short irradiations (0.05 s) were alternated with pauses (0.95 s). The radiolytic products from the vapor–gas mixture were condensed at 16°C. This condensate was considered as the end product.

A propane–butane mixture (30 wt% propane, 40 wt% isobutane, and 30 wt% n-butane), methane, and the superheated steam, supposed as sources of hydrogen and also helium, were applied as carrier gases at a flow rate of 80–120 cm³/kg/s. The end products were analyzed using a chromatograph–mass spectrometer.

Characteristics of the end products are shown in Table 2.16 and Figure 2.88.

The distillation curves describing fractional composition of the cracking end products are presented in Figure 2.89. The end product of TC contained more low-boiling components than the end product of RTC. At the same time, TC was three times slower; it yielded the lowest conversion of feedstock to end product (about 37 wt%) and the higher content of unsaturated HCs (about 16 wt%). An effect of a carrier gas on composition of TC end product was not observed.

The alkenes detected in the composition of the end product are the foregone cracking products. However, high alkene concentrations in the end product are undesirable from a viewpoint of making stable synthetic oil from bitumen. The end products of RTC contained much less alkenes (about 3–7 wt%) due to the efficient addition of radiolytic intermediates to unsaturated bonds. Such an addition becomes more probable at high concentrations of intermediates hence it may be stimulated by a higher dose rate.

TABLE 2.16

Average Characteristics of the End Products:

Refraction Index n_d^{20}, Density ρ^{20},

Concentration of Unsaturated Hydrocarbons N

Mode	Carrier Gas	n_d^{20}	ρ^{20}, kg/dm³	N, wt%
TC	$C_3H_8 + C_4H_{10}$	1.4855	0.897	16.1
RTC A	$C_4H_8 + C_4H_{10}$	1.4962	0.887	7.1
RTC B	CH_4	1.4941	0.881	5.8
RTC B	He	1.4970	0.877	7.0
RTC B	H_2O	1.4887	0.872	3.2

Source: Bludenko, F.V. et al., Mendeleev Commun., 17, 227, 2007.

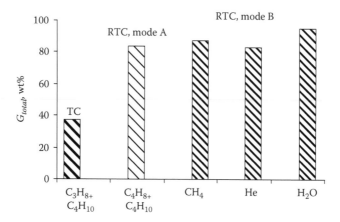

FIGURE 2.88 Total yields G_t of fraction boiling below 350°C for different carrier gases. (Plotted using the data by Bludenko, F.V. et al., *Mendeleev Commun.*, 17, 227, 2007.)

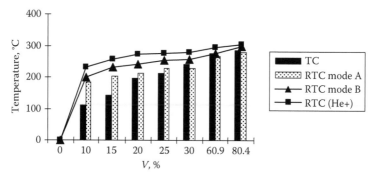

FIGURE 2.89 Curves of atmospheric distillation for end products of TC and RTC. (Plotted using the data by Bludenko, F.V. et al., *Mendeleev Commun.*, 17, 227, 2007.)

The radiation-initiated alkylation is a variety of addition reactions. A free alkyl radical adding an alkene or an alkyne can form a larger radical before the synthesis of a large compound (Cserep et al. 1985). The alkylation may occur as the chain radical process (Woods and Pikaev 1994):

$$\cdot RH + RHC{=}CHR' \rightarrow (RH)_2 C\dot{C}\, HR' \qquad (2.44)$$

$$(RH)_2 C\dot{C}\, HR + RH_2 \rightarrow (RH)_2 CCH_2R + \cdot RH \qquad (2.45)$$

The presence of unsaturated bonds in feedstock molecules can simultaneously initiate both alkylation and phenylation, resulting in an increase in the content of large molecules in the end products (Woods and Pikaev 1994). In particular, the end product of RTC had a higher average molar mass and, accordingly, higher boiling points (Figure 2.89).

Aromatic components also protect bitumen against radiation damage but owing to efficient dissipation of absorbed beam energy. At the same time, the yield of radiolytic decomposition of alkylbenzenes by scission of bonds in the alkyl groups is increasing as these groups are becoming larger. Besides, aromatic compounds can act as efficient scavengers of electrons and radicals. In particular, the reactions of hydrogen addition to a benzene ring with the formation of alkylhexacyclodienyl radical

$$\cdot H + C_6H_5R \rightarrow \cdot C_6H_6R \qquad (2.46)$$

provide a rather low yield of hydrogen. There is a high probability of thermal decomposition of an alkylhexacyclodienyl radical with the formation of an aromatic molecule and an alkyl radical (Woods and Pikaev, Cserep et al.).

Aromatic compounds (Ar) can also act as scavengers of a positive charge from the cations of alkanes:

$$RH^+ + Ar \rightarrow RH + Ar^+ \qquad (2.47)$$

Charge-transfer reactions (2.47) proceed when the ionization potential of RH is higher than the ionization potential of Ar. The rate constants of reactions (2.47) between cations of alkanes and aromatic scavengers (Cserep et al. 1985, Woods and Pikaev 1994) at ambient temperature are 10^9–10^{11} dm^3/mol/s. Regeneration in the reactions of charge neutralization is also characteristic of aromatic compounds:

$$Ar^+ + Ar- \rightarrow Ar * + Ar \qquad (2.48)$$

The aromatic compounds containing sulfur can also protect the dissolved components against radiation damage. For example, polyphenylene sulfides are considered as compounds with high radiation resistance (Woods and Pikaev 1994). In spite of the specified protection effects, the essential decomposition accompanied by elimination of large pendent groups, hydrogenation, and alkylation of aromatic rings has been detected in the aromatic compounds of bitumen.

In addition, RTC produces more branched HCs than TC, which results in the difference of octane numbers of their end products. Aliphatic and aromatic final compounds apparently originate by the dimerization and disproportion of HC radicals. Radicals in the RTC processing are formed mainly due to the action of radiation. It is known (Woods and Pikaev 1994) that the radiolytic decomposition of an organic molecule produces mainly a radical having an unpaired electron in the central part of a carbon skeleton. In turn, TC results in radicals having an unpaired electron at the end of a carbon backbone chain. The secondary and tertiary radicals also originate as a result of addition (2.11) of small radicals to unsaturated compounds. Therefore, the probability of radiolytic formation of branched molecules increases essentially because of the prevalence of the recombination of secondary and tertiary radicals. Octane numbers for the branched HCs are more than those for linear homologues. Similar dominance of the branched alkanes in the end products of RTC of other bitumens was observed earlier at the electron dose rate of 1.5 kGy/s in the study (Zaikin and Zaikina 2004b).

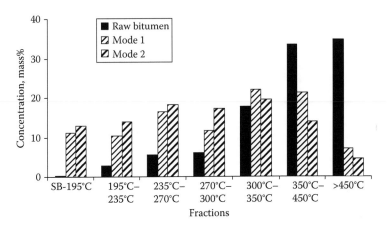

FIGURE 2.90 Fractional contents of synthetic oil produced by RTC of Athabasca bitumen at 400°C. Mode 1: $T=390°C$, $P=3.7$ kGy/s, $D=1.6$ kGy. Mode 2: $T=400°C$, $P=3.4$ kGy/s, $D=2.5$ kGy.

The composition of the end product of RTC in mode B depends on the nature of a carrier gas. The use of a propane–butane mixture produces an end product with a lower average boiling point than the use of helium or methane. In the presence of butanes, the end product is enriched in HCs formed mainly by recombination of *tert*-butyl radicals with organic radicals from bitumen. For example, *tert*-butylbenzene, *tert*-butylcyclopentane, and 2,2,3,3-tetramethylbutane were accumulated. Butanes, and especially isobutane, are the most responsive to irradiation among gaseous alkanes, while the yield of radiolytic decomposition of methane is the smallest (Makarov et al. 2007).

It was noted (Bludenko et al. 2007) that the presence of water in bitumen or the use of water vapor as a carrier gas results in the formation of alcohols and carbonyl compounds, decreasing long-term stability of the end product as slow polycondensation reactions are provoked.

Essential advantages of radiation initiation of TC have been revealed earlier at lower dose rates of an electron beam and γ-radiation. However, the identification of features of the cracking at extreme parameters of electron beam irradiation is very important. Industrial accelerators are acting at a beam current of about 0.1–1 A and at an average beam current density of 100 μA/cm². The results of the study (Bludenko et al. 2007) show that the radiation initiation increases the productivity of a TC even at extremely high dose rates. Unproductive thermal losses in RTC are lower than those in TC. The end product of RTC has both higher knock characteristics and smaller concentrations of sulfur and unsaturated compounds. At the same time, the concentration of low-boiling components in the end product in the experimental conditions of the study (Bludenko et al. 2007) was minimal at the high dose rates.

In the studies of RTC of bitumen (Zaikin and Zaikina 2004b, 2006), bitumen radiation processing was directed to production of upgraded oil with properties that answer demands for oil commodities and ecological requirements.

The modes of RTC remarkable for high yields of light fractions are illustrated in Figure 2.90 on the example of Athabasca bitumen (API 10.5°).

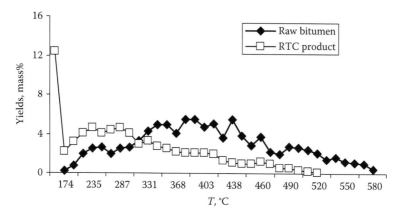

FIGURE 2.91 Fraction distribution in Shilikty bitumen before and after radiation-thermal processing.

HC molecular mass distributions in bitumen from Shilikty field, Western Kazakhstan (density ~ 1.02 g/cm³), before and after radiation-thermal processing are shown in Figure 2.91. Similar to the RTC of Athabasca bitumen, radiation processing of Shilikty bitumen resulted in high yields of light fractions. The yield of the liquid RTC product was 77.5 mass%. Concentration of fractions boiling below 350°C increased by 2.2–2.4 times compared with that in the original feedstock.

Comparison of light fraction yields obtained by application of different methods for deep bitumen processing is shown in Figure 2.92 on the example of bitumen from Mortuk field (Western Kazakhstan). Figure 2.92 shows that the one-stage bitumen radiation processing provides almost twofold reduction in the average molecular mass of the liquid fraction that allows production of upgraded oil with high contents of the gasoline–diesel fraction. RTC yields of light fractions are higher than those obtained by other methods, including TCC and such effective, though rather expensive, technology as combination of ozonolysis with subsequent thermal processing at 350°C.

Fractional contents of upgraded oil produced from Tubkaragan bitumen, Kazakhstan, are shown in Figure 2.93 with respect to the number of carbon atoms in a molecule. This sort of bitumen contained a considerable amount of water. During irradiation in the mode of distillation under an electron beam, water partially evaporated. However, the presence of water assisted at a considerable increase in the yields of light fractions boiling below 220°C (up to 17 mass% in the liquid product of bitumen radiation processing).

Figure 2.94 shows the fraction yields of RTC and TCC with respect to fraction boiling temperatures. Conversion of the heavy residue with the start of boiling at 450°C is two times higher in the case of bitumen radiation processing.

Typically, yields of upgraded oil from RTC of different types of bitumen reached 82–86 mass%; yields of the coking residue (feedstock for coke production) were up to 10 mass%; and yields of gases (H₂, CH₄, C₂H₄, etc.) were 4–8 mass%. The average density of the upgraded oil was 0.86 g/cm³.

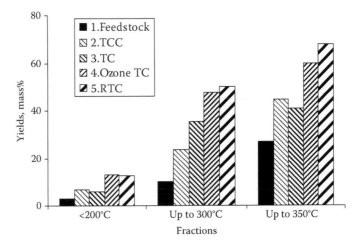

FIGURE 2.92 Radiation-thermal cracking of Mortuk bitumen (Kazakhstan). 1. Original feedstock (natural bitumen). Fractions were thermally distillated by a conventional method (Kamyanov et al. 1994); 2. TCC—thermocatalytic cracking using facility (Musaev et al. 1994) for bitumen processing in the temperature range of $T=400°C–500°C$; clay and sand particles in the rock served as catalysts; 3. TC—thermal cracking (Kamyanov et al. 1994); 4. Ozone TC—preliminary bitumen ozonization by introduction of 34 g of O_3 per 1 kg of feedstock and subsequent thermal processing at 350°C (Kamyanov et al. 1994); 5. RTC—radiation-thermal cracking (Zaikin and Zaikina 2004b).

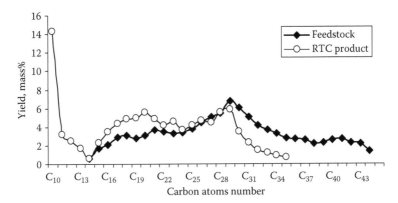

FIGURE 2.93 Fraction distribution with respect to the number of carbon atoms in Tyubkaragan bitumen before and after radiation-thermal processing.

Figure 2.95 provides comparison of the fractional contents of upgraded oil obtained by TCC and RTC of natural bitumen having different API gravity (7.2°, 12.6°, and 13.2°) (Zaikin and Zaikina 2006).

Similar to the experiment on fuel oil radiation processing described earlier, higher yields of light fractions were obtained from heavier sorts of bitumen. This result shows that efficiency of heavy oil radiation-thermal processing depends not on oil density but rather on oil availability of highly branched aromatic structures.

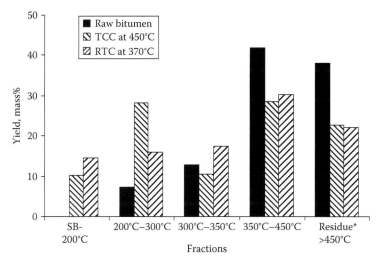

FIGURE 2.94 Fractional contents of upgraded oil produced by thermocatalytic technology (TCC) and by means of radiation processing (RTC) of bitumen from Tyubkaragan field. Heavy residue boiling above 450°C includes RTC coking residue and a residual part of the liquid RTC product.

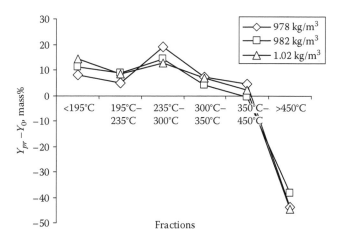

FIGURE 2.95 Fractional contents of liquid RTC products from different sorts of bitumen. $Y_{pr} - Y_0$ is the difference in the fraction yields from the liquid RTC product and from the feedstock.

The data on HC compositions of the gasoline fractions from upgraded oil obtained by TCC and RTC of the three sorts of bitumen are compared in Table 2.17. The table shows that the content of gasoline produced by RTC essentially differs from that obtained by a thermocatalytic method.

Concentration of isoparaffins after TCC was almost two times lower than that of n-paraffins due to the highly pronounced effect of radiation-induced isomerization.

TABLE 2.17

Hydrocarbon Content of Gasoline Fraction Produced by Bitumen Processing

Field, Type of Processing	Hydrocarbon Contents, Mass%					Octane Number
	n-Alkanes	Iso-Alkanes	Aromatic	Naphthene	Unsaturated	
Mortuk, TCC*	29.0	17.0	22.6	13.4	19.0	68
Mortuk, RTC	17.9	33.4	21.8	13.8	10.4	75
Shilikty, TCC*	28.2	17.3	22.5	12.7	19.3	66
Shilikty, RTC	20.6	36.1	15.5	13.0	14.8	72
Tyubkaragan, TCC*	30.1	16.4	22	13.5	18.0	68
Tyubkaragan, RTC	20.0	40.9	11.6	19.1	8.4	76

In the case of RTC, iso-alkanes, concentration was almost two times higher compared with n-alkanes, concentration in RTC gasoline and isoparaffin concentration in the TCC product. Octane numbers were correspondingly higher in the case of RTC gasoline.

Another characteristic feature of the gasoline fraction obtained by RTC of bitumen is a higher concentration of paraffin HCs both of the linear and of the branched structure and, therefore, lower concentrations of aromatics and unsaturated HCs compared with gasoline produced by a thermocatalytic method. Since alkyl substituents in the aromatic rings are the main source of aliphatic compounds, predominance of paraffins in HC contents of light RTC products is the evidence of a more efficient substituent detachment during RTC.

HC contents of the aromatic part of gasoline fraction generally correspond with the original molecular structure of bitumen (Evans 1964). The latter was characterized by a number of polycyclic and benzol aromatic HCs C_{21}–C_{42} with the maximum of molecular mass distribution in the region of C_{31}. There were many cycloalkyl structures that could be either substituents of other HCs, for example, cycloalkylbenzols, or bicyclic naphthene structures with cycloalkyl groups divided by long chains. Together with alkyl substituents, many compounds had cycloalkyl and phenylalkyl substituents.

Most of the components observed in the aromatic part of gasoline fraction after RTC, such as isooctane (2,2,4-trimethylpetane) and cycloolefins, could be formed as a result of the destruction of substituent substructure during radiation-thermal processing. However, higher absolute yields of aromatics compared with a thermocatalytic process did not exclude appearance of monoaromatic HCs as a result of direct radiation-induced destruction of polyaromatic compounds.

The general bitumen characteristic is the deficiency of hydrogen that limits yields of light fractions after any type of processing. In bitumen radiation processing, water was used as an additional source of hydrogen (Zaikin and Zaikina 2004b). Water addition allows considerable increase in the yields of synthetic oil, even up to a level higher than the original feedstock mass.

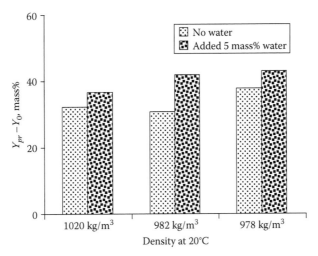

FIGURE 2.96 Changes in the yields of RTC fraction boiling below 350°C from different sorts of bitumen due to water addition. $Y_{pr} - Y_0$ is the difference in the fraction yields from the liquid RTC product and from the feedstock.

Fractional contents of upgraded oil produced from different types of bitumen with water addition (Zaikin and Zaikina 2006) are shown in Figure 2.96. Although water addition increased the yields of light fraction ($T_b < 350°C$) only by 3%–5%, it should be taken into account that original bitumen could contain a considerable amount of bound water.

Increase in the total product mass due to water addition and observed changes in its chemical composition are the evidence of the radiation-induced reactions with water participation.

Apparently, most chemically active particles in these reactions are hydrogen atoms H and radicals OH and HO_2 that can initiate a great number of reactions, for example,

$$RH + H \rightarrow R + H_2$$

$$HO_2 + RH \rightarrow ROOH + R$$

$$HCOOH + H \rightarrow H_2 + COOH$$

$$HCOOH + OH = H_2O + COOH, \text{ etc.}$$

Typical material and elemental balances of bitumen radiation-thermal processing are shown in Tables 2.18 and 2.19. Bitumen feedstock usually contains a considerable amount of water (up to 30 mass%) that must be taken into account in the mass and elemental balances of bitumen processing as well as in the forecast of the yields of light fractions.

In a common practice, heavy oils and bitumen are classified by their density (or API gravity) and viscosity. The data on RTC of two sorts of Athabasca bitumen of different API gravity, AB1 having API 13° and AB2 having API 10°, are shown in Figures 2.91 through 2.95.

TABLE 2.18

Yields of Synthetic Oil in Material Balances of Products Obtained by Processing of Bitumen from Tyubkaragan Field by Means of TCC and RTC

| Product | Yields, Mass% | |
	TCC at 450°C	RTC at 370°C
Synthetic oil	82	92
Coke	6	2
Gases	12	6

TABLE 2.19

Typical Elemental Balance of the Overall Product Produced by Radiation-Thermal Cracking of the Organic Part of the Bitumen Rock

Products and Their Yields, Mass%	H	C	S	N+O	C/H
Feedstock	84.480	9.000	1.700	4.820	9.400
Bitumen+6% H_2O	79.410	9.120	1.600	9.800	8.700
70%—light gas oil	57.290	7.130	0.635	4.945	8.000
25%—heavy gas oil	20.880	1.780	0.850	1.490	11.700
5% gases	1.240	0.210	0.115	3.435	5.900

The experiments were conducted in the mode of under beam distillation.

Figure 2.97 represents the material balance in the RTC of bitumen AB2 (API 10°). The bitumen samples were irradiated with 2 MeV electrons at a temperature of 400°C and a dose rate of 3.5 kGy/s. A twice-increased irradiation dose leads to a lower conversion of heavy residue and lower yields of light fractions.

The dose dependence of gaseous product yields in the RTC of bitumen (API 10°) obtained in a series of experimental runs under similar conditions is shown in Figure 2.98. It demonstrates nearly linear gas yields up to the dose of 46 kGy.

For comparison, a material balance in the RTC of lighter bitumen (API gravity 13°) irradiated in the same conditions at the dose rate of 4.5 kGy/s is shown in Figure 2.99. In both cases of lighter and heavier bitumen, a degree of the heavy residue (C20+) conversion was about 50%. However, in heavier bitumen characterised by a higher content of the heavy residue, the yields of liquid light fractions were considerably higher. It indicates that the main source of light fractions in the RTC of bitumen is the aliphatic part of the branched polynuclear aromatic system.

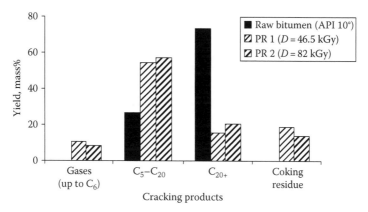

FIGURE 2.97   Products yields in RTC of bitumen (API gravity 10°).

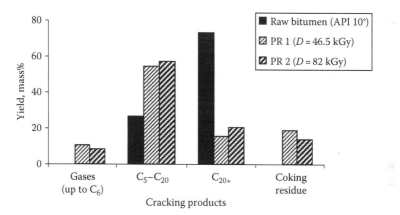

FIGURE 2.98   Dose dependence of the gas yields in RTC of bitumen (API gravity 10°).

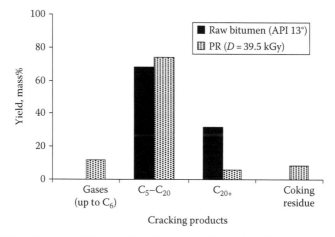

FIGURE 2.99   Products yields in RTC of bitumen (API gravity 13°).

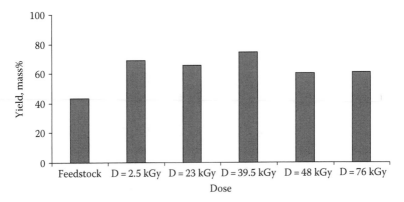

FIGURE 2.100 Yields of fractions (SB-350°C) produced by RTC of bitumen (API 13°) for different doses of electron irradiation. $P = 5$ kGy/s; $T = 410$°C.

The yields of fractions C_5–C_{20} boiling below 350°C in the liquid RTC product of bitumen (API 13°), calculated from the data of gas–liquid chromatography, are shown in Figure 2.100. Lower yields of light fractions at high doses rates are associated with bitumen's tendency to radiation-induced polymerization and adsorption of the light fractions by the polymerized residue.

Comparison of the dose dependences of liquid fraction yields in the RTC of the two types of bitumen (Figure 2.101) shows higher yields of the light product for all doses of electron irradiation. It confirms a conclusion that the most part of light products was obtained from the branched aromatic system in the heavy fractions of bitumen.

The effect of water addition on the RTC of bitumen was studied in a series of experiments. Water was added to bitumen at a temperature of about 50°C with

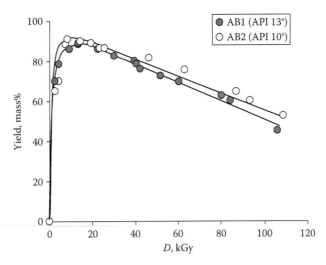

FIGURE 2.101 Yields of the liquid RTC product versus irradiation dose for two sorts of bitumen.

continuous stirring. Then bitumen with added water was slowly heated under an electron beam up to the reaction temperature of 390°C. The preheated bitumen samples were irradiated with 2 MeV electrons in the mode of under beam distillation at the dose rate of 3 kGy/s. The absorbed irradiation dose was 26 kGy.

A step at the temperatures of 100°C–120°C observed in a monotonous dependence of bitumen temperature on time of heating testified to partial water evaporation. However, addition of even small amounts of water considerably raised a yield of liquid RTC products (Figure 2.102). In the experiments with an addition of 2.5 mass% water, the average yields of liquid condensates were 81.5 mass%, that is, higher by 6%–8% than that without water addition.

Water addition to bitumen had a pronounced effect on the HC contents of the gasoline fraction of a liquid RTC product (Figure 2.103). Gasoline fractions produced from bitumen with added water were characterised by higher concentrations of iso-alkanes and aromatic compounds and lower concentrations of naphthenes. Figure 2.104 shows that higher concentrations of added water caused increase in the molecular weight of aromatic compounds in gasoline.

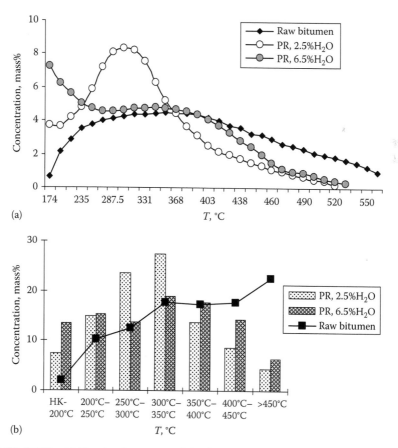

FIGURE 2.102 (a) Fractional contents of the liquid RTC product of bitumen with added water and (b) concentrations of narrow fraction in the liquid product.

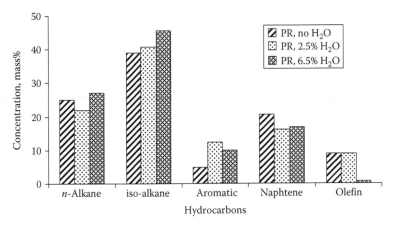

FIGURE 2.103 HC group contents of gasoline fraction distilled from the liquid RTC product of bitumen with added water.

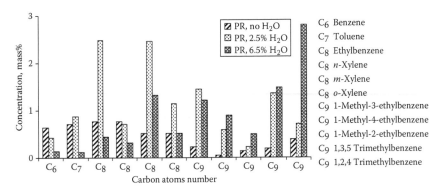

FIGURE 2.104 Mass-molecular distribution of arenes in gasoline fraction distilled from the liquid RTC product of bitumen with added water.

Addition of water to bitumen in concentrations higher than 6.5 mass% did not affect the liquid product yields and did not improve its HC contents.

As an approach to the evaluation of synthetic oil's commercial value and development of methods for utilization of heavy coking residue (if available in the target product), distributions of vanadium-containing compounds in products of bitumen radiation-thermal processing were determined. Properties of the coking residue after radiation-thermal processing were studied in different irradiation modes (Zaikin and Zaikina 2006).

Data on concentrations of vanadium compounds in products of bitumen processing obtained using different experimental methods were compared and analyzed. These experiments were intended to find out how conditions of RTC affect chemical conversion of vanadium-containing molecules, including such compounds of high commercial value as vanadium porphyrins.

It was shown that concentration of the quadrivalent vanadium in the coking residue after RTC varies in the range 800–1200 g/ton, while its content in the original feedstock was 160 g/ton. This result testifies to intense radiation-induced vanadium oxidation and can be compared with the available data on vanadium accumulation in coke during tar thermo-destruction (Kurdumov 1999). In this work, oil residua were subjected to thermal destruction at temperatures higher than 400°C in the presence of particles (carriers) of different nature. Asphalt-pitch and metal-containing compounds precipitated on the surface of the carriers and transformed into coke. In such a way, they were removed from the products formed. Together with thermopolycondensation of asphaltenes and pitches, this process was accompanied by formation of up to 70% distillate fractions.

To achieve higher vanadium extraction, solid particles of iron ore concentrate were used as carriers in the coking process. Oxidation of coked carriers was studied at 575°C (Kurdumov 1999). Oxidation time was 30–90 min; rate of air supply was 2–3 mL/s per 1 g of carrier. In these experiments, maximum observed vanadium concentration in coke particles was $17 \cdot 10^{-2}$ mass%, that is, 1700 g/ton that is about 10 times higher than that in the original feedstock.

Thus, without the application of any additional technique, radiation bitumen processing provided practically the same efficiency of vanadium transfer to coke as in specially developed methods using oxidation with different additives. This result shows that intense bitumen oxidation that accompany RTC causes not only sulfur transformation and its predominant concentration in the heavy residue of processing but also metal transfer to the coking residue.

Radiation facility for heavy oil and bitumen radiation processing with the production rate of 200 kg of feedstock per hour (Zaikin and Zaikina 2006) is shown in Figure 2.105. The tests have shown its stable and reliable work and high efficiency in the processing of heavy oil and the organic part of bitumen.

FIGURE 2.105   Experimental facility for radiation-thermal oil processing.

Generally, radiation-thermal processing of bitumen allows obtaining high yields of synthetic oil together with the reliable control of its fractional and HC contents more efficiently compared with conventional methods of bitumen refining.

2.4.4 COAL RADIATION PROCESSING

A series of experiments was conducted on RTC of coal (Mitsui and Shimizu 1981, Kuztetsov et al. 2005, Torgashin 2009).

In the study by Mitsui and Shimizu (1981), RTC of Taiheiyo coal mixed with asphalt was carried out in the presence of hydrogen at 400°C with the dose rate of 0.9 Gy/s in a static batch-type autoclave of 100 mL capacity and with the TC under the same conditions.

The elemental contents of coal and asphalt are shown in Table 2.20.

The initial hydrogen pressure was 20 atm at 30°C. Tetralin and benzene were used as solvents. The yields of gas, nonvolatile BS residue, and benzene-insoluble (BI) residue were measured while the remaining products were defined as oil. The results of the cracking without solvent showed that decomposition of the component, most difficult for destruction in coal, was accelerated by gamma irradiation. At the same time, the components easily decomposed in the conventional process were less affected. In the case of cracking in tetralin, it was considered that gamma rays accelerate cracking at its early stage through the same mechanism as that of the cracking without solvent.

Decomposition of coal and asphalt was accelerated by gamma irradiation. The main gaseous products were methane and carbon dioxide in all the reaction systems studied. The time dependence of the yields of the gaseous products showed that formation of the gaseous HCs was accelerated by gamma irradiation while formation of carbon dioxide and carbon monoxide were almost independent of irradiation.

The product yields resulting from RTC and TC of coal without solvent are shown in Figure 2.106. The effect of gamma irradiation on the product yields becomes apparent at the doses higher than 10 kGy.

TABLE 2.20
Elemental Composition of Taiheiyo Coal and Straight Asphalt

Material	C	H	N	S	O	Ash, wt%
	\multicolumn{5}{c	}{wt%}	
Coal	76.8	7.4	1.2	0.2	14.4	11.0
Asphalt	85.2	11.0	0.7	3.1	0.0	0.06

Source: Mitsui, H. and Shimizu, Y., *Radiat. Phys. Chem.*, 18(3–4), 817, 1981.

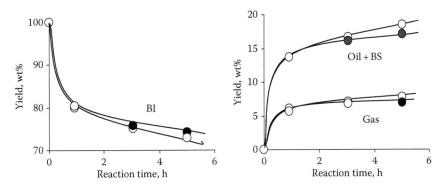

FIGURE 2.106 Time dependence of product yields (based on dry coal) from RTC (○) and TC (●) of coal in the presence of hydrogen at 400°C. (From Mitsui, H. and Shimizu, Y., *Radiat. Phys. Chem.*, 18(3–4), 817, 1981.)

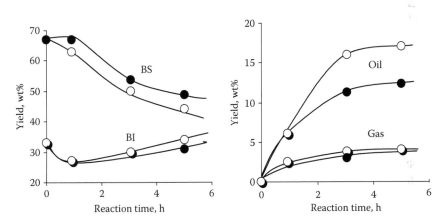

FIGURE 2.107 Time dependence of product yields (based on dry coal and asphalt) from RTC (○) and TC (●) of coal in asphalt in the presence of hydrogen at 400°C. (From Mitsui, H. and Shimizu, Y., *Radiat. Phys. Chem.*, 18(3–4), 817, 1981.)

In the case of coal cracking in tetralin, gamma irradiation enhances BI decomposition even at the early cracking stage. The yield of oil considerably increases under gamma irradiation while that of BS tends to slightly decrease (Figure 2.107).

In the case of RTC and TC of coal in asphalt, the increase in BI yields was attributed to the asphalt decomposition. It was suggested that coal decomposition proceeds as the main reaction at the early stage of cracking while asphalt cracking becomes predominant at a later stage. Both coal and asphalt decompositions are accelerated by gamma irradiation.

As shown in Figure 2.108, the main gaseous products were methane and carbon dioxide. No remarkable effects of gamma irradiation on gaseous product distribution were detected.

Time dependence of gas yields from RTC and TC without solvents and that in tetralin and asphalt, respectively, is represented in Figures 2.109 through 2.111.

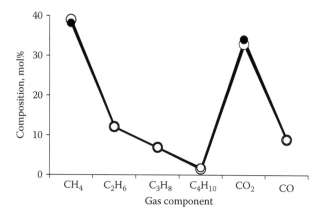

FIGURE 2.108 Distribution of gaseous products from RTC (○) and TC (●) of coal in tetra-lin in the presence of hydrogen at 4000°C; reaction time 3 h. (From Mitsui, H. and Shimizu, Y., *Radiat. Phys. Chem.*, 18(3–4), 817, 1981.)

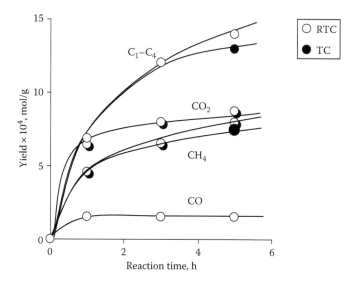

FIGURE 2.109 Time dependence of the yields (based on dry coal) of gaseous products from RTC (○) and TC (●) of coal in the presence of hydrogen at 400°C.

Figures 2.109 through 2.111 show that gamma irradiation increases the yields of gaseous HCs while the yields of carbon dioxide and carbon monoxide are little affected by irradiation.

In the work (Torgashin 2009), brown coals were irradiated with 1 MeV electrons at the dose rates of 1.9–3.3 kGy/s. The irradiation dose was varied from 100 to 2000 kGy. It was stated that electron irradiation of brown coal stimulated its chemical conversion accompanied by the evolution of volatile fractions. At the dose of 2 MGy, the level of radiation-enhanced coal decomposition into volatile products reached 15.4%. Methane (27.9%–65.9%) and *n*-butane (11.8%–23.1%) predominated in the composition of HC

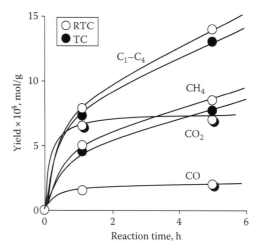

FIGURE 2.110   Time dependence of the yields (based on dry coal) of gaseous products from RTC (O) and TC (●) of coal in tetralin in the presence of hydrogen at 400°C. (From Mitsui, H. and Shimizu, Y., *Radiat. Phys. Chem.*, 18(3–4), 817, 1981.)

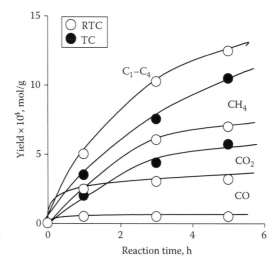

FIGURE 2.111   Time dependence of the yields (based on dry coal and asphalt) of gaseous products from RTC (O) and TC (●) of coal in asphalt in the presence of hydrogen at 400°C. (From Mitsui, H. and Shimizu, Y., *Radiat. Phys. Chem.*, 18(3–4), 817, 1981.)

gases. HCs C_5–C_8 were also formed in a smaller amount. At the dose rate of 500 kGy and a temperature of 60°C–70°C, evolution of volatiles was inconsiderable.

The form of the dependence of swelling coefficient in tetralin on absorbed dose differed in the regions of high and low doses (Figure 2.112). At the doses of 10–50 Mrad (100–500 kGy), the observed increase in the coal swelling capacity reflected the prevalence of partial depolymerization of the organic mass of coal. It was

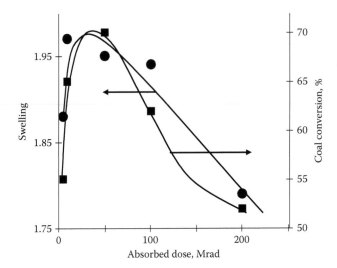

FIGURE 2.112 Changes in the coefficient of coal swelling in tetrahydrofuran in the process of coal hydrogenation in tetralin at the temperature of 380°C versus absorbed irradiation dose. (From Torgashin, A.S., Effects of mineral components and modifying processing on permolecular organization and reaction ability of brown coals, PhD dissertation, Institute of Chemistry and Chemical Technology, Krasnoyarsk, Russia, 2009.)

accompanied by a considerable increase in the swelling rate; the power exponent ($n = 0.75$) approached the value characteristic of diffusion due to relaxation oscillations of macromolecules that testified to the increase in the molecular fragment mobility. Further increase in the irradiation dose stimulated crosslinking processes.

Conversion of irradiated coals in the process of their hydrogenation in the tetralin medium at the temperature of 380°C depended on the irradiation dose and conditions of radiation processing. Both conversion of irradiated coal to liquid HCs and its swelling capacity had a pronounced maximum at the dose of 500 kGy, which corresponds to 3% of the total coal heat content (Figure 2.112). The effect of electron irradiation was more considerable in the presence of organic solvents. The conversion of coal pretreated with ionizing irradiation in the presence of tetralin or ethyl alcohol increased by 1.5–1.6 times compared with the conversion of the original coal. Increase in the coal conversion occurred owing to liquid HCs without noticeable changes in the yields of gaseous products.

The effects of coal activation using different chemical processes (de-cation processing in HCl solvents or solvation with a polar solvent, such as ethyl alcohol), mechanochemical treatment, and irradiation by accelerated electrons were compared in the work (Torgashin 2009).

It was stated that compared with chemical and mechanochemical treatments, irradiation by accelerated electron beams is the most efficient method of brown coal activation. It allowed raising the yield of liquid HCs up to 79%, which is 1.7 times higher than that observed for the original coal (Figure 2.113).

Since the beginning of the 1990s, researchers were focused on the radiation processing of heavier types of oil feedstock (Nadirov et al. 1994a,b, 1995, 1997a,b,

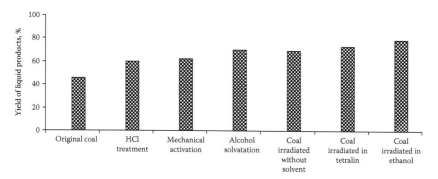

FIGURE 2.113 Effect of different types of brown coal activation pretreatment on the yield of liquid products in the process of coal hydrogenation in tetralin. (Plotted using the data from Torgashin, A.S., Effects of mineral components and modifying processing on permolecular organization and reaction ability of brown coals, PhD dissertation, Institute of Chemistry and Chemical Technology, Krasnoyarsk, Russia, 2009.)

Zaikina et al. 1997, 2002a,b, 2004, Zaikin et al. 1999a,b, 2001, 2003, 2004a,b, Zaikin and Zaikina 2004a–e).

Systematic studies of radiation-chemical transformations in complex HC mixtures and technological approaches to radiation-thermal processing have shown that radiation technologies are most advantageous and promising for large-scale industrial application when they are applied to the processing of high-viscosity or high-paraffin crude oils, bitumen, wastes of oil extraction, used oil products, and heavy residua of oil primary processing. Processing of such types of oil feedstock by traditional methods is not economical or comes across considerable technological difficulties. In this connection, physical and chemical aspects of radiation processing of heavy HC feedstock are of special technological interest. The large-scale application of high-efficiency and environmentally friendly radiation methods for heavy oil processing could be a radical solution of many acute problems of oil extraction, transportation, and refining (Zaikin et al. 1999c, Zaikina et al. 2002b, 2005, Mirkin et al. 2003).

In contrast to light crude oil, the specificity of self-sustaining radiation-chemical transformations at heightened temperatures in heavy HC feedstock of complex chemical composition displays itself in strong synergetic effects that accompany RTC. Such effects are provoked by redistribution of radiation energy between the original components of a complex HC mixture and a great number of intermediate reaction products. These phenomena facilitate side reactions of nondestructive character that considerably affect the rate of feedstock conversion and the contents of final products (Zaikina and Ailyev 2001, pp. 114–120; Zaikin et al. 2001, 2004).

The two most important synergetic effects, revealed and investigated in experiments with the problem sorts of high-paraffin and high-viscous oil, are the phenomena of radiation-induced isomerization and radiation-induced polymerization in conditions of RTC. Statement of conditions and evaluation of the intensity of these effects are important for obtaining high yields and quality control of the commodity oil products produced as a result of radiation processing of heavy oil feedstock.

2.4.5 RADIATION-INDUCED ISOMERIZATION IN THE PROCESS OF RADIATION-THERMAL CRACKING OF HIGH-VISCOUS OIL

Isomerization by-effects in conditions of induced cracking were noted in the works on radiolysis and photolysis of light oil fractions (Topchiev and Polak 1962, Dolivo et al. 1986). The phenomenon of strong radiation-enhanced isomerization in the process of RTC was first observed in the experiments on radiation-thermal processing of high-viscous oil from Karazhanbas field, Kazakhstan (Zaikin et al. 2001, Zaikin and Zaikina 2004c). General characteristics of Karazhanbas oil are given in Section 2.3. Together with the considerable paraffin component, this sort of oil is characterised by high concentrations of resins and heavy aromatic compounds.

The main products obtained by radiation-thermal processing of Karazhanbas oil (Zaikin et al. 2001) were a liquid gas oil fraction with boiling temperature from 60°C to 350°C, a heavy coking residue (pitches, asphaltenes, and solid coke particles), and the gaseous fraction (4–7 mass% hydrogen, 35–40 mass% methane, 18–21 mass% ethane, 10–12 mass% butane, 10–12 mass% ethylene, and 8–12 mass% propylene and other gases).

Oil samples were irradiated by electrons having an energy of 2 MeV from the linear electron accelerator ELU-4. The time-averaged dose rate of electron irradiation was varied in the range 0.5–1.5 kGy/s.

Figure 2.114 shows the typical dose dependences of G-values for gas oil fractions with boiling temperature up to 350°C obtained as a result of the RTC of Karazhanbas oil in two different modes. Mode 1 provides conditions for intense molecular destruction, while mode 2 is more favorable for isomerization reactions.

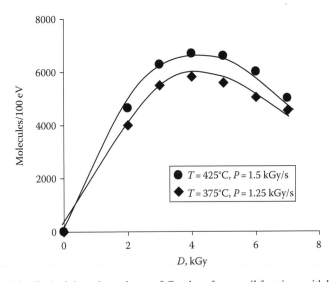

FIGURE 2.114 Typical dose dependence of G-values for gas oil fractions with boiling temperature up to 350°C obtained by RTC. (Plotted using the data from Zaikin, Y.A. et al., *Radiat. Phys. Chem.*, 60, 211, 2001.)

The maximums of *G*-values corresponding to the yields of liquid fractions of 80 and 52.5 mass% in modes 1 and 2, respectively, were observed at the same dose values of about 6 kGy. The further dose increase makes *G*-values of the liquid RTC products lower due to the effect of polymerization reactions. Intramolecular isomerization stabilizes alkyl radicals, increases activation energy necessary for their disintegration, and, therefore, also contributes to the reduction of the rate of radiation-induced chemical conversion. As irradiation dose grows, alterations in structure and length distribution of paraffin chains cause changes in the share of radicals stabilizing in a unit of time, and this affects the rates and the yields of products in radiation-induced reactions.

Contents of liquid fractions obtained in the two RTC modes were determined from GLC data (Table 2.21). The maximum yield of gasoline fractions C_6-C_{10} (boiling temperature up to 180°C) in mode 1 characterized by the heightened rate of heavy fraction disintegration was 32.2 mass% (with respect to the feed mass), while in mode 2 it attained only 15 mass%. Chromatography data show that the gasoline yield from the condensate of light fractions in mode 2 is higher than that in mode 1 for the dose of 2 kGy. However, the lower yield of liquid fractions in mode 2 makes the resulting yield of the gasoline fraction with respect to the feed mass lower than that in mode 1 (Table 2.21). The gasoline fractions produced in the two modes

TABLE 2.21

Hydrocarbon Contents (Mass%) of Gas Oil Fraction with Boiling Temperature from 70°C to 350°C Obtained by RTC of Karazhanbas Crude Oil

Dose, kGy	Mode	Concentration of the Gas Oil Fraction (70°C < T_{boil} < 350°C) in the Total Product	Hydrocarbons[a]			
			C_6-C_{10}	$C_{11}-C_{13}$	$C_{14}-C_{18}$	$C_{19}-C_{21}$
0	—	26.8	17.9	19.8	24.6	37.7
			4.8	5.3	6.6	10.1
2	1	65.0	13.2	37.1	43.4	6.3
			8.6	24.1	28.2	4.1
	2	45.0	15.0	42.7	34.6	7.7
			6.7	19.2	15.6	3.5
6	1	80.0	40.2	33.6	23.7	2.5
			32.2	26.9	19.0	2.0
	2	52.5	28.5	45.7	24.6	1.2
			15.0	24.0	12.9	1.0

Source: Zaikin, Y.A. et al., *Radiat. Phys. Chem.*, 60, 211, 2001.

[a] Numerator—mass% with respect to the analyzed fraction. Denominator—mass% with respect to the total product.

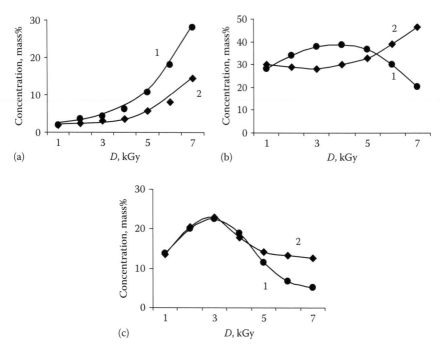

FIGURE 2.115 Dose dependence of (a) *n*-alkane, (b) iso-alkane, and (c) olefin concentrations in the gasoline fractions with the start of boiling at 100°C obtained by RTC of Karazhanbas oil in two irradiation modes: 1, mode 1; 2, mode 2. (From Zaikin, Y.A. et al., *Radiat. Phys. Chem.*, 60, 211, 2001.)

essentially differ in their HC contents due to different mechanisms of radiation-induced reactions.

The yields of different HC groups (*n*-alkanes, iso-alkanes, and olefins) are shown in Figure 2.115 versus irradiation dose for gasoline fractions (start of boiling at 100°C) produced by RTC of Karazhanbas oil in the two modes. Figure 2.115 shows that the amount of *n*-alkanes in the gasoline fraction grows rapidly in mode 1, while the yields of iso-alkanes and olefins reach maximums and then decrease as the dose rises. Such behavior of the yield curves testifies to the high degree of HC disintegration and the prevailing role of polymerization reactions in limitations of unsaturated HC accumulation.

The unusually high yields of isoparaffins in the RTC modes characterized by relatively low values of temperature and dose rate were attributed to the effects of energy transfer from paraffin to aromatic components of the HC mixture.

Decrease in the reaction temperature and the dose rate in mode 2 displays the "protective" role of heavy aromatic compounds that take the excess radiation energy away from big alkyl radicals. It leads to increase in the lifetime of alkyl radicals, which becomes sufficient to transform primary alkyl radicals to more thermodynamically stable secondary (tertiary) radicals. Figure 2.115 demonstrates that concentration of *n*-alkanes in the gasoline fraction produced in mode 2 increases slowly with irradiation dose compared with mode 1 that is accompanied by the cymbate increase in iso-alkane yields.

Taking into account rather considerable decrease in RTC product yields in mode 2 compared with mode 1 (nearly proportional to decrease in the dose rate), it was assumed that the fraction of alkyl radicals stabilizing due to intramolecular isomerization grows as the dose increases and contributes to some limitation of the rate of destructive conversion, together with polymerization.

These results can be compared with the data on isomerization processes observed during oil thermal processing without irradiation. Studies of cracking mechanisms and kinetics in HCs (see Chapter 1) have shown that isomerization of primary radicals plays an important role in thermal disintegration of high-molecular alkanes (C_6 and higher). Activation energy for isomerization can be much lower than the energy necessary for radical disintegration that would influence the equilibrium between primary and secondary isomer forms.

Decrease in temperature provokes intensification of isomerization processes. It was evaluated (Stepukhovich and Ulitsky 1975) that equilibrium concentration of secondary butyl radicals is about nine times higher than the concentration of primary radicals in the case of conventional TC (900 K), while in the case of initiated cracking (700 K) this factor is estimated as 16. According to the calculations (Stepukhovich and Ulitsky 1975), concentrations of tertiary isobutyl radicals exceed concentrations of primary isobutyl radicals by a factor of 16 in the case of TC and by a factor of 35 in the case of initiated cracking.

The absolute rates of HC isomerization during TC are very low, and isomer concentrations in an HC mixture usually do not exceed 1%–2% after processing. To reach higher isomer concentrations, catalytic cracking is usually applied. Three temperature intervals characteristic of catalytic isomerization were registered (Farkhadova et al. 1987). The lower temperature range (up to 600°C) is characterized by predominant isomerization of the components originally present in the mixture. At higher temperatures, increase in the rates of carbon cation reactions leading to isomerization of the appearing decomposition products was observed; isoparaffin concentrations increased to 26–28 mass%. Further temperature rise (above 800°C) caused drastic fall of isoparaffin HC yields.

Isomerization rates and isomer yields after radiation-induced cracking are much higher than those in the case of standard TC. It was stated (Topchiev and Polak 1962) that the isobutane yield after RTC of the light gasoline fraction with boiling temperature up to 140°C was 0.09 mass% for the cracking temperature of 350°C. At the same time, no measurable amounts of TC products were detected for reaction temperatures below 500°C. The isobutane yield observed after radiation-induced cracking at 500°C was 1.4 mass%, that is, 200 times higher than that in the case of TC.

Radiation-induced cracking opens unique opportunities for the intensification of intramolecular isomerization due to combination of high rates of radical generation and rather low rates of thermal processes, all of them regulated in wide ranges.

Isomerization intensity observed in mode 2 is defined not only by RTC conditions but also by specific HC contents of the oil feedstock characterized by heightened concentrations of pitch-aromatic components. The latter are the compounds of high radiation resistance that absorb a considerable part of radiation energy and thus can lower the rate of radiation-induced chemical conversion and provoke isomerization processes.

This conclusion is supported by the results of RTC of the light oil of Kok-Zhide field, Kazakhstan, in similar irradiation conditions (Zaikin et al. 2001). This feedstock is characterized by higher paraffin contents (51.5 mass% in the gasoline fraction) and lower concentrations of naphthenes and arenes in light fractions (32.2 mass% in the gasoline fraction compared with 45.2 mass% for Karazhanbas oil). The yields of n-alkanes, iso-alkanes, and olefins obtained in the result of Kok-Zhide oil processing in conditions of mode 2 are more characteristic for mode 1, which is less favorable for isomerization. So, reactions of polymerization and isomerization can be considered as the main factors limiting the rates of radiation-induced chemical conversion and essentially affecting the yields and the HC contents of RTC products. The relative contributions of these processes depend on RTC conditions and feed contents.

Thus, a high isomerization rate in the RTC of HC compositions with high concentrations of heavy aromatics was explained by availability of the lower temperature and dose rate limits for noticeable RTC reactions. Transfer of the excess radiation energy to an alkyl radical in the excited paraffin composition assists its disintegration and impedes intramolecular isomerization. On the contrary, the addition of heavy aromatics allows combination of rather high dose rates and temperatures with favorable conditions for isomerization. Aromatic compounds, known for high radiation resistance, can absorb the excess energy of a considerable part of radiation-generated radicals. In this case, many alkyl radicals can have enough time to stabilize their electron structure and to form isomers before their disintegration or recombination.

This interpretation was confirmed by experiments on bitumen, which are characteristic for higher concentrations of heavy aromatics and, therefore, still more pronounced effects of enhanced isomerization (Zaikin and Zaikina 2004b).

Experimental data have shown that the most part of iso-alkanes is concentrated in the gasoline fraction of the liquid product of radiation processing. Maximal iso-alkane yields were observed at the lowest values of dose rate and temperature sufficient for noticeable chain reaction. At the given dose and dose rate, they increase as the density of the feedstock becomes higher.

Maximal yields of isomers in gasoline fraction obtained by RTC of HC mixtures with very high concentrations of aromatic compounds, such as bitumen and highly viscous oil, can be approximately estimated using the following empiric equation (Zaikin et al. 2001, Zaikin and Zaikina 2004c):

$$Y_{i-par}^{par} = \frac{1040}{PT} \frac{[(1+190/T)\tilde{\rho} - \rho_0]}{\rho_0} \tag{2.49}$$

where

Y_{i-par}^{par} is the isoparaffin concentration in the paraffin part of the gasoline fraction

P is the dose rate, kGy

T is the absolute temperature, K

$\tilde{\rho}$ is the feedstock density at 20°C, kg/m³

ρ_0 is the density of the gasoline fraction (in the following calculations, ρ_0 was taken as equal to 780 kg/m³)

TABLE 2.22

Isoparaffin Concentration in the Paraffin Part of Gasoline Fraction of Synthetic Oil Produced by Radiation Processing of Bitumen and High-Viscous Crude Oil

				Iso-Alkane Concentration, %		
Feedstock	Reaction Temperature T, K	Dose Rate, P, kGy/s	Feedstock Density $\tilde{\rho}$, kg/m³	In Gasoline Fraction $\left(Y_{i-par}^{gas}\right)$, %	In Paraffin Part of Gasoline $\left(Y_{i-par}^{par}\right)\cdot$calc	In Paraffin Part of Gasoline $\left(Y_{i-par}^{par}\right)$ exp
Fuel oil	673	1.1	939	46.7	76.4	75
Heavy oil	648	1.25	942	35.5	72.1	67–73
AB 1	683	1.3	970	38.2	68.9	65–68
KB 1	683	1.5	976	33.4	60.3	59.6
KB 2	693	1.5	998	36.1	63	61–65
AB 2	693	1.5	1020	40.2	66.7	66–68

Note: AB 1,2—two types of Athabasca bitumen; KB 1,2—two types of Kazakhstan bitumen.

Table 2.22 shows a good agreement of experimental and calculated concentrations of iso-alkanes in the RTC products after processing different types of feedstock.

Formula (2.46) was based on the following nonrigorous consideration in a paper (Zaikin and Zaikina 2004c), where energy transfer from light paraffin to heavy aromatic molecules was considered in the hypothetical two-component HC mixture. It was assumed that the paraffin component can be partially or completely formed by radiation-induced detachment of alkyl substituents from the aromatic structure. It was also supposed that an alkyl radical can form an isomer only in the case when the light paraffin fraction transmits a part of its excess energy to aromatic molecules.

The isomer yield from the feedstock, Y_i^{feed}, can be written in the form

$$Y_i^{feed} = \int_{\rho_{min}}^{\rho_{hf}} \frac{P_2(\rho_{hf})-P_1(\rho)}{P_1(\rho)} f(\rho,T,P)C_{hf}\,C(\rho)d\rho \qquad (2.50)$$

where $f(\rho)$ is the probability density of intramolecular isomerization of an alkyl radical in the absence of energy transfer from the fraction characterized by "partial density" (concentration) $\rho(\rho=M/V$, where M is the molecular mass of the fraction and V is the molecular volume); $P_1(\rho)$ is the probability of excitation energy absorption by a molecule of the light fraction characterized by "partial density" ρ; $P_2(\rho_{hf})$ is the probability of excitation energy absorption by the heavy aromatic fraction; C_{hf} and $C(\rho)$ are atomic concentrations of the "heavy" and "light" fractions, respectively.

For the sake of simplicity, it was supposed that $f(\rho)$ is the δ-function:

$$f(\rho,T,P)=\frac{\alpha}{PT}\delta(\rho-\rho_0) \tag{2.51}$$

where α is the constant; ρ_0 is the fraction where isomers are concentrated; in the case of RTC product of bitumen, it is gasoline and partially kerosene fraction.

Substitution of $f(\rho)$ to expression (2.50) yields

$$Y_i^{feed}=\frac{\alpha}{PT}\frac{P_2(\rho_{hf})-P_1(\rho_0)}{P_1(\rho_0)}C_{hf}\,C(\rho_0) \tag{2.52}$$

In suggestion that isomerization proceeds only in the fraction with the average density ρ_0 that transfers its excess energy to the remaining part of material with density ρ_{hf},

$$C_{hf}=1-C(\rho_0)$$

and

$$Y_i^{feed}=\frac{\alpha}{PT}\frac{P_2(\rho_{hf})-P_1(\rho_0)}{P_1(\rho_0)}[1-C(\rho_0)]C(\rho_0) \tag{2.53}$$

In supposition that probability of energy absorption by each of the two components of the HC mixture is proportional to their "partial densities,"

$$P(\rho)\sim\rho \tag{2.54}$$

Equation 2.53 can be rewritten as follows:

$$Y_i^{feed}=\frac{\alpha}{PT}\frac{\rho_{hf}-\rho_0}{\rho_0}C(\rho_0)[1-C(\rho_0)] \tag{2.55}$$

Proportionality of the probability of excitation energy transfer to the substance density corresponds to the quantum mechanical solution of the problem when static dipole–dipole interaction is considered in the case of the ideal resonance of two oscillators (Topchiev and Polak 1962). The linear losses of radiation energy that can be considered as a measure of substance ability to accumulate excitation energy are also proportional to the substance density.

Introduction of the average feedstock density

$$\tilde{\rho}=\rho_0 C(\rho_0)+\rho_{hf}[1-C(\rho_0)] \tag{2.56}$$

allows reduction of Equation 2.55 to the simple expression

$$Y_{i-par}^{par}=\frac{Y_i^{feed}}{C(\rho_0)}\approx\frac{\alpha}{PT}\frac{\tilde{\rho}-\rho_0}{\rho_0} \tag{2.57}$$

Thus, an expression derived is of the same form as Equation 2.49. Formula (2.49) shows that the degree of paraffin isomerization during RTC is determined by transfer of excess excitation energy to the more dense and radiation-resistant medium. The factor $(1 + 190/T)$ can be considered as a correction for different thermal expansions of the "heavy" and "light" fractions, and $1040/PT = 1$ for the characteristic mode of radiation ($P = 1.5$ kGy, $T = 420°C$) favorable for paraffin isomerization in the gasoline fraction of bitumen.

If linear energy losses for excitation energy transfer are proportional to substance density, then volume energy losses should be proportional to $\rho^{3/2}$. Based on the analysis of experimental data on HC radiolysis inhibition by different chemical additions, it was stated (Topchien and Polak 1962) that inhibitor capacity for energy absorption is proportional to $\rho^{3/2}$; in this case, expression $\tilde{\rho} - \rho_0/\rho_0$ in formula (2.56) should be replaced with $(\tilde{\rho} - \rho_0/\rho_0)^{3/2}$. However, the available experimental data on isomer yields can be satisfactorily described by the linear correlation (2.57).

The favorable conditions for isomerization are lowered dose rates of ionizing irradiation and lowered temperatures. Therefore, the effect of radiation-enhanced isomerization in the presence of heavy aromatics should be considerable in the case of low dose rate gamma irradiation of aromatic-rich HC mixtures at lowered temperatures.

For experimental verification of this conclusion, special experiments were conducted on radiation processing of low-octane gas-condensate gasoline at ambient temperature. Gasoline was irradiated with the bremsstrahlung x-rays from the 2-MeV electron beam. The experiments have shown that radiation-enhanced isomerization became especially pronounced when heavy aromatic HCs were added to relatively lighter feedstock, and this mixture was irradiated with moderate x-ray doses. In these experiments, heavy residua of bitumen radiation processing were used as additional agents for initiation of isomerization.

Figure 2.116 summarizes the observed changes in HC contents of gasoline extracted from gas condensate and demonstrates the effect of aromatics addition on

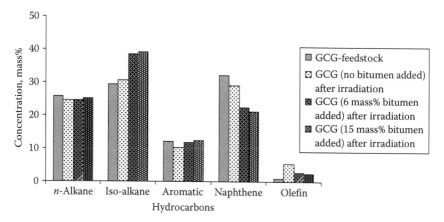

FIGURE 2.116 HC contents of gas-condensate gasoline after bremsstrahlung x-ray irradiation. GCG—gas-condensate gasoline.

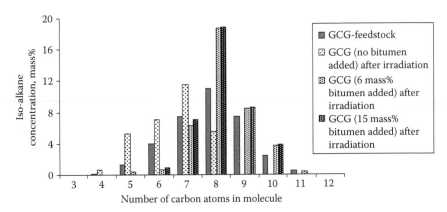

FIGURE 2.117 Molecular mass distribution of isoparaffin concentrations in mixtures of gas-condensate gasoline: GCG—gas-condensate gasoline.

radiation-induced isomerization. Experimental data of Figures 2.116 and 2.117 show that the effect of paraffin isomerization becomes pronounced in the presence of heavy aromatic compounds as a result of mixture radiation processing. Mixing gas-condensate gasoline with 15 mass% residue of bitumen vacuum distillation and subsequent x-ray irradiation at room temperature provides increase in iso-alkane concentration by 33.8% and increase in gasoline octane number from 54 to 67 without any chemical additions.

2.4.6 RADIATION-INDUCED POLYMERIZATION IN CONDITIONS OF RADIATION-THERMAL CRACKING OF HIGH-PARAFFIN OIL

This effect, most clearly pronounced in oils with high contents of high-molecular paraffins, was observed in experiments on processing of crude oil from Kumkol field, Kazakhstan, with accelerated electron beams and described in the papers (Zaikina and Aliyev 2001, Zaikin et al. 2004a,b, Zaikin and Zaikina 2004c). This type of oil is characterised by high contents of C_{15}–C_{22} HCs of paraffin series with solidification temperature 10°C–44°C with their total concentration of 60%–77%. The contents of polycyclic aromatic HCs and asphalt-pitch substances are very low.

High temperatures of crude oil solidification are due to high concentrations of heavy paraffins. The contents of heavy aromatic compounds in fractions 400°C–450°C and 450°C–500°C are 5%–7%, while the total concentration of paraffin and naphthene HCs is 80%–90%. Considerable amount of paraffins with high melting point affects the rheological properties of oil.

Oil samples were irradiated (Zaikin et al. 2004a) by 2 MeV electrons from the electron accelerator ELU-4 using current densities from 1 to 3 μA/cm². Different conditions of radiation-thermal processing were provided by variation of main processing parameters: temperature in the range 340°C–450°C, irradiation dose in the range 1–4 kGy, and dose rate in the range 1–4 kGy/s.

The results of these experiments were unexpected. Kumkol oil is highly paraffin, that is, it contains HCs mostly subjected to decomposition under irradiation.

High concentration of high-molecular paraffins in this feedstock testifies to great hydrogen reserves that principally assumes high yields of light fractions. However, experiments have shown that radiation-thermal processing of mixtures of light and heavy paraffins at moderate temperatures and dose rates does not result in the expected high yields of light fractions.

The main products were the liquid gas oil fraction with boiling temperature from 60°C to 350°C (70–80 mass%), the gaseous fraction (8–10 mass%) and the heavy paraffin residue (15–20 mass%) with the strong tendency to polymerization. Product yields varied depending on conditions of RTC.

According to the GLC data, the gasoline fraction separated from the feedstock contained 38.0 mass% normal alkanes, 27.0% iso-alkanes, 29.0% naphthenes, 3.5% aromatics, and 1.8% unsaturated HCs. After radiation-thermal processing of Kumkol oil in different conditions, 30–35 mass% n-alkanes, 28%–33% isoparaffins, 25%–30% naphthenes, 5%–7% aromatics, and 2.5%–4.2% olefins were identified in the HC composition of the gasoline fraction (C_4–C_{10}) from the RTC product. Cyclic olefins and dienes were also present, but their concentrations did not exceed 1%.

Dependence of the yields of gasoline and kerosene fractions on the electron dose is shown in Figure 2.118. It is obvious that radiation-thermal processing of the mixture of heavy and light paraffins in these conditions does not raise distillation yields of motor fuels compared with those from the feedstock ($D=0$). Gasoline originally available in such feedstock is partially converted into gases and partially absorbed by the excited heavy paraffin substance. As a result, yields of light fractions are much lower than those observed after RTC of highly viscous Karazhanbas oil under the same processing conditions (Figures 2.118 and 2.119).

Figure 2.119 shows dependence of gasoline and diesel yields on dose rate. It demonstrates that increase in the dose rate up to 2 kGy/s intensifies destruction processes in the gasoline fraction and leads to a decrease in the yields of light fractions. When the dose rate is higher than 2 kGy/s, the light fractions increase in their yields due to heavy paraffin destruction.

Radiation-chemical yields of n-alkanes in the gasoline fraction decrease with increasing dose (Figure 2.120). However, there is a noticeable trend to an increase in

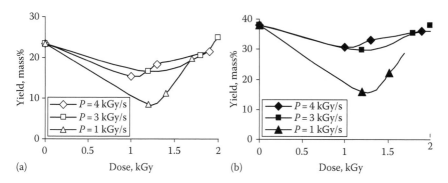

FIGURE 2.118 Dose dependence of the yields of (a) gasoline and (b) diesel fractions after RTC of Kumkol crude oil; $T=400°C$. 1—4kGy/s; 2—3 kGy/s; 3—1kGy/s. (From Zaikin, Y.A. et al., *Radiat. Phys. Chem.*, 69(3), 229, 2004a.)

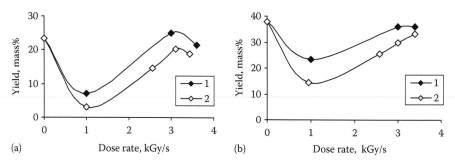

FIGURE 2.119 Dependence of (a) gasoline and (b) diesel yields after RTC of Kumkol crude oil on dose rate; $T = 400°C$. 1—D = 2kGy; 2—D = 1.4 kGy. (From Zaikin, Y.A. et al., *Radiat. Phys. Chem.*, 69(3), 229, 2004a.)

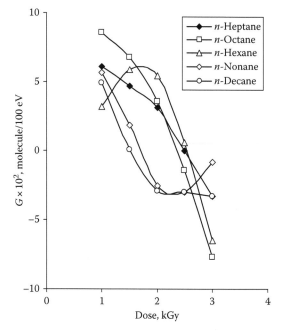

FIGURE 2.120 Dose dependence of n-alkane (C_7–C_{10}) G-values after RTC of Kumkol crude oil. $P = 3$ kGy/s; $T = 400°C$. (From Zaikin, Y.A. et al., *Radiat. Phys. Chem.*, 69(3), 229, 2004a.)

the relative contribution of heavier paraffins (n-pentane and n-hexane for low doses, and n-nonane and n-decane for heightened doses). Figure 2.121 shows higher rates of the yields of heavy alkanes with the increasing dose rate.

Generally, increase in the molecular weight of light fractions with an increase in electron dose rate was the characteristic of RTC of the paraffin Kumkol oil. Increase in the dose rate leads to increase in concentrations of gasoline fractions with a higher boiling temperature.

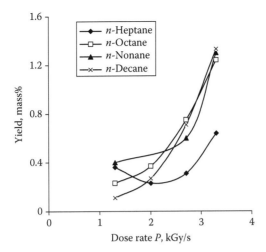

FIGURE 2.121   Dependence of *n*-alkane yields on dose rate after RTC of Kumkol crude oil. $D = 1.5$ kGy; $T = 400°C$. (From Zaikin, Y.A. et al., *Radiat. Phys. Chem.*, 69(3), 229, 2004a.)

The effect of the carbon chain length on the yields of *n*-alkane radiolysis products was studied in the work (Mirkin et al. 2003). It was shown that increase in the number of C–C bonds in a molecule (from 6 to 16) leads to excitation energy redistribution over the greater amount of bonds that diminishes the efficiency of C–C bond breaks. This causes increase in the yields of parent radicals and reduces the yields of fragmentary radicals.

The increase in the molecular weight of the gasoline fraction observed in the paper (Zaikin et al. 2004a) could be explained by increased destruction of the paraffins in the middle of molecules as the dose rate increased. It raises the probability of alkyl radical recombination with subsequent formation of paraffin molecules lighter than the molecule destroyed but heavier with respect to the gasoline fraction.

Compared with the HC contents of gasoline fractions produced as a result of RTC of high-viscous Karazhanbas oil (see previous section), the RTC of the gasoline of Kumkol oil contained lower concentrations of isoparaffins. The study of Karazhanbas oil has shown that irradiation modes characterized by low values of temperature and dose rate provided a high level of paraffin isomerization. It was reasonable to suppose that a share of alkyl radicals stabilizing per unit of time due to intramolecular isomerization was higher in these conditions. It caused a steady increase in the iso-alkane concentration in the gasoline fraction with increasing dose.

Figure 2.122 shows that in similar irradiation conditions ($P = 1$ kGy/s), isomerization observed during RTC of Kumkol oil is much less pronounced. The maximum increase in iso-alkane concentration is about 7% compared with 15% in the case of Karazhanbas oil. When the dose is higher than 1.5 kGy, iso-alkane concentration decreases with increasing dose, while its monotonous growth is observed for Karazhanbas oil in similar conditions.

These observations show the synergetic nature of the isomerization processes in complex HC mixtures when isomerization rate depends not only on conditions of

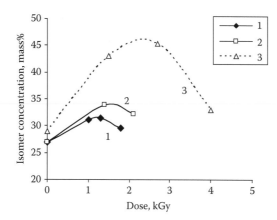

FIGURE 2.122 Dose dependence of isomer contents in the paraffin part of RTC products of Kumkol crude oil (1, 2) and Karazhanbas crude oil (3): $T = 400°C$. 1—P = 4 kGy/s; 2—P = 1 kGy/s; 3—P = 1.2 kGy/s. (From Zaikin, Y.A. et al., *Radiat. Phys. Chem.*, 69(3), 229, 2004a.)

radiation-thermal processing but also on the specific HC contents of the feedstock. Higher concentrations of pitch-aromatic components in the case of Karazhanbas oil are responsible for the higher yield of isomerization. In the case of Kumkol oil with very low concentrations of aromatic HCs, this yield of isomerization cannot be achieved even at low dose rates.

Important characteristics of RTC in the paraffinic oil can also be drawn from the analysis of the yields of unsaturated HCs.

Dose dependence of olefin concentrations in the gasoline fractions separated from liquid products after RTC of Kumkol oil is shown in Figure 2.123 in comparison with those in Karazhanbas oil. It shows that unsaturated HCs that appear during radiation processing are practically entirely concentrated in the gasoline fractions, that is, they are mainly low-molecular compounds.

Figure 2.123 shows that the maximum olefin concentration in the gasoline fraction after RTC of Kumkol oil does not exceed 4%–5%, that is, it is three to five times lower than that in Karazhanbas oil. These values are very low for RTC gasoline fractions. Another characteristic feature of the results shown in Figure 2.124 is the low dose of electron irradiation corresponding to the olefin maximum (0.8–1.5 kGy).

Such composition of cracking products and its dose dependence are characteristic neither of the RTC of relatively light paraffins (Brodskiy et al. 1961, Topchiev and Polak 1962, Lavrovskiy 1976,) nor of the RTC of high-viscous oil with high concentrations of aromatic compounds (Zaikin et al. 2001, Zaikin and Zaikina 2004b). Note that similar RTC conditions in the case of Karazhanbas oil provide gasoline containing 20%–25% olefins with the dose of maximum olefin yields being about 2.5 kGy.

Radiation-chemical conversion of the model paraffin HCs $C_{16}H_{34}$ and $C_{17}H_{36}$ in different aggregate states was studied in works (Panchenkov et al. 1981, Farkhadova et al. 1987, Zaikin and Zaikina 2004a). According to the data by Földiàk (1981), the rate of $C_{16}H_{34}$ decomposition is practically constant up to about 523 K. Further increase in temperature causes a sharp increase in the reaction rate as a result of its transformation to a chain process (RTC).

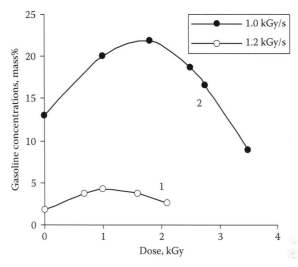

FIGURE 2.123 Olefin concentrations in gasoline fraction after RTC of Kumkol crude oil (1) and Karazhanbas crude oil (2); $T=400°C$. (From Zaikin, Y.A. et al., *Radiat. Phys. Chem.*, 69(3), 229, 2004a.)

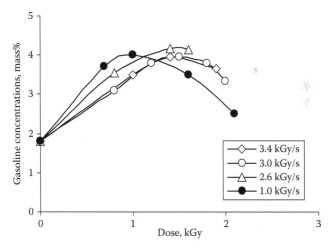

FIGURE 2.124 Olefin concentrations in gasoline fraction after RTC of Kumkol crude oil; $T=400°C$. (From Zaikin, Y.A. et al., *Radiat. Phys. Chem.*, 69(3), 229, 2004a.)

These studies have shown that the main products of radiolysis of these HCs in the liquid phase are dimers forming mainly as a result of secondary radical recombination. Our studies of high-paraffin Kumkol oil show that the increase in molecular mass of the paraffin feedstock gives a considerable rise to polymerization during and after RTC.

The observed regularities of olefin accumulation in the gasoline fraction of high-paraffin oil can be qualitatively interpreted in frames of the simplest model of RTC kinetics (Zaikin et al. 2004a) where this process is reduced to disintegration of paraffin molecules with the given average molecular mass that leads to olefin accumulation limited by reactions of polymerization. Paraffin disintegration and olefin polymerization are initiated by radiation generation of radicals, their quasi-stationary concentration being calculated in the approach of the square-law chain break and assumed to be independent of irradiation dose in this rough model.

In accordance with this simplified standard mechanism,

$$\frac{d[\text{Par}]}{dt} = -k[\text{R}][\text{Par}] \tag{2.58}$$

$$\frac{d[\text{Ol}]}{dt} = k[\text{R}][\text{Par}] - k_p[\text{R}][\text{Ol}]$$

where
[Par] and [Ol] are concentrations of paraffins and olefins, respectively
[R] is the concentration of carbon-centered radicals

The rate constants of olefin formation k and polymerization k_p are defined by the usual exponential relations:

$$k = k_0 \exp\left(-\frac{E}{k_B T} \right)$$

$$k_p = k_{op} \exp\left(\frac{E_p}{k_B T} \right)$$

where
E, E_p are activation energies of the processes
k_0, k_{op} are pre-exponential factors
k_B is the Boltzmann constant

Radical concentration [R] can be found from the equation

$$[\text{R}] = \left(\frac{G\dot{D}}{k_r} \right)^{1/2} \tag{2.59}$$

where
$\dot{D}$ is the dose rate
G is the radiation-chemical yield for radicals
k_r is the radical recombination rate constant

Solution of the system of Equations 2.58 with the initial condition $[Par]_{t=0}=[M]$ leads to the following equation for olefin concentration:

$$[Ol]=k[R][M]\cdot\frac{\exp(-kt)-\exp(-k_p t)}{k_p - k} \qquad (2.60)$$

The maximum olefin concentration

$$\frac{[OL]_{max}}{[M]}=\frac{1}{(k_p/k)-1}\left[\exp\left(\frac{1}{1-(k_p/k)}\right)\ln\left(\frac{k_p}{k}\right)-\exp\left(\frac{k_p/k}{1-(k_p/k)}\right)\ln\left(\frac{k_p}{k}\right)\right] \qquad (2.61)$$

should be observed at the time t_0 corresponding to irradiation dose D_0:

$$D_0=\dot{D}t_0=\frac{\dot{D}}{k[R]}\varphi\left(\frac{k_p}{k}\right)=\frac{\dot{D}^{1/2}}{\left(G\dot{D}/k_r\right)^{1/2}k_0\exp\left(-\dfrac{E}{k_B T}\right)}\varphi\left(\frac{k_p}{k}\right) \qquad (2.62)$$

where

$$\varphi\left(\frac{k_p}{k}\right)=\frac{1}{(k_p/k)-1}\ln\left(\frac{k_p}{k}\right) \qquad (2.63)$$

The plot of the dependence (2.63) of olefin concentration in the cracking product on the ratio of constants of paraffin scission and olefin polymerization is shown in Figure 2.125.

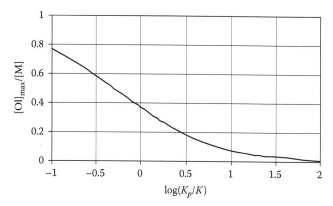

FIGURE 2.125   Dependence of maximum olefin concentration on ratio K_p/K. (From Zaikin, Y.A. et al., *Radiat. Phys. Chem.*, 69(3), 229, 2004a.)

Figure 2.125 shows that the maximum olefin concentration in cracking products increases nonlinearly as the ratio k_p/k decreases. In the simple model under consideration, this effect does not depend on the dose rate. According to formula (2.60), the dose corresponding to maximum olefin yield is proportional to $P^{1/2}$ and essentially dependent on the ratio k_p/k.

These regularities indicate considerable increase in the polymerization constant during the RTC of high-paraffin Kumkol oil compared with Karazhanbas oil with high contents of aromatic compounds. Qualitatively, Equations 2.60 through 2.63 describe the observed dose dependence of olefin yields.

For the rough quantitative estimation, the following values for reaction constants characteristic of HCs were used: $E=35$ kJ/mol; $k_p=k_r=3 \cdot 10^{-19}$ m^3/molecule/s; $G=3.1$ radicals/J. The average molecular mass of the feedstock and its density were assumed as $M=0.35$ kg/mol and $\rho=800$ kg/m^3, respectively. For the values of constants given earlier, Equation 2.62 can be written in the form

$$D_0\left(\text{Gy}\right)=\dot{D}\,t_0=8.3*10^5\,\dot{D}\,\varphi\left(\frac{k_p}{k}\right) \tag{2.64}$$

where the $\dot{D}$ dimension is Gy/s.

In the case of Karazhanbas oil, the maximum olefin concentration of 20% corresponds to the value $k_p/k \cong 3$ (Figure 2.125); in the case of Kumkol oil, the maximum olefin concentration of 4%–5% corresponds to the value $k_p/k \cong 20$, which testifies to the increase in the polymerization rate constant by six to seven times.

Substituting these values of k_p/k in to formula (2.64), we obtain the values of 0.6 kGy for Karazhanbas oil and 2.0 kGy for Kumkol oil for the doses corresponding to maximum olefin yields. These estimations approximately agree with the observed alterations in characteristics of dose dependence of olefin concentrations in different types of oil feedstock.

Correlation of polymerization rate in the RTC process with the average length of the paraffin molecule in the original mixture was considered in the paper (Zaikina and Aliyev 2001). The values of k_p/k were calculated using experimental data on maximal olefin concentrations in gasoline fractions after RTC of crude oils and oil products with different average molecular masses of the paraffin component (Zaikin et al. 2004). The value of k_p/k can be easily determined from the graph in Figure 2.125. The results of calculations presented in Figure 2.126 quantitatively characterize the decrease in the olefin polymerization constant with the densification of paraffin mixtures subjected to RTC (Table 2.23). If the ratio k_p/k is known, formula (2.61) can be used for the calculation of the irradiation dose, D_0, characteristic of the maximum olefin yield.

Calculated and experimental values of the irradiation dose, D_0, corresponding to the maximum olefin yields are given in Table 2.23 for various types of oil. Comparison of the data allows estimation of the applicability of the simplified model used and, therefore, reliability of the correlation shown in Figure 2.126. It shows that for the most part of the samples of crude oil and oil products, calculated and experimental values of the dose for the maximum olefin yield are in a satisfactory agreement.

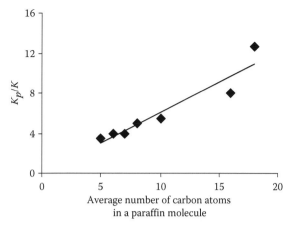

FIGURE 2.126 Dependence of the ratio of the rates of olefin polymerization and formation during RTC on the length of a paraffin chain.

TABLE 2.23

Calculations of the Dose Corresponding to the Maximum Olefin Yields for Petroleum Crude Oil of Different Original Composition

Type of Crude Oil	Average Number of Carbon Atoms in the Paraffin Chain	Maximum Olefin Concentration [OL], Mass%	Dose Corresponding to the Maximum Olefin Yield, D kGy	
			Experimental Value	Calculated Value
High paraffinic oil 1	16	4.2	2	1.6
High paraffinic oil 2	18	2.9	1.0	1.2
Heavy oil 1	10	7.3	2.2	1.8
Heavy oil 2	8	6.5	2.0	2.1
Heavy oil 3	7	8.4	2.5	1.8
Heavy oil 4	6	9.9	2.2	2.0
Fuel oil	5	11.5	2.5	3.0

Intense polymerization of RTC products was also observed in earlier experiments with such wastes of oil extraction as asphalt-pitch-paraffin sediments (APPS) (Zaikin et al. 1999a,b). This type of feedstock contains high concentrations of heavy paraffins and, hence, great reserves of hydrogen. Therefore, it has high potential yields of light fractions. However, reactions of polymerization and chemical adsorption limited the maximum yields of light RTC products to 40%.

A specific feature of APPS radiation-induced decomposition is the formation of a considerable amount of the reactive paraffinic residue. The products of APPS destruction are unstable and maintain the tendency to polymerize in a few days if kept in contact. In the absence of stirring, a distinct interface was formed between light fractions and heavy paraffin residue. As a result of interaction with the polymerizing residue, the light liquid fractions were completely absorbed and the heavy residue got denser and solidified after several days of exposure at room temperature. When separated, the light fraction was stable, but the residue continued increasing its molecular weight.

Thus low olefin contents in RTC products and the high yield of their polymerization are characteristic features of the RTC of highly paraffinic oils. The effects observed can be explained taking into account the known features of heavy paraffin radiolysis. The G-value of radical generation in n-alkanes is almost independent of molecular mass, being usually about $5 \cdot 10^{17}$ radicals/J. Nevertheless, distribution of radiolysis products as a function of molecular mass differs significantly for different HCs (Gäuman and Hoigne 1968). In particular, as we pass from n-butane to n-heptane, the yields of methyl and ethyl radicals drop down by nearly an order of magnitude, and the yield of radicals containing more than six carbon atoms makes more than 80% of the heavy and light alkyl radicals that appear during n-heptane radiolysis.

According to the radical chain cracking theory (Topchiev and Polak 1962, Saraeva 1986), the excited n-alkane molecules disintegrate as a result of C–C or C–H bond breaks. In the latter case, there is a high probability of formation of the secondary carbon valence while a hydrogen atom of excess energy participates in the fast abstraction reactions forming another radical of this type.

According to Kossiakoff–Rice theory (Kossiakoff and Rice 1943), it is probable that the big alkyl radical formed in RTC, for example, a primary radical $C_{10}H_{21}$, can convert to equilibrium state through an intermediate (e.g., six member) secondary alkyl radical before it disintegrates.

At 300°C–500°C, there are the well-documented chain reactions of long-chain alkanes. Of course, the degradation reactions compete with the reverse reactions of chain addition of alkyl radicals to the olefins. Also the secondary radicals, in the hypothesis of radical–radical recombination, yield branched alkanes (Kossiakoff and Rice 1943, Topchiev and Polak 1962, Földiàk 1981).

Thus, formation of high concentrations of relatively long-living alkyl radicals that cannot directly be the RTC chain carriers should cause the observed intensification of polymerization and isomerization in heavy paraffin fractions. These reactions result in the limitation of the yields of light RTC products, low olefin concentrations in light fractions, and increase in their molecular weights as the dose rate grows.

A considerable increase in polymerization rate for heavy paraffins in RTC conditions is analogous to the well-known phenomenon of acceleration of radiation-induced polymerization as the polymer molecular mass grows due to decrease in the mobility of polymer radicals (Pikaev 1987).

Generally, the characteristic features of RTC that should be noticed in oil with high contents of heavy paraffins are a low level of isomerization in light RTC fractions, a high polymerization rate and low olefin contents in RTC products, relatively low yields of light fractions at low irradiation dose rates, and increase in the molecular weight of the gasoline fraction as the irradiation dose rate grows. These observations are to be attributed to the behavior of heavy alkyl radicals that initiate polymerization and isomerization in heavy paraffin fractions.

Radiation-induced polymerization always limits the yields of light products of RTC of heavy HC feedstock. In high-paraffin oils, the effect is so strong that efficient oil upgrading by means of RTC becomes difficult. The problem of paraffin oil radiation processing was resolved in PetroBeam technology (Chapter 5), where the optimal processing conditions at lowered temperatures take into account the structural state of the paraffin feedstock.

Dependence of the synergetic effects, described earlier, on HC contents of the original feedstock and irradiation conditions was used in the new technological approaches for the control of yields and quality of the target products of heavy oil radiation processing.

The earlier-described approaches to oil processing based on the RTC of HCs are remarkable for such unique advantages as high efficiency of energy transfer to the feedstock processed, ability of radiation to create high concentrations of reactive particles for initiation of various chemical reactions at any temperatures, relative easiness of control of radiation-chemical reactions, and minimum energy consumption.

The methods for RTC of oil feedstock allow economically, technologically, and environmentally overcoming many acute problems of the oil industry. However, chain propagation in RTC is still thermally activated; therefore, RTC should be conducted at heightened temperatures (350°C–420°C in the case of heavy oil feedstock). Therefore, further research was directed to achieve radical decrease in the temperature of oil radiation processing (Chapters 4 and 5).

3 Methods for Petroleum Processing Based on Radiation-Thermal Cracking

3.1 RADIATION METHODS FOR HIGH-VISCOUS OIL AND BITUMEN PROCESSING

Very shortly after the discovery of radiation-thermal cracking (RTC) of hydrocarbons, Topchiev et al. (1959) discussed the prospects of its commercial applications. Most of the successful technological approaches to oil radiation processing (RP) developed later were based on this phenomenon.

Comberg (1988) developed a method for the extraction of liquid hydrocarbon products from fossil resources such as oil shales, tar sands, heavy oil, and coals. The method comprised the mixing of a donor solvent and the exposure of the mixture to gamma irradiation. The donor solvent supplied hydrogen for combination with molecules whose bonds were broken by the irradiation process. The synergetic effect of hydrogen donors and ionizing irradiation was demonstrated in the following examples.

Two 50 g samples of granulated oil shale were each mixed with 50 cm^3 *n*-heptane designed to serve as a hydrogen donor solvent. One sample was subjected to 1 MGy Co-60 irradiation. No protective cover was used; the sample was exposed to air. Irradiation was carried out for 6 days at the temperature of 40°C and at atmospheric pressure. The solvent was then drained into open beakers of known weight and evaporated at 80°C. The yields were measured by reweighing the beakers.

The solvent-only run (without irradiation) yielded 0.1 g of a dull, thin, hard plastic-like material coated firmly and nonuniformly distributed over the bottom and the walls of the container. The shale irradiated in solvent yielded 0.75 g of a brown liquid of moderate viscosity, which flowed slowly, like honey, at room temperature. The same shale samples were then each mixed with fresh 50 cm^3 supplies of solvent, and the process including irradiation repeated.

The control sample produced no additional measurable yield while the irradiated sample produced an additional 0.3 g of a lighter less viscous liquid. The total yield from the irradiated sample was 1.05 g, 10 times the solvent-only (control) production. The elemental analysis had demonstrated increase in the H/C ratio and

TABLE 3.1

Elemental Contents of Shale and Products of Its Processing

	Control Sample (Material Scrapped from the Wall of the Beaker)	First Irradiation	Second Irradiation
H/C	1.72	1.99	1.99
N, %	0.31	0.08	0.033
S, %	6.1	1.78	0.39

Source: Comberg, H.J., Extraction and liquefaction of fossil fuels using gamma irradiation and solvents, US Patent 4,772,379, 1988.

considerable decrease in the nitrogen and sulfur contents as a result of shale RP (Table 3.1). Together with the higher yield of relatively light fraction, the irradiated samples were characterised by higher clarity and low ash content.

The next series of irradiation experiments was conducted with the medium volatile bituminous coal (atomic H/C = 0.57). Four samples consisting of 9 g each of coal were mixed with 9 mL of the donor solvent, tetrahydrofuran. One sample was exposed to 1 MGy of Co-60 radiation at ambient temperature and pressure, a second was exposed to 2 MGy, and the last two retained as control.

After irradiation, the donor solvent was removed by evaporation at about 125°C. The remaining solid was then processed with pyridine; the control samples received the same treatment without irradiation. The extracted material was freed of pyridine at 130°C and then analyzed for carbon, hydrogen, and nitrogen. The results are represented in Table 3.2.

The primary effect of the radiation was the increased yield of pyridine-soluble hydrocarbons. A higher H/C ratio and reduced nitrogen content demonstrated increased hydrogen transfer from the donor solvent. It was proposed to use the method developed for kerogen extraction and shale oil upgrading.

The advantage of the method is an opportunity of feedstock processing at ambient temperature and pressure. However, low-rate radiolysis reactions involved in the process result in very high irradiation doses, low yields of the target product, and

TABLE 3.2

Effect of Radiation Processing on Coal Elemental Contents

Dose, MGy	C	H	N	H/C	Yield, g
1	78.01	7.01	1.55	1.08	0.22
Control	84.88	6.92	2.88	0.98	0.07
2	76.6	7.45	1.25	1.17	0.45
Control	81.1	5.98	2.53	0.88	0.08

low production rate. The methods based on RTC at heightened temperatures have demonstrated much higher efficiency.

Based on the specific features of the RTC mechanism in highly viscous oil, described in Chapter 2, technological approaches were developed for RP of these types of petroleum feedstock.

Putilov et al. (1981) discussed the opportunities of high-temperature nuclear reactor applications in radiation-thermal processes of oil chemistry in the example of ethylene production from crude oil. Specifically, estimations were made for a nuclear reactor having a heat power of 300–400 MW and radiation power of 2000 kW. It was estimated that at the temperature of helium coolant of 650°C, absorbed radiation dose of 500 kGy, and the product radiation-chemical yield of 10^4 molecules per 100 eV of absorbed radiation energy, the reactor capacity for ethylene production can reach 1 million tons per year.

Trutnev et al. (1998) proposed a design of a nuclear power plant for distillation and RTC of crude oil in the production of motor fuels and other oil products. The plant comprised a unit for crude oil heating and a separation column with a sump for residual fuel oil. The sump was designed in the form of an outer circular shell of the reactor core region. Together with the outer perimeter of the active zone, the shell formed a reactor volume where the residual fuel oil was subjected to heating and ionizing irradiation from the active zone to initiate RTC of fuel oil with the subsequent thermal fractionation of the product in the separation column. The outer circular shell of the active reactor zone could be realized in the form of an elliptic paraboloid. To the authors' opinion, application of this production scheme would allow more rational use of fuel oil and other heavy oil fractions for motor fuel production by the realization of a highly productive cracking process at atmospheric pressure.

Melikzade et al. proposed radiation-thermal methods for gasification and coking of oil bitumen (Melikzade et al. 1982, Mustafaev et al. 1987) and tars (Bakirov et al. 1984). The main products of these processes are fuel gases H_2, CO, and CH_4. In the temperature range of 400°C–500°C, pressure of gasifying agents up to 0.5 MPa, and dose rate of gamma radiation of 20 kGy/h, radiation-chemical yields of fuel gases reached 800–1000 molecules/100 eV with a hydrogen concentration of above 60 mass%. It was shown that the addition of catalysts NiCl, $KMnO_4$ ($\leq$5 mass%) leads to an increase in the yield of target products by 20%–25% (Yakubov et al. 1983).

Melikzade et al. (1980) developed a method for the high-molecular olefin production by RTC of the oil fraction boiling in the temperature range of 162°C–400°C under an electron beam from a high-current accelerator. Physical and chemical parameters of the process were varied in a wide range: the dose range was 0.36–21.6 kGy, the dose rates were 0.34–1.03 Gy/s, and the temperature range was 400°C–550°C. It was shown that in optimal conditions, the total yield of olefins reached 52.5 mass%. The ratios of olefin concentrations in alpha- and trans-positions were determined by means of IR spectroscopy. In the most favorable conditions, the yield of alpha-olefins reached 33%.

A method for treating heavy hydrocarbon fractions using the combination of different types of ionizing radiation was described by Pokacalov (1998). In this method, heavy hydrocarbon material was treated simultaneously with different accelerated particles so that at least one type of the particles was positively and the other was negatively charged, for example, accelerated electrons and protons. Feedstock was processed in the RTC temperature range (300°C–600°C).

It was reported that combined gamma–electron–proton irradiation resulted in the 24 vol.% yield of light fractions from crude oil of 11°API gravity compared with the 4 vol.% yield in the case of the sole thermal treatment. It was stated that gamma–proton–electron irradiation provides higher yields of light fractions at lower absorbed doses than gamma–electron irradiation. The light fraction yield increased with the absorbed dose. It was suggested that the role of accelerated electrons was predominantly breaking molecular bonds and generation of free radicals, while protons compensated free valence bonds and prevented recombination of radicals, which might lead to the formation of even bigger hydrocarbon chains.

Sorokin (2002) offered to supplement the apparatus design in Pokacalov's patent with an ultrasonic generator or pressure pulsator installed to the feed running system. The synergetic action of ionizing irradiation and cavitation effects was intended to increase the light fraction yields.

Gafiatullin et al. (1997b) proposed a method for processing liquid and solid hydrocarbons where the feedstock was dispersed with 10%–40% water to form uniform ultradisperse foam and/or aerosol having a density of 5–350 g/L. This mixture was subjected to ionizing irradiation at temperatures of 90°C–150°C with continuous stirring in the steam flow. A mixture of the surplus steam and low-boiling product formed in the process was withdrawn from the reaction zone and separated. The end product contained low-molecular hydrocarbons, alcohols, and ethers.

The best results in processing condensed hydrocarbons under ionizing irradiation were obtained in the case when feedstock was mixed with water and when this mixture was transformed into an ultradisperse state. Importance of low-boiling product removal from the irradiation zone was emphasized.

The method offered was tested on the example of RP of Arlan crude oil, Russia, which contained no mineral impurities. At the temperature of 95°C, oil was intensely mixed with 35 mass% boiling water; the mixture was foamed by blowing with oil-well gas in the amount of 1 L of gas per 1 L of feedstock. The formed ultradisperse mixture was poured into the steel reactor and irradiated with 6 MeV electrons at the temperature of 135°C and foam density of 100 g/L. During irradiation, the foamed mixture was continuously bubbled with steam. Live steam and fresh foam were continuously supplied to the reactor to maintain a constant foam level. The target product was withdrawn in the steam flow and condensed for the organic phase separation.

The condensate obtained in these experiments contained 19% hydrocarbons, 34% monohydric alcohols, 26% polyhydric alcohols, and 3% ethers. The yield of the target product was 2.2 kg per 1 kW h of absorbed electron energy. The average octane number of the product was 89 at 100% feedstock conversion. The yield of the target product exceeded feedstock consumption by 35 mass% due to water participation in the formation of alcohols and ethers. A similar effect was observed in the work (Zaikin and Zaikina 2004b) on bitumen RP.

The same procedure was applied to processing oil fraction with the start of boiling at 400°C, crude oil from Grozny field (Chechnya, Russia), fuel oil M-100, and tar. The process conditions and the results obtained are shown in Table 3.3.

TABLE 3.3
Composition and Properties of the Products of Heavy Oil Radiation Processing

Process Conditions

Water concentration, wt.%	25	35	15	40
Temperature, °C	125	140	110	140
Density of foam/aerosol, g/L	210	16	115	240
Feedstock	Oil fraction $T_b = 400°C$	Grozny crude oil	Fuel oil M-100	Tar $T_m = 70°C$

Product Contents, wt.%

Hydrocarbons	15	24	20	14
Monohydric alcohols	37	35	30	35
Glycols	28	27	30	33
Polyhydric alcohols	16	9	17	14
Ethers	4	5	3	4

Product Properties

Boiling range, °C				
Start/end	28/295	25/275	27/300	24/280
Octane number	94	88	90	95
Density, kg/L	0.77	0.78	0.80	0.78
Yield, kg/kW h	2.6	2.1	3.4	3.7

Source: Gafiatullin, R.R. et al., A method for processing condensed hydrocarbons, Patent of Russia RU 2087519, 1997b.

At the process parameters (temperature, water concentration, and foam/aerosol density) beyond the recommended optimal range, oil conversion was still high but, at the same time, such undesirable products as pitches, aldehydes, ketones, and acids appeared in the product composition.

In the method proposed by Pruss (1999), oil irradiation was combined with its mechanical processing in a special device (master spray tower). After cleaning and desalting, oil passed to the master spray where it was sprayed and whirled. It was supposed that at this stage, hydrocarbon molecules were already partially cracked. The subsequent irradiation with high-energy electrons from an electron accelerator provided a higher degree of oil conversion. The irradiation and flow conditions were to be specified in each specific case of oil RP. It was suggested but not shown in any examples that "cold cracking," that is, rapid chain cracking reactions at lowered temperatures, was possible in such dispersed feedstock. To enhance the reaction rate and to increase conversion under irradiation, the author proposed application of a reaction camera with a platinum substrate as a catalyst, imposition of the electric field to platinum plates, and also using fullerenes as an additional catalyst.

The method for heavy oil and oil residua RP (Nadirov et al. 1994b, 1997d) allowed expanding the raw material base for the production of motor fuels by involvement of oil fractions with $T_b > 450°C$ into RP with the increased yield and improved quality together with the reduction of energy consumption compared to the conventional thermal cracking (TC).

Heavy oil fractions were processed under the beam of accelerated electrons with energy of 1–4 MeV. The process was carried out in experimentally determined optimal processing modes that provided the maximal approach to industrial conditions and the highest yield of motor fuels (gasoline, kerosene, straw distillate, and light gas oil fractions with $T_b < 350°C$).

An electron accelerator of ELU type was used as a source of ionizing irradiation. Oil feedstock of practical importance for industrial upgrading and refining was used as the object of processing. Application of RTC allowed improvement of the technological conditions for destructive processing of oil feedstock with $T_b > 400°C$ by reduction of the process temperature and pressure, increase in the yield of the target product (fractions with $T_b < 350°C$) by 2.5 times, and improvement of the product quality (considerable decrease in the sulfur content and better anti-detonation properties) compared with thermal and thermocatalytic cracking.

In this method, hydrocarbon feedstock was subjected to heating with the simultaneous effect of ionizing irradiation at nearly atmospheric pressure. The specific features were that (1) an electron accelerator with an electron energy of 1–4 MeV was used as a radiation source, (2) the feedstock was subjected to ionizing irradiation to absorbed doses in the range from 1 to 20 kGy, (3) the dose rate necessary for RP was in the range from 1 to 94 kGy/s, and (4) the process temperature was 380°C–410°C.

The reason for application of the electron accelerators was that, unlike the isotope sources, they guarantee high production rate, high reliability, and safety in the continuous operations: any possibility of radioactive pollution and necessity of radioactive waste disposal are completely excluded when using electrons in the energy range of 1–4 eV.

The selection of the lower dose limit was caused by a considerable decrease in the yield of light fractions with $T_b < 350°C$ when RTC was conducted with doses below 1 kGy. No considerable increase in the yield of the target product was observed for doses higher than 20 kGy that indicated the upper dose limit.

The lower limit of the dose rate was also associated with a sharp decrease in the target product yield at the dose rates below 1 kGy/s. The upper dose rate was the maximal dose rate accessible for the authors because of the accelerator's engineering restrictions.

RTC of oil fractions with $T_b > 450°C$ under the beam of accelerated electrons did not intensely proceed in the feedstock tested at the atmospheric pressure when the process temperature was below 380°C, which defined the lower temperature limit. However, further studies (Zaikin and Zaikina 2008a, Zaikin 2013a) have shown that the temperature threshold for RTC of heavy oils can be considerably lowered. The upper temperature limit came from the fact that a further increase in the process temperature required higher energy consumption without any essential increase in the yield of the target product (oil fractions with $T_b < 350°C$).

The test subject was fuel oil M-40 produced at the Atyrau refinery, Kazakhstan, by the straight-run distillation of crude oil. This type of feedstock was a mixture of heavy oil fractions with $T_b > 400°C$ and flash temperature $T_f = 170°C$. Water content was 2 mass% and sulfur concentration was 1.4 mass%. Kinematic viscosity of the fuel oil was 32.86 cSt at 80°C; its density was 0.95 g/cm^3.

The experiments were conducted in two different modes of electron irradiation: the stationary mode where fuel oil was enclosed in the finite volume and in flow conditions where the feedstock was continuously running through the radiation zone of the reactor. The 4 MeV electron beam current was about 600 µA; the current density varied in the range of 0.5–5.0 µA/cm^2. The stationary irradiation of fuel oil in batches was conducted in a cylindrical metal reactor with a standard cooler for separation of light oil fractions by distillation.

The liquid-phase RTC of fuel oil in flow conditions was studied using the apparatus shown in Figure 3.1.

A 2 L batch of fuel oil was poured into the feed tank. The flow rate of the liquid feedstock was set up by variation of nitrogen pressure in the feed tank and using valves for its fine regulation. The volume rate of fuel oil was varied in the range of 1–2 mL/s. The vertical furnace provided fuel oil heating up to the temperatures 400°C–500°C. The temperature of the walls of reactor 10 was kept at the level of 400°C ± 20°C. Pressure and temperature of oil in the reactor were measured and controlled.

After passing irradiation zone, the product flow was transported to a one-plate fractionation device supplied with an electric heater in its lower part and with a cooler in its upper part. The condensed vapors entered the cooler and then passed to the gas separator. The condensed products were collected in the lower part of the gas separator while the gaseous products passed to the gas collector.

The fractional contents of the liquid product of fuel oil RTC is shown in Figure 3.2 for two dose rates of electron irradiation. Increase in the dose rate leads to increase in the yield of light products.

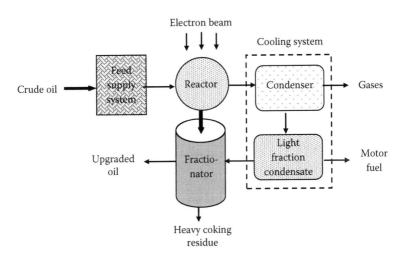

FIGURE 3.1 Layout of the flow RTC facility.

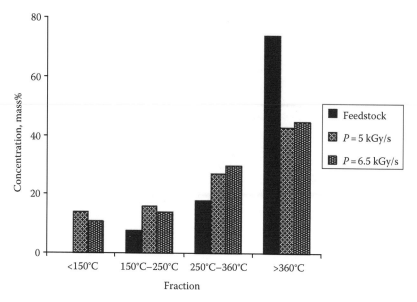

FIGURE 3.2 Fractional contents of fuel oil and liquid RTC product ($T = 400°C$, $D = 4$ kGy).

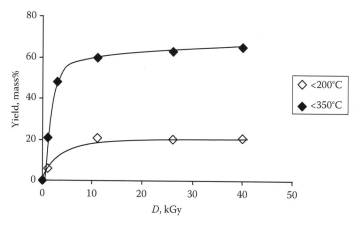

FIGURE 3.3 Yields of liquid fractions obtained by RTC of fuel oil versus dose of electron irradiation for fraction $T_b < 350°C$ and fraction $T_b < 200°C$.

The yields of the target product (oil fractions with $T_b < 350°C$) after RTC of oil fractions with $T_b > 450°C$ are shown in Figures 3.2 and 3.3 as functions of dose and dose rate, respectively. They illustrate an experimental observation that the yield of the target product rapidly increased as the dose increased from 1 to 20 kGy at the constant dose rate of 1 kGy/s.

Further increase in the irradiation dose resulted in a slow increase in the product yield (Figure 3.3). A linear dependence of the yields of light fractions with

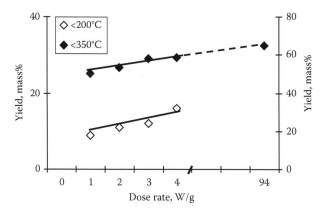

FIGURE 3.4 Dependence of the yields of RTC fractions $T_b < 350°C$ and $T_b < 200°C$ on dose rate of electron irradiation; $D = 1$ kGy.

$T_b < 350°C$ on the dose rate was observed at the given irradiation dose (Figure 3.4). In the dose rate range studied, the dose rate dependence of the product yields was nearly linear. However, there is only one experimental point in the plot of Figure 3.4 at the dose rate as high as 94 kGy/s. More detailed studies have shown that the dose rate dependence of the RTC reaction rate and of the yields of RTC products is more complicated (see Chapter 1).

At the given temperature, the production rate of the RTC facility depends on the beam power and irradiation dose:

$$Q\,(\text{kg/s}) = \alpha\,\frac{N(\text{kW})}{D(\text{kGy})} \qquad (3.1)$$

where
  N is the beam power
  D is the dose
  α is a part of the beam power consumed for the process

In the case of 100 kW beam power of an electron accelerator and the dose of 1 kGy/s, the maximal production rate calculated from formula (3.1) will be 100 kg/s = 360 ton/h ≈ 54,000 bbl/day.

Fractional contents of the products obtained by RTC of fuel oil and quality characteristics of gasoline, solar, and gas oil fractions were determined using the methods of optical spectroscopy, measurements of iodine numbers, gas chromatography, and chemical analyses. The maximal yields of the target product in flow conditions were close to those obtained in static irradiation experiments.

Characteristics of the products obtained by TC and RTC of oil fractions with $T_b > 450°C$ are shown in Figure 3.5 and Table 3.4.

The gasoline fraction of the RTC product was characterised by high octane numbers of 76–80 and low sulfur concentration (35 times lower than that in the original feedstock), that is, RP leads to considerable desulphurization of light fractions.

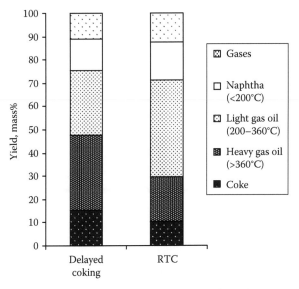

FIGURE 3.5 Fractional contents of delayed coking and RTC products.

TABLE 3.4
Quality Characteristics of Gasoline Fraction Extracted from RTC Product

Hydrocarbon Contents and Other Characteristics	Mass%		
	Feedstock	TC	RTC
Paraffins	60	35	25.4
Unsaturated hydrocarbons	<5	45	43.5
Aromatic hydrocarbons	18	15	21.2
Naphthenes	17	5	9.9
Density at 20°C, g/cm³	0.95		0.743
Octane number by motor method		70–72	76–80

Comparison with the hydrocarbon content of the original feedstock showed that RTC products predominantly resulted from paraffin decomposition. However, concentrations of aromatic and naphthenic hydrocarbons in RTC products were noticeably higher compared with those after conventional TC.

Another group of researchers (Dolgachev et al. 2008) studied RTC of fuel oil under electron irradiation in similar conditions at the temperature of 400°C. The fractional content of the liquid RTC product is shown in Figure 3.6.

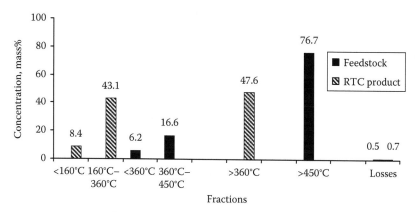

FIGURE 3.6 Fractional contents of the total product obtained by RTC of fuel oil. (Plotted using the data from Dolgachev, G.I. et al., *Voprosy Atomnoi Nauki I Tekhniki (Prob. Atom. Sci. Technol.) Ser. Nucl. Fusion*, 1, 57, 2008.)

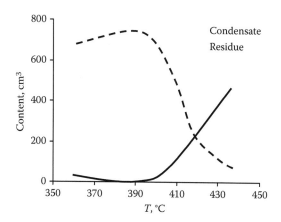

FIGURE 3.7 Volume yield of liquid fractions boiling below 360°C obtained by RTC of fuel oil. (From Dolgachev, G.I. et al., *Voprosy Atomnoi Nauki I Tekhniki (Prob. Atom. Sci. Technol.) Ser. Nucl. Fusion*, 1, 57, 2008.)

Figure 3.7 shows temperature dependence of the liquid condensate ($T_b < 360°C$) yield obtained in the work (Dolgachev et al. 2008).

Generally, radiation methods based on RTC allow efficient high-rate processing and upgrading of high-viscosity oil at moderate temperatures and near atmospheric pressure.

3.2 RADIATION METHODS FOR LUBRICATING OIL PROCESSING

Since the early 1960s, generous efforts of the researchers were applied to development of the radiation methods for lubricating oil production and upgrading.

The resistance of various types of lubricating oils to ionizing irradiation and methods for improvement of their radiation resistance are described in the papers (Haley 1961, McDaniel 1964, Potanina et al. 1976, 1977, Arakawa et al. 1987).

A method for preparing multipurpose lubricant additives by ionizing irradiation was proposed by Lucchesi and Long (1961).

It is known that one of the desirable properties of lubrication oils is a high-viscosity index, which means that the oil is subject to less change in viscosity with temperature changes. Other desirable properties of lubricants are their high resistance to shearing action or shear stability and sufficiently low pour point so that they would function effectively at relatively low temperatures. These requirements are particularly important in the lubrication of internal combustion engines.

It was found that normally solid paraffin constituents, usually removed from lubricating oils during their preparation, can be converted to obtain potent lubricating oil additives by high-intensity ionizing irradiation.

The process developed by Lucchesi and Long (1961) comprised irradiating a paraffin wax fraction containing at least 80 wt% of paraffin hydrocarbons having from 21 to 32 carbon atoms with mixed ionizing irradiation comprising neutrons. The recommended intensity of both gamma and neutron radiation was above 10^5 R/h and the dose rate was at least 1 Gy/s. Under these conditions, high conversions of the paraffin fraction, usually over 90%, were obtained. Irradiation was preferably carried out in the presence of a porous solid hydrocarbon conversion catalyst. The product obtained by this treatment was a relatively rubbery material at room temperature, soluble in oil.

In a demonstrational experiment, a paraffin distillate boiling in the range of 370°C–480°C was used as a feedstock. The wax melted at 52°C and contained 99% n-paraffins. This feedstock was processed in an atomic pile in the presence of two different types of hydrocarbon conversion catalysts, namely, a silica–alumina cracking catalyst and a platinum/alumina hydroforming catalyst. The thermal neutron flux in the reaction zone was $2.5 \cdot 10^{12}$ cm^{-2} c^{-1} and the gamma ray flux was $1.6 \cdot 10^6$ R/h. It was shown that 500 cm^3 of catalyst and 500 cm^3 of paraffin wax were irradiated in a vented aluminum container on a horizontal aluminum sled. Under these conditions, more than 95 wt% of paraffin wax was converted to a rubbery polymer. The rubbery material began to liquefy in a nitrogen atmosphere at 288°C and was completely liquid at 343°C.

Rubbery material of 3 wt% was added to lubricating oil, which boiled in the range of 70°C–538°C and had a pour point of −29°C, viscosity index of 114, and viscosity of 6.6 cSt at 99°C. As a result, the viscosity index of the lubricant increased up to 120–122 and its pour point fell to −31.7°C. Addition of the rubbery material produced by wax irradiation with a silica–alumina catalyst to a lubricating oil base stock resulted in the increase in viscosity index from 113 to 120 and decrease in the pour point from −9.4°C to −23.3°C.

Chester and Lucchesi (1962) proposed a process for producing lubricating oils from select hydrocarbon mixtures by irradiation. Distillate hydrocarbon mixtures, comprising predominantly paraffinic hydrocarbons and a minor amount of unsaturated straight-chain hydrocarbons, were subjected to neutron and gamma irradiation from a nuclear reactor. The RP was carried out at a temperature below 316°C until a dose above 100 MR had been absorbed. In this manner, a high-quality lubricating oil amounting to about 50–90 wt% was obtained. The oil had a viscosity index above 125.

The hydrocarbon mixture used as a feedstock boiled in the range of 149°C–371°C. It contained 70–95 wt% of essentially saturated hydrocarbons, 5–40 wt% unsaturated or olefinic straight-chain hydrocarbons, less than 2 wt% of aromatics, and less than 25 wt% of cycloparaffins or naphthenes. It was essential that the olefin content was within the range indicated earlier. It was found that in the ranges higher than this, the yield and quality of the lubricating oil fell off. With smaller amounts of olefins than specified, the process approached that of irradiating n-hexadecane, wherein the tendency to be overconverted was present and the yield greatly fell off.

The recommended dose rate was at least 1 MR/h with the total dosage of at least 100 MR that corresponds to the absorbed dose of 84 kGy. Of the energy received, it was preferred that at least 30% was obtained through neutrons while the rest might come from the associated gamma rays. The temperature was recommended to be below 316°C to avoid any predominance of cracking reactions and, therefore, yield loss.

A distillate boiling in the range of 316°C–593°C obtained from a mixture of Venezuelan, South Louisiana, and West Texas crude oils was used in the experiments. It was converted in a fluid catalytic cracking at the temperature of 490°C with silica–alumina–type catalyst. A fraction boiling in the range of 177°C–260°C and amounting to 9.4% on total cracked products was segregated from the catalytically cracked material. This fraction was treated with 125 vol.% sulfur dioxide at a temperature of –37°C to remove aromatics. The feedstock contained 60 vol.% aromatics and the raffinate from this extraction contained about 7 vol.% aromatics. The sulfur dioxide–treated material was then further contacted with 170 wt% of silica gel. The extracted raffinate, amounting to about 30 wt% of the cracked fraction, contained 87 wt% saturates and 13.2 wt% olefins. It had a naphthene content of 19.4 wt% and a paraffin content of 67.4 wt%.

This material was exposed to irradiation in a vented aluminum container in the graphite-moderated atomic pile. The pressure during irradiation was substantially atmospheric and the temperature was about 150°C. The flux of thermal neutrons, having energy less than 0.6 MeV, was $2.5 \cdot 10^{12}$ n/cm^2 s in the reaction zone. The flux of fast (above 0.6 MeV) neutrons was $1.2 \cdot 10^{12}$ n/cm^2 s. The gamma ray flux was $1.6 \cdot 10^6$ R/h. The irradiation was continued for 10 days until about 1400 equivalent megaroentgens had been absorbed.

About 95 wt% of the original material was recovered, the rest having distilled off during irradiation. The fractional content of the irradiated product is shown in Figure 3.8.

Since olefins polymerize easier than paraffins, it was expected that the olefin unsaturated mixture would have gone to solid easier than does cetane. However, the product boiling at temperatures above 420°C was liquid and its yield amounted up to 77 wt%.

A light lubricating oil was made by blending equal volumes of 260°C–370°C, 370°C–420°C, and >420°C cuts. The resultant oil had a boiling range of 260°C–430°C, viscosity of 8.5 cSt at 99°C and 45 cSt at 38°C, and 136 viscosity index. For comparison, a lubricant made from pure cetane under the same conditions boiled in the range of 370°C–480°C and had a viscosity of 4 cSt at 99°C and 14.5 cSt at 38°C, and viscosity index of 150–156. Thus a considerably higher yield of a high VI material was obtained from the catalytically cracked and extracted material.

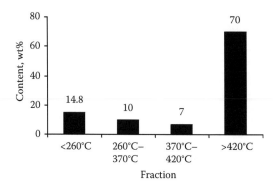

FIGURE 3.8 Fractional contents of the irradiated lubricant fraction. (Plotted using the data from Chester, L.R. and Lucchesi, P.J., Producing lubricating oils by irradiation, US Patent 3043759, 1962.)

TABLE 3.5

Properties of the Base Stock Blended with 3% Product Boiling above 420°C

Characteristics	Base Stock	Base Stock + 3 wt% Additive
Viscosity, cSt at 99°C	7.2	8.0
Pour point, °C	−9.4	−15
Viscosity index	113	120

Source: Chester, L.R. and Lucchesi, P.J., Producing lubricating oils by irradiation, US Patent 3043759, 1962.)

The product boiling above 420°C was blended in 3% concentration in a lubricant base stock with the results shown in Table 3.5.

A fraction of 10 wt% boiling above 420°C having a viscosity of about 1300 cSt at 99°C was mixed with a base stock. The base stock consisted of 95 wt% of a neutral, 5 wt% of a bright stock, and a small amount of a pour point depressant. The results are shown in Table 3.6.

These examples illustrate the thickening and viscosity index-improving power of the lubricant fractions obtained by RP when added to conventional lubricant base feedstock.

Hartzband et al. (1960) patented a continuous process for producing high-quality lubricating oils from paraffinic feedstocks through the use of high-energy ionizing irradiation. In this process, a predominantly paraffinic feedstock boiling in the range of about 38°C–370°C was irradiated at the temperature in the range above the pour point of the feedstock (about 10°C) to 370°C, until an irradiation dose of at least

TABLE 3.6

Properties of the Base Stock Mixed with

10% Fraction Boiling above 420°C

Characteristics	Base Stock	Base Stock + 10 wt% Additive
Viscosity, cSt at 99°C	6.5	8.5
Viscosity, cSt at 43°C	32.4	51.2
Viscosity index	114	120

Source: Chester, L.R. and Lucchesi, P.J., Producing lubricating oils by irradiation, US Patent 3043759, 1962.)

40 kGy had been absorbed to obtain a conversion of at least 1 mass%. After irradiation, the material was then separated to recover an intermediate boiling-range (about 400°C/510°C) lubricating fraction having a viscosity in the range of 3–10 cSt at 100°C and a viscosity index about 120. The material boiling below this intermediate boiling range was recycled to the radiation zone.

The product obtained by irradiating a paraffinic feed material showed a rather unique variation of viscosity index with viscosity, as compared with conventional lubricating distillates. As expected, the irradiated material increased in viscosity as the boiling point increased. However, the viscosity index for irradiated materials reached a maximum at some intermediate viscosity. This is not characteristic for conventional lubricating distillates as they show a relatively flat but decreasing viscosity index with increasing viscosity. This finding was incorporated into a continuous process typified by recovery of products with high viscosity indexes.

Pure *n*-hexadecane was irradiated from an atomic pile in a vented aluminum container at the temperature of 127°C and at the pressure of about 1 atm. The flux in the container was about $3 \cdot 10^{12}$ n/cm^2 s and 1.6 MR/h of gamma radiation (1 MR is equivalent to the absorbed dose of about 7.9 kGy). The irradiation was continued for about 8 days until the dose of 2.4 MGy had been absorbed. As a result of RP, the viscosity index of the material reached a maximum of about 5 cSt at 99°C. The irradiated material was separated to obtain a fraction boiling in the range of about 370°C–480°C. This fraction had a viscosity of 4 cSt at 99°C, a viscosity index of 145, and 4.5°C pour point. It could be dewaxed to −32°C pour point by removing 3.5 wt% wax and did not break down under shear.

Irradiation of *n*-hexadecane was also carried out in flow conditions. The approximate flux was $0.2 \cdot 10^{12}$ n/cm^2 s of fast neutrons, $1 \cdot 10^{12}$ n/cm^2 s of thermal neutrons, and $1 \cdot 10^6$ R/h of gamma rays. The pressure was about atmospheric, and the temperature was 93°C–204°C, varying with the length of the reactor. The material from the reactor was passed through a condenser at 18°C. The liquid product was returned to the storage drum for recycle, and the gas was metered and vented.

The yields after 3140 h of operation are shown in Figure 3.9.

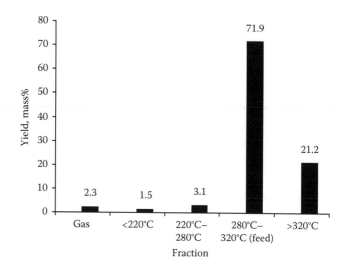

FIGURE 3.9 Product yields after *n*-hexadecane reactor irradiation in flow conditions. (Plotted using the data from Hartzband, H.M. et al., Preparing lubricating oils using radiation, US Patent 2951022, 1960.)

The results of continuous *n*-hexadecane irradiation, shown in Figure 3.10, demonstrate an unusual relationship between viscosity and viscosity index in the irradiated material.

Stoops and Day (1960) treated lubricating petroleum oil with an olefin in the presence of ionizing irradiation. Such a mixture was employed as a coolant and/or moderator for an atomic reactor. At the same time, it was treated by exposure to reactor radiation. A portion of lubricating oil mixed with 22 mass% ethylene was placed in a high-pressure container and exposed to gamma irradiation of an average intensity $5.6 \cdot 10^6$ R/h for 18.02 h, for a total dosage of 10^8 R. The treatment was performed at ambient temperature of about 27°C. Characteristics of the irradiated material are summarized in Table 3.7.

In Table 3.7, sample 1 is a sample of untreated oil, while sample 2 is a sample of the same oil after irradiation. Samples 3 and 4 were derived from sample 2 by centrifugal separation. The liquid portion (sample 3) was high-quality lubricating oil having a much-improved viscosity index. The other portion was a soft wax or jelly resembling petroleum useful as grease for lubrication of bearings.

Humphrey and McGrath (1964) developed a method for preparing an improved lubricating oil having a viscosity in the range between 7.7 and 28.3 cSt at 99°C and a low trend to deposit coke. For this purpose, lubricating oils having a viscosity in the earlier-said range and which had been treated with hydrogen were subjected to ionizing irradiation with absorbed doses between 3.6 and 36,000 kJ/kg.

A hydrogenated residue having a viscosity of 20 cSt at 99°C and a nonhydrogenated residue of the same viscosity were compared before and after irradiation. The irradiation was carried out by pumping the oil sample through a glass tube exposed to a beam of 3 MeV electrons from a Van de Graaff accelerator at a rate of 4–5 gal/min until the necessary dose was absorbed.

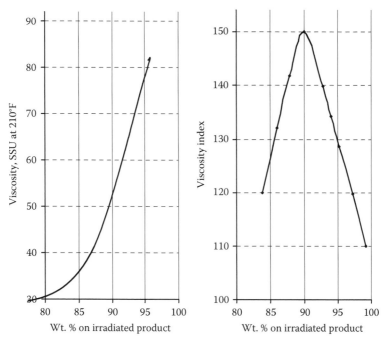

FIGURE 3.10 Viscosity and viscosity index behavior in irradiated *n*-hexadecane. (From Hartzband, H.M. et al., Preparing lubricating oils using radiation, US Patent 2951022, 1960.)

TABLE 3.7

Characteristics of Irradiated Mixture of Lubricating Oil and Ethylene

	1	2	3	4
			Irradiated	**Irradiated**
Sample	**Unirradiated**	**Irradiated**	**Centrifuged Liquid**	**Centrifuged Jelly**
Specific gravity	0.8645	0.8692	0.8645	
Viscosity cSt				
At 38°C	99.88	384.5	130.7	
At 99°C	39.88	45.17	42.68	
Viscosity index	100	<0	121	
Drop point				76°C

Source: Stoops, C.E. and Day, J.M., High VI oil and process for preparing same, US Patent 2954334, 1960.

Table 3.8 demonstrates improvement in the lubricant properties due to the electron irradiation.

To obtain lubricating oil having higher viscosity indices and other improved characteristics, such as viscosity, color, etc., Natkin et al. (1961) subjected the oils to gamma irradiation without application of any additives or special refining methods.

TABLE 3.8

Effect of Electron Irradiation on Properties of Oil Residue

Irradiation Dose, kGy	Hydrotreated Residue				Non-Hydrogenated Residue	
	0	100	1,000	10,000	0	10,000
Gravity, °API		29.8	29.1	25	27.2	27.1
Viscosity, cSt						
At 38°C	243	246.8	324.8	14.904	244.5	311.9
At 99°C	20.81	21.03	25.79	583.5	19.91	23.8
Viscosity index	107	106	108	121	101	103
Color D 1500	1.5	2	3.5	5.5	3	3
Flash, °C	>288	>288	268	216	296	>316
Fire, °C			316		327	>316
Pour, °C	−15	−15	−17.8	−6.7	−17.8	
Total acid no. D664	0.04	0.12	0.06	0.46	0.01	0.08
Panel coking test (371°C, 8 h)						
Coke deposit, mg	1238	84	374	95	90	362
260°C oxidation test						
Increase in 38°C viscosity, %	110		129		179	236
Increase in total acid no	3.48		3.11		3.87	3.7
Weight change of metals, mg/cm^2						
Mg	0.03		−0.06		0.04	0.04
Al	0		0.01		−0.21	0
Cu	−0.01		0		−0.01	−0.02
Steel	0.03		0.01		0.03	0.03
Ag	0.01		0.01		0.03	0.01
Ti	0.01		−0.01		0.03	−0.01

Source: Humphrey, E.L. and McGrath, J.J., Irradiation of lubricating oils, US Patent 3153622, 1964.

Various types of lubricating oil were subjected to gamma irradiation from a Co-60 source at ambient temperature and atmospheric pressure:

1. *Lubricating oil A*—an acid-treated naphthenic-type oil having viscosity of 198.8 cSt at 38°C and 12.5 cSt at 99°C, 36.1 VI, API gravity of 22.4°, a flash point of about 227°C, and a pour point of about −26°C.
2. *Lubricating oil B*—a paraffinic-type oil having a viscosity of 857 cSt at 38°C and 39.6 cSt at 99°C, 87.5 VI, API gravity of 22.0°, a flash point of about 293°C, and a pour point of about −4°C.
3. *Lubricating oil C*—a synthetic ester comprising di-2-ethylhexyl sebacate having viscosity of 12.7 cSt at 38°C and 3.4 cSt at 99°C, and 156 VI.
4. *Lubricating oil D*—a phenol-treated naphthenic-type oil having viscosity of 268.3 cSt at 38°C and 14.9 cSt at 99°C, 34.2 VI, API gravity of 22.2°, a flash point of about 227°C, and a pour point of about −18°C.

5. *Lubricating oil E*—a phenol-treated naphthenic-type oil having viscosity of 20.5 cSt at 38°C and 3.7 cSt at 99°C, 54.3 VI, API gravity of 28.9°, a flash point of about 171°C, and a pour point of about −46°C.

6. *Lubricating oil F*—a paraffinic-type oil having viscosity of 86.0 cSt at 38°C and 10.0 cSt at 99°C, 104.8 VI, API gravity of 30.2°, a flash point of about 252°C, and a pour point of about −7°C.

7. *Lubricating oil G*—a synthetic oil comprising C_{13} Oxo formal having viscosity of 16.0 cSt at 38°C and 3.60 cSt at 99°C.

The observed changes in the properties of lubricating oils after gamma irradiation are summarized in Table 3.9. The radiation intensity was $2.2 \cdot 10^6$ R/h.

The data in Table 3.9 show that gamma irradiation of lubricating oils resulted in marked changes in their physical properties. The paraffinic-type mineral lubricating oils were less susceptible to change in certain properties by irradiation than the naphthenic-type mineral lubricating oils, particularly the phenol-treated naphthenic oils, while the synthetic-type lubricating oils were the most affected. However, the synthetic lubricating oils, as represented by sample C, showed a very large increase in their neutralization numbering in contrast to the mineral lubricating oils, which did not show any change at all in this respect.

The opportunities of radiation technology application to recycling used automotive lubricating oil were investigated by Scapin et al. (2007). Used automotive lubricating oil was treated by the ionizing radiation for metal removal and degradation of organic compounds. A Co-60 isotope facility was used as a source of gamma radiation. The samples were irradiated with the doses of 100 and 200 kGy.

Determination of the elements before and after irradiation was accomplished using x-ray fluorescence technique, and the organic profile was obtained by infrared spectroscopy. Irradiation of the used lubricants was performed with and without the addition of hydrogen peroxide.

TABLE 3.9
Effect of Gamma Irradiation on Lubricating Oils

| | Viscosity, cSt | | | | | | | |
| | At 38°C | | At 99°C | | VI | | Neut. No. | |
Ludricating Oil	Before	After	Before	After	Before	After	Before	After
A	198.8	250.2	12.5	14.4	36.1	36.3	0.04	0.08
B	857	1165	39.6	49.2	87.5	90.2	0.19	0.19
C	12.7	21.4	3.4	4.8	156.1	159	0.12	0.21
D	268.3	344	14.9	17.5	34.2	38.6	0.03	0.02
E	20.5	25.5	3.7	4.3	54.3	61	0.06	0.02
F	86	112.5	10	12.3	104.8	107.2	0.06	0.12
G	16.14	20.4	3.6	4.23	120.3	129.3	0.03	30.53

Source: Natkin, A.V. et al., Radiation of lubricating oils, US Patent 2990350, 1961.

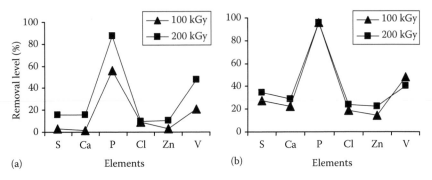

FIGURE 3.11 Removal level percentage of used oil sample (a) without and (b) with H_2O_2. (From Scapin, M.A. et al., *Radiat. Phys. Chem.*, 76, 1899, 2007.)

As a result of RP, a considerable amount of impurities were precipitated, and their concentration in the purified lubricants was reduced to a level shown in Figure 3.11. Addition of H_2O_2 noticeably increased the efficiency of the impurity removal (Figure 3.11).

The infrared spectra (FTIR) of a sample recycled by a conventional process (CRP) and that of the used automotive lubricating oil sample irradiated with doses of 100 and 200 kGy without and with H_2O_2 addition are shown in Figure 3.12.

The lubricating oil irradiated with the doses of 100 and 200 kGy, with and without H_2O_2, did not show noticeable alterations in the main groups of organic compounds, such as aldehydes, ketones, and aliphatics. It was interpreted as evidence that the experimental conditions used did not degrade these compounds.

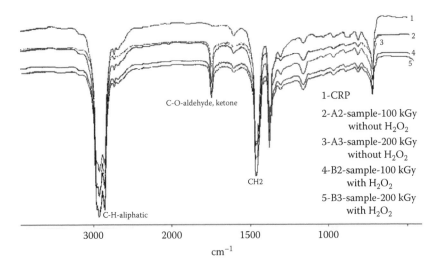

FIGURE 3.12 FTIR spectra of the used automotive lubricating oil irradiated with the doses of 100 and 200 kGy. (From Scapin, M.A. et al., *Radiat. Phys. Chem.*, 76, 1899, 2007.)

The hydrocarbon degradation by gamma irradiation of the waste automotive lubricating oil at higher absorbed doses was studied by Scapin et al. (2009).

The waste automotive oil was collected in a Brazilian oil recycling company. To the used lubricants, 50 and 70 vol.% water were added. Each sample was irradiated with the doses of 100, 200, and 500 kGy using a Co-60 gamma source with 5×10^3 Ci total activity. Gas chromatography–mass spectrometry was used to identify the degraded organic compounds.

The samples irradiated with the doses of 200 and 500 kGy with H_2O addition showed the most significant changes in their composition. Such compounds as 2-(2-hydroxyethyl-methyl-amino)ethanol($C_5H_{13}NO_2$); 1-octyn-3-ol,4-ethyl($C_{10}H_{18}O$); 1-ethoxy-2-(2-ethoxyethoxy) ethane ($C_8H_{18}O_3$); 1,4-pentanediamine, N1-diethyl ($C_9H_{22}N_2$); 1,4,7,10,13-pentaoxacyclopentadecane ($C_{10}H_{20}O_5$); ethylene glycol ether monovinyl ($C_4H_8O_2$); ethanol,2-(1-methylethyl) amino ($C_5H_{13}NO$), diptych (boroxazolidin); and B-ethyl ($C_6H_{14}BNO_2$) compounds were identified in the composition of the samples irradiated to high irradiation doses.

The UV/VIS absorption spectra for the control and irradiated samples are shown in Figure 3.13.

The spectra in Figure 3.13 show lubricant clarification promoted by irradiation. Sample A1 (control) shows the greatest absorbance value (at 260 nm) while sample C_4 shows the lowest absorbance demonstrating the highest clarification of the used oil lubricants.

A group of researchers from the University of Maryland proposed a method for industrial transformer oil cleaning from polychlorinated biphenyls (PCBs) (Schmelling et al. 1998, Chaychian et al. 2002, Jones et al. 2003).

Although the use of PCBs has been tightly restricted, PCBs still remain in use in a variety of industrial and commercial applications when circumstances permit.

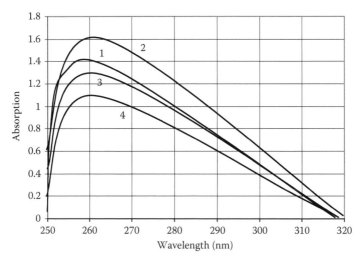

FIGURE 3.13 UV/VIS absorption spectra for control and irradiated lubricant samples: 1, pure mineral oil; 2, control sample; 3, –200 kGy; 4, 500 kGy. (From Scapin, M.A. et al., *Radiat. Phys. Chem.*, 78, 733, 2009.)

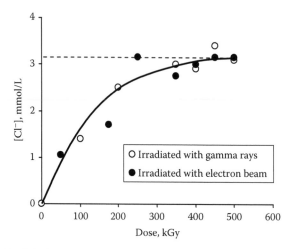

FIGURE 3.14 Radiolytic dechlorination of PCBs in transformer oil containing ≈95 µg/g PCB and mixed with TEA (9.1 v/v), irradiated with gamma (O) or electron beam (●). The dotted line indicates the level of complete dechlorination.

The concern for environment contamination of PCB-laden transformer oil today is that of overheating of electric components, such as electric transformer or capacitor oil containing PCBs that can produce emissions of vapors or expansion of the oil, leading to small spills, leaks, or airborne releases into the environment. In addition, accidental spills are a problem as possibly illegal disposal.

It was shown that irradiation of transformer oil by a 7 MeV electron beam or gamma-irradiation in the presence of special additives, such as triethylamine (TEA) or KOH mixed with 2-propanol (2-PrOH), induces the complete PCB radiolytic degradation (Figure 3.14).

Gamma irradiations were carried out using Co-60 isotope sources with dose rates 6.40 and 6.84 kGy/h. Electron irradiations were carried out with a Varian electron accelerator, using 3-µs pulses of 7 MeV electrons. The solutions were irradiated with sequences of 10,000 pulses at a time, followed by a 2 min cooling period. The absorbed dose was 10 Gy/pulse.

Additional solvent is necessary to prevent oil solidification. The mechanism for this process is based on PCB (ArCl) dechlorination by reacting with solvated electrons:

$$e_{solv}^- + ArCl \rightarrow Ar^* + Cl^-$$

However, the oil contains various aromatic hydrocarbons, such as biphenyl (Ph_2), that can react with electrons to form radical anions:

$$e_{solv}^- + Ph_2 \rightarrow Ph_2^{*-}$$

PCBs are dechlorinated also by electron transfer from aromatic radical anions:

$$Ph_2^{*-} + ArCl \rightarrow Ph_2 + Ar^* + Cl^-$$

These reactions explain why PCBs can be dechlorinated despite the presence of aromatic hydrocarbons in the oil and despite the formation of biphenyl as a radiolysis product that reacts rapidly with solvated electrons.

Similar results were obtained using PCB photolytic (UV) treatment. For the photolysis experiments, solutions (5 mL aliquots) were placed in Pyrex (cutoff ≈ 300 nm) cylindrical cells and irradiated with a 300 W UV Xenon arc lamp with continuous stirring. The output of the lamp is nearly flat throughout the visible range, and it drops in the UV range to about 10% at 220 nm. The duration of the photolytic processing is about 30 times lower than that of the radiolytic process. But this factor can be gained by the lower cost of equipment and operation of UV sources as compared with electron beam and gamma sources.

Radiolytic or photolytic PCB treatments are not too different from the conventional base-catalyzed decomposition or the polyglycoxide methods in terms of treatment materials, but they are advantageous in terms of energy consumption because they involve a chain reaction that is merely initiated by radiation. PCBs can be dechlorinated also by reaction with strongly reducing metals and by catalytic hydrogenation. These methods do not involve a strong base but use a highly reactive metal that must be rereduced for recycling or specific catalysts and high pressures of hydrogen, all of which involve costly materials and equipments. This method was extended to the problem of PCB elimination in aquatic sediments. Sediment treatment with ionizing and UV radiation leads to effective PCB degradation.

The behavior of lubricating greases under ionizing irradiation was discussed in the paper (Kobzova et al. 1978). It was noted that, whereas radiation effects on oils show up mainly in increasing the kinematic viscosity, the main effect of irradiation on lubricating greases is a gradual breakdown of the structural skeleton of the grease, producing changes in its rheological properties. The breakdown of structural bonds in grease usually begins at certain absorbed doses causing chemical changes of the dispersion medium or thickening agent. The initial effect of radiation on lubricating grease is manifested in a drop in its yield stress and viscosity. As the absorbed dose is further increased, the grease, after changes in its rheological parameters generally corresponding to a fluid state, begins to harden. The final stage of change in all lubricating greases is solidification, which is related to radiation polymerization of the dispersion medium.

The authors (Kobzova et al. 1978) adopted, as a general criterion of grease radiation resistance, that it must retain its plastic properties. They referred to the presence of a yield stress and a dependence of apparent viscosity on shear rate, as well as to the absence of any separation of solid and liquid phases. The greases were irradiated in a flux of gamma quanta from a cobalt source of 6 Gy/s. Measurements of the thermal oxidative stability of the greases before and after irradiation indicated significant drops in this property index as a result of the radiation. For example, the greases studied, when tested at 250°C without previous irradiation, hardened in 2000 and 700 h. When tested after irradiation, they hardened in 700 and 1200 h.

Application of radiation methods for lubricant production from heavy oil fractions was discussed in the paper (Zaikina and Zaikin 2003).

The previous experiments on RTC of heavy hydrocarbon mixtures, such as raw heavy oil, fuel oil, and petroleum residua (Zaikina et al. 1996, 1998), have shown that optimal conditions for lubricant production from this type of feedstock imply a rather high degree of a degradation-type conversion necessary to increase the content of the lubricant-containing fraction lighter than the original heavy mixture. It also leads to a decrease in the cyclicity of aromatic and naphthene hydrocarbons, and promotes the dealkylation process. However, in contrast to the modes optimal for motor fuel production, an important role of lubricant production in radiation technologies is performed by reactions of radiation-induced polymerization making mono-olefin contents in the lubricant-containing fraction much lower and attenuating its trend to oxidation. The heavy polymer deposit formed during processing is of a high adsorption capacity. The intense olefin polymerization combined with radiation-induced adsorption causes efficient clearing of the lubricant-containing fraction from pitches, asphaltenes, and mechanical impurities, if available, and allows further easy separation of the purified lubricants. A combination of the high rates of destruction and olefin polymerization is provided by RP at temperatures higher than the temperature value characteristic for the intense RTC.

Such a combination can be much easier achieved in radiation oil processing than in traditional methods of oil refining. In conventional thermal processes based on thermal activation of chain cracking reactions, any rise in temperature always leads to a rise in the rate of molecule destruction. An advantage of RP is its ability to control nondestructive thermally activated processes by temperature variation without catalyst application, while the destruction rate is independently controlled by variation of the dose rate. In particular, it allows exclusion of many stages of the conventional technology for lubricant production and obtaining higher yields of lubricant-containing fractions by radiation-thermal processing of high-viscosity oil.

The experimental studies were conducted both in stationary and in flow conditions. A lubricant-like material was produced from heavy fuel oil produced by the Atyrau refinery, NW Kazakhstan.

Physical and chemical properties of the lube fraction produced by radiation initiation of complex degradation and nondestructive reactions in fuel oil are shown in Table 3.10 in comparison with the product of conventional delayed coking.

Figure 3.15 demonstrates IR spectra for the original fuel oil (curve 1), for the gas oil fraction with boiling temperature from 60°C to 3500°C obtained by fuel oil RTC (curve 2), and for the basic lubricant produced from the same feedstock (curve 3). Curves 2 and 3 are relevant to different irradiation modes characterized by different sets of values for temperature, dose, dose rate, and other operation parameters.

R-spectroscopy data show that, as distinct from the original feedstock (curve 1), the products of prevailing destructive conversion (curve 2) and products of complex radiation-induced reactions (curve 3) contain plenty of saturated hydrocarbons of C_5H_{12}–$C_{12}H_{26}$ series and partially higher molecular mass hydrocarbons of aliphatic series with normal and branched structure. That follows from the presence of significant absorption in the valence oscillations of the spectrum region over C–H bonds in the range of 1500–1300 cm^{-1}.

Saturated hydrocarbons with branched chains, aromatic hydrocarbons, and a small quantity of cyclic hydrocarbons are also available in the substance of the basic lubricant; it was concluded from comparison of the bands' position and intensity in the deformation oscillation region. The degree of alkane branching is higher in

TABLE 3.10

Characteristics of Basic Lubricant Produced by Radiation Processing and Conventional Delayed Coking of Fuel Oil

Characteristics	Fuel Oil	Lubricant-Containing Fraction ($T_b > 350°C$)	
		Radiation Processing	Delayed Coking (Meraliev et al. 1996)
Density at 20°C, g/cm³	0.9185	0.8687	0.882
Viscosity at 50°C, cSt	92	17.7	18.7
Viscosity index	—	120	52
Pour point, °C	34	15	28
Flash point, °C	180	208	76
Sulfur, mass%	1.0	0.2	0.3
Acid number	—	0.02	0.40

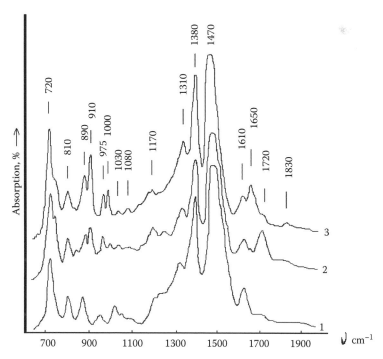

FIGURE 3.15   IR spectra of products of radiation processing.

the case of conditions favorable for radiation-induced cracking (absorption bands [a.b.] 720, 1380, 1610 cm^{-1}). According to IR spectra, considerable growth of mono-aromatic structures was observed in both cases. In general, the basic lubricant (curve 3) was characteristic of a much wider spectrum of adsorption bands (a.b. 870, 890, 910, 975, 1000, 1080, 1170, 1650 cm^{-1}).

High quality and chemical stability of the basic lubricant produced were provided by the sufficiently high level of olefin polymerization and high contents of polycyclic aromatic compounds without side chains that lead to lower lubrication oxidizability and corrosion activity, and inhibits oxidation of the naphthene compounds. High-molecular sulfuric compounds concentrated in the lubricant-based oil fraction after RP increase the lubricant viscosity at room temperatures, prevent viscosity anomalies, and improve anticorrosion properties of the lubricant.

The problem of regeneration and refining of used oil products in the industrial scale was studied in the papers (Zaikina and Zaikin 2003, Zaikina et al. 2005).

Environment contamination by ecologically dangerous components of used oil products and, first of all, by lubricants is the problem of a global scale. Annually about 6 millions tons of oil products are introduced into the biosphere all over the world, including more than 50% used lubricants. Their toxic components spread in the atmosphere, water, and soil fall into food circuits and meals. Used oil products are not regenerated in sufficiently large scales because of the complicacy and low economic efficiency of existing technologies.

Zaikina et al. (1999b) proposed a universal method and an apparatus for reprocessing of any type of used and residual mixtures of oil products. The method is remarkable for a high production capacity and high yields of different types of commodity products, such as base lubricants and motor fuels. The utilization of the waste oil products contributes to environment protection against oil contaminators.

The existing lubricant production technology consists of three main stages (Evdokimov et al. 1992): (1) production of lubricant-containing fractions; (2) extraction of basic hydrocarbon components from lubricant-containing fractions; and (3) mixing (compounding) of these components and, if necessary, introduction of appropriate additional agents into basic lubricant to produce commodity products.

Technology of basic lubrication production includes a series of processes enabling the extraction of some hydrocarbon groups and compounds undesirable in lubricants (asphalt-pitch compounds, polycyclic aromatic hydrocarbons with low viscosity indexes, solid paraffin hydrocarbons). Application of the radiation method allows replacing all these complex operations with a one-stage process, combining RTC and nondestructive radiation-induced reactions.

The main difference of the technology for used oil product reprocessing from other methods of oil upgrading, including those for lubricant production from heavy oil fractions, is that used oil products do not require deep conversion. Therefore, the most favorable conditions for their regeneration are those combining slight hydrocarbon destruction with deep olefin polymerization. In this case, the role of the polymerization process can be explained by comparison with a well-known radiation technology for the purification of sewage water in the chemical industry (Ivanov 1988). The conditions necessary for regeneration of the used oil products are provided by the performance of RP at temperatures lower than that for the start of the intense reaction of RTC. The rate and the degree of the radiation-induced destruction are considerably lower than those in the case of lubricant production from heavy fuel oil, and application of the polymerization effect requires a use of heightened irradiation doses.

In this method, after cleaning from water and mechanical impurities, used oil products were subjected to the action of an accelerated electron beam or gamma irradiation having an energy of 0.25–8.00 MeV in the temperature range of 240°C–450°C at the dose rates of 1–60 kGy/s and the doses of 1–80 kGy. As a result of simultaneous proceeding of such reactions as decomposition, oxidation, polymerization, and chemical adsorption stimulated by ionizing irradiation, effective cleaning and regeneration of the used oil mixtures were observed. The commodity products (regenerated lubricants, motor fuels, and basic lubricants) were produced by fractionation of the overall product of RP.

A single-step processing of various oil-containing mixtures with high-energy electrons or gamma rays from a source of ionizing radiation was carried out according to the scheme shown in Figure 3.16.

In the case when oil products form resistant water emulsions, radio-frequency or gamma irradiation can be use for emulsion destruction and more efficient separation of water and mechanical impurities.

The technology was tested for different types of used oil products, such as

- Mixtures of diesel and heavy gas oil fractions from precipitation tanks (MFD 1);

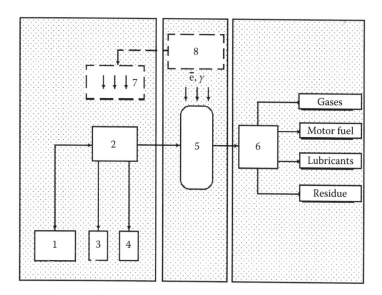

FIGURE 3.16 Layout of the device for processing used and residual oil products with electron or gamma irradiation *Note*: (1) Feed tank, (2) reactor for feed preparation (centrifuge), (3) water accumulator, (4) mechanical impurity accumulator, (5) radiation-chemical rector, (6) separator, (7) microwave irradiator (switched to centrifuge 2 if hardly separated impurities are available), (8) electron accelerator..

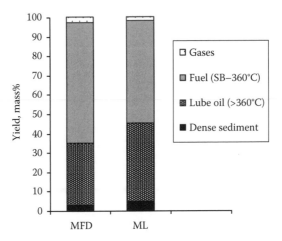

FIGURE 3.17 Material balance of used oil product radiation processing. MFD, mixture of used diesel fuel and detergents; ML, mixture of used lubricants.

- Mixtures of used diesel fuels and detergents from railway stations (MFD 2); and
- Mixtures of used motor fuels collected from cars and trucks, and used transformer oils (ML).

The material balance of the process is shown in Figure 3.17 in the example of the two types of used oil products processed according to the scheme described earlier (Figure 3.16)

Temperature and dose dependences of the regenerated product yields obtained by radiation-thermal processing of different types of used and residual products are shown in Figures 3.18 and 3.19.

Figure 3.18 shows that increase in the process temperature above 450°C is inexpedient because it does not lead to increase in the yield of target products. Besides, oil RP

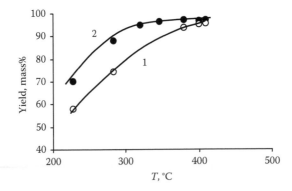

FIGURE 3.18 Temperature dependence of the overall product yield (mixture of basic lube and diesel fractions) in radiation processing of used oil products: 1, mixture of used diesel fuel and detergents (MFD); 2, mixture of used lubricants (ML).

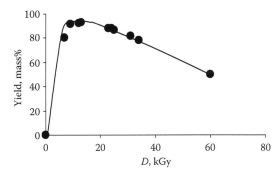

FIGURE 3.19   Dose dependence of target product yield (basic lube and diesel fractions) from mixtures of used diesel fuel and detergents (MFD) and used lubricants (ML).

at temperatures as high as 450°C is accompanied by intense reactions of hydrocarbon decomposition undesirable in the case of used oil regeneration. At the temperatures below 240°C, the yields of target products sharply fall (Figure 3.18).

Figure 3.19 shows that the range of absorbed radiation doses characteristic of a high yield of target products is about 8–30 kGy. Further increase in the absorbed dose leads to decrease in the yields of basic products. At the dose rates below 1 kGy/s, the rates of the reactions responsible for regeneration and refining of used and residual oil products become considerably lower.

In the experiments conducted, the processing rate was varied in a wide range up to the values characteristic of the industrial scale (from 1 to 200 kg/h). The temperature of RP was in the range 340°C–380°C, the dose rate was 1–2 kGy/s, and the dose varied from 6 to 10 kGy (Zaikin 2005). Yields of the basic products of RP, that is, purified motor fuels (fraction "start of boiling—360°C") and basic lubricants (the residue of processing with boiling temperature higher than 360°C), are given in Table 3.11.

The percentages of diesel and lube fractions in used oil products after RP substantially depend on the type of the processed mixture. Processing of oil mixtures with a

TABLE 3.11

Yields of Regenerated and Refined Products after Radiation Processing

Product of Radiation-Thermal Processing	Yields in mass% for Different Types of Used Oil Products as Feedstock	
	MFD	**ML**
Refined product (liquid fractions)	94–96	92–94
Coking residue	3–4	4–5
Gases and other losses	1–2	2–3

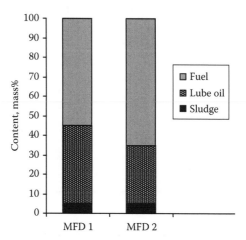

FIGURE 3.20 Fractional contents of the refined mixtures of used oil products: MFD 1, mixture of diesel fuel and detergents from precipitation tanks; MFD 2, mixture of diesel fuels and detergents from railway stations.

higher percentage of diesel fractions leads to higher yields of fractions boiling below 350°C, that is, motor fuels.

Figure 3.20 shows fractional contents of two refined mixtures of used oil products; a mixture of diesel and heavy gas oil fractions from precipitation tanks (MFD 1) and a mixture of diesel fuels and detergents from railway stations (MFD 2). RP with 2 MeV electrons was conducted at the temperature of 380°C and the dose rate of 2–4 kGy/s. The absorbed radiation dose was 6 kGy. Concentrations of the refined diesel fraction were 55 and 65 mass% in products of MFD 1 and MFD 2 processing, respectively.

Regeneration and refining mixtures of used motor lubricants (ML) required higher process temperatures (up to 410°C). Two sorts of used oil lubricants were processed at 410°C and at the same other process parameters as those used for processing diesel fuel mixtures (MFD). Figure 3.21 shows that gas yields were nearly the same at temperatures of 380°C and 410°C; however, increase in temperature resulted in a lower concentration of heavy residue in the overall product of used lubricant RP.

Characteristics of the "lubricant" fraction with boiling temperature higher than 360°C produced by RP of different types of used oil product mixtures are given in Table 3.12. The product characterized in Table 3.12 satisfies standard requirements for basic lubricants. Commodity lubricants can be produced by special additions to basic lubricants.

The necessary conditions for a large-scale industrial application of radiation technology for regeneration and refining of used oil products are the availability of the continuous feed sources (a great car fleet, the developed net of railways, and other industrial objects), and well-organized collection of used oil products. Provided these preconditions are available, application of even a not-very-powerful electron accelerator, for example, 20 kW beam power, would allow processing up to 70,000 tons of used oil products annually. If decentralized regeneration of lower volumes of used oil products (20,000 tons a year) is desirable, application of less expensive gamma-isotope sources (Co-60, Cz-137) could be expedient.

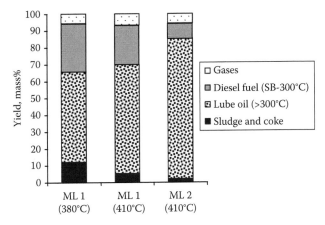

FIGURE 3.21 Fractional contents of the used lubricant mixtures after radiation processing: ML-1, mixtures of engine lubricants used in trucks; ML-2, mixtures of engine lubricants used in passenger cars.

TABLE 3.12

Characteristics of Basic Lubricants after Radiation Processing

	Types of Used Oil Products			
Characteristics	**MFD**	**After RP**	**ML**	**After RP**
Density at 20°C, g/cm³	0.8895	0.8764	0.928	0.782
Viscosity, cSt				
At 50°C	4.2	28.2	20.0	NA
At 100°C	1.8	6.4	37.1	7.6
Index of viscosity	NA	90	NA	92
Pour Point, °C	NA	−15	NA	−17
Flash Point, °C	50	206	140	209
Sulfur, wt.%	NA	0.3	NA	0.2
Acid number	NA	0.015	NA	0.02
Solid particles	0.1%	—	0.14%	—
Water	0.6%	—	>10%	—

3.3 RADIATION PROCESSING OF HIGH-PARAFFINIC OIL AND OIL WASTES AT HEIGHTENED TEMPERATURES

Efficiency of high-paraffinic oil RP in the temperature range of 370°C–410°C characteristic of RTC, to a great extent, depends on the relationship of paraffin and aromatic hydrocarbon concentrations in heavy oil fractions. In Chapter 2, the difficulties in RP of such sorts of oil caused by suppression of hydrocarbon decomposition due to intense radiation-induced polymerization were demonstrated on the example of crude paraffinic oil from Kumkol field, Kazakhstan. At higher ratios of aromatic to paraffin concentrations in asphalt-pitch–paraffin sediments (APPS), RTC becomes more efficient.

A series of experiments were conducted on RTC of crude oil from Akshabulak field, Kazakhstan. This sort of oil has a density of 0.82 g/cm^3, pour point of +10°C, and concentrations of pitches and paraffins in fraction boiling above 400°C of 14.38 and 8.44 mass%, respectively. For comparison, Kumkol crude oil had about the same density and pour point, but concentrations of paraffins and pitches in the same fraction were 16.5 and 6.67 mass%, respectively. Thus the ratio of paraffin to pitch concentrations was 1.7 in Akshabulak oil compared with 2.5 in Kumkol oil. A higher concentration of pitch-aromatic compounds in Akshabulak oil resulted in a decreased-rate radiation-induced polymerization of paraffins, considerable increase in the yield of light fractions, and a higher stability of the product of oil RP.

Akshabulak oil was irradiated with electrons having an energy of 2 Mev at the temperature of 380°C and the dose rate of 1 kGy/s. The absorbed radiation dose was 4 kGy. The yield of the gaseous products was 9.2 mass%. The fractional content of the liquid RTC product is shown in Figure 3.22. As a result of RTC, the yield of light fraction having $T_b < 195$°C increased by 9.5 mass% with the considerable decrease in the average molecular mass of the overall product.

Remarkable changes were observed in the gasoline fraction of the RTC product. Figure 3.23 shows that RP leads to a shift of the maximum in hydrocarbon distribution in the gasoline fraction from C_8 to C_6.

Increase in the paraffin component was observed in the hydrocarbon content of the gasoline fraction (Figure 3.24); increase in the iso-alkane concentration was greater than that of n-alkanes. Most considerable was increase in the concentrations of iso-alkanes C_7–C_8 (Figure 3.25). The changes observed in the composition of the gasoline fraction show that higher concentrations of heavy aromatic compounds in Akshabulak oil promoted intense alkane isomerization in the process of RTC.

As the temperature of oil RP was lowered to 360°C, the condensate yield at the same dose (4 kGy) and dose rate (2.5 kGy/s) decreased to 55 mass% while the concentration of light fractions in the condensate was still high (Figure 3.26).

Changes in the distribution of paraffin hydrocarbons in the gasoline fraction were nearly the same as those observed for RTC temperature of 380°C. Together with a small increase in the paraffin concentration, a small decrease was observed in the concentrations of arenes and cycloparaffins in the gasoline fraction. Olefin concentration did not exceed 4 mass% decreasing with the dose increase. Such olefin behavior testifies to the intense polymerization in the heavy residue of the product. However, light fractions separated from the heavy residue demonstrated sufficiently high stability.

Dose dependence of the yield of the liquid RTC product of Akshabulak oil at the process temperature of 360°C is shown in Figure 3.27. The optimal dose of electron irradiation at this temperature was about 10 kGy.

An important ecological problem of the oil industry in oil-exploring countries is the environment pollution by oil extraction wastes and residua of oil refining, the scales of contamination being tremendous. In particular, a number of pills of the

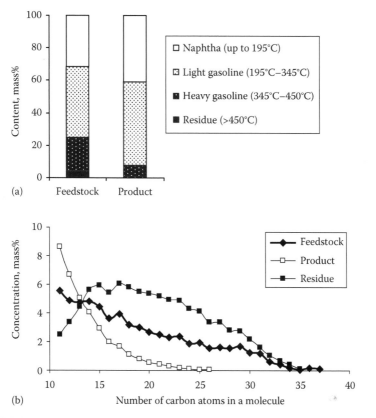

FIGURE 3.22 (a) Fractional contents of the liquid product obtained by RTC of Akshabulak crude oil by boiling temperature and (b) by number of carbon atoms in a molecule.

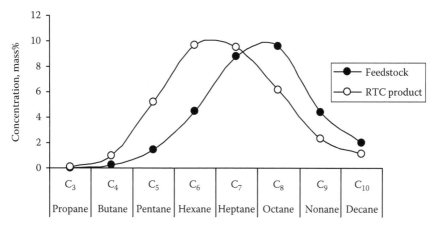

FIGURE 3.23 *n*-Alkane distribution in RTC gasoline fraction of Akhabulak crude oil.

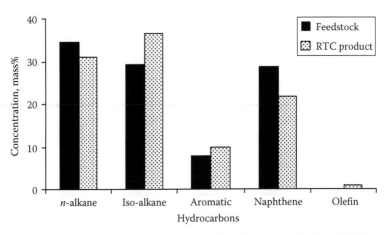

FIGURE 3.24 Hydrocarbon contents of the gasoline fraction in the liquid RTC product obtained by RTC of Akshabulak crude oil.

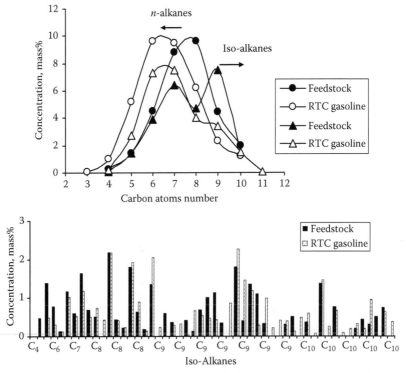

FIGURE 3.25 Distribution of *n*- and iso-alkanes in the gasoline fraction of the liquid RTC product of Akshabulak crude oil (*T* = 380°C).

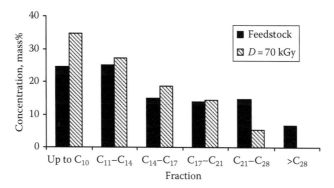

FIGURE 3.26   Fractional contents of the liquid product after RTC at 360°C.

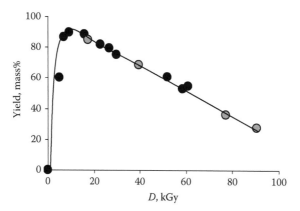

FIGURE 3.27   Dependence of the yield of light fraction condensate obtained by RTC of Akshabulak crude oil on the dose of electron irradiation at 360°C (black dots $P=3$ kGy/s; grey dots $P=2.5$ kGy/s).

so-called lake oil formed as a result of failures and damages in oil extraction and transportation are spread over the world. Since efficient methods for oil lake processing are not available, these ponds are annually overfilled with additional oil wastes.

Additional sources of environment pollution are APPS. Annually up to 150,000–200,000 tons of APPS are accumulated all over the world without any utilization. Partially APPS are buried in swamps, mucks, holes, barns, and collectors and partially burned. Anyway it causes contamination by oil wastes and contravenes environmental equilibrium (Zaikin et al. 1999a,b)

The technology used today for paraffin sediment processing includes the stage of feed preparation (removal of water, light oil fractions, and solid impurities), refining by means of traditional oil-chemistry methods using clays and sulfuric acid, propane desulfurization, crystallization in acetone–benzene, etc. In this technology, the product of APPS processing is ceresin with a boiling temperature up to 80°C.

A selection of the technology for APPS processing is defined by APPS composition and properties that are very different for different oil fields. Generally, the existing methods for APPS processing are characterized by the low degree of feed conversion and low economic efficiency; they do not solve the environmental problems associated with APPS accumulation.

A promising approach to oil waste processing is application of radiation technologies (Zaikin et al. 2004b) characterized by flexibility with respect to the type of oil wastes processed. To make radiation technology advantageous, specific features of radiation-induced reactions in this type of feedstock are to be taken into account. APPS and lake oils contain extremely high concentrations of pitches, asphaltenes, and heavy paraffins; to a great extent, they are contaminated with different impurities.

High concentrations of water and mechanical impurities in the original feedstock should be taken into account in the technologies of oil processing. In particular, redundant water in the samples of APPS and lake oil upsets the normal behavior of radiation-thermal conversion and causes entrainment of unreacted products by water steam. Therefore, one of the requirements to the technology of oil waste RP is their predewatering and cleaning from mechanical impurities. However, it does not exclude using water as an additional source of hydrogen and its intentional introduction into heavy oil feedstock under controlled conditions.

A specific feature of such types of feedstock as APPS and lake oil is their considerable trend to coking during radiation-thermal processing. It should be also taken into account in the facility design, in the selection of irradiation conditions, and in the rate of the feedstock flow through the pipelines of the given cross section.

The samples of lake oil and APPS were irradiated in flow and stationary conditions by 2–4 MeV electrons.

As a result of radiation-thermal processing of lake oil, the following products can be obtained: 40 mass% liquid fractions with boiling temperature up to 360°C; 22% lube fraction with boiling temperature higher than 360°C and solidification temperature $T > 23°C$; 25.8% pitchy coking residue; and 12.2% gases. Physicochemical properties and elemental contents of the original lake oil and RTC products are given in Tables 3.13 and 3.14.

The maximum yield of light fractions after lake oil irradiation was 60%–65% for irradiation temperatures of 360°C–390°C, the dose rate of 1 kGy/s, and doses of 3–5 kGy (1 Gy = 1 J/s). The fractional contents of products separated from RTC products ($D = 3$ kGy; $T = 380°C$) are shown in Table 3.15.

The EPR data have shown that concentration of paramagnetic centers decreased by almost two orders of magnitude in the course of RP of lake oil. Concentration of free radicals in the heavy residue ($T_b > 450°C$) was about 1.5 times higher than that in a lighter fraction of the liquid product. It indicated considerable decomposition of the original feedstock under irradiation. Concentration of the bivalent iron in the heavy fraction of the product of lake oil RP decreased almost by two orders of magnitude and was not detected in the lighter fraction. Obviously, iron partially concentrated in coke and partially transformed to another valence state.

The process kinetics is complicated with the concomitant polymerization and other intermediate reactions. However, it is predominantly defined by the kinetics of the basic chain RTC reactions as described in Chapter 1.

TABLE 3.13

Physical and Chemical Properties of the Original Lake Oil and Products of Its Radiation-Thermal Processing

		Viscosity, cSt		Flash	Solidification
	Density at	At	At	Temperature	Temperature
Sample	20°C, g/cm³	50°C	95°C	T, °C	T, °C
Feedstock	0.85	27.0	—	27	±31
Product 1					
Oil fraction	0.83	3.68	1.89	70	—
(SB-360°C)					
Product 2					
Paraffinic oil fraction	0.86	6.86	3.0	118	—
(>360°C)					

TABLE 3.14

Elemental Contents of Lake Oil and Products of Its Radiation-Thermal Processing

	Concentration, Mass%				
Sample	C	H	N+O	S	C/H
Feedstock	85.90	13.18	0.46	0.45	6.51
Product 1					
Oil fraction (SB-360°C)	85.07	14.02	0.35	0.56	6.07
Product 2					
Paraffinic oil fraction (>360°C)	85.33	13.81	0.53	0.33	6.17
Pitchy coking residue	86.43	12.19	0.68	0.70	7.09

TABLE 3.15

Fractional Contents of Product 1

Fractional Contents of Product		Fraction Yield, Mass%
Up to C_{10}	(SB-180 °C)	7.60
C_{10}–C_{13}	(180°C–240 °C)	34.50
C_{13}–C_{18}	(240°C–320°C)	40.11
C_{18}–C_{22}	(320°C–360°C)	17.80

Note: $D = 3$ kGy, $P = 5$ kGy/s, $T = 380$°C.

The most noticeable characteristic feature of APPS RTC is the availability of the intensive competing processes of polymerization of hydrocarbon destruction products. Polymerization limited the maximum yield of light fractions that did not exceed 40%. Unlike the final products of lake oil processing, the products of APPS destruction were unstable and tended to polymerize after RP. As a result of interaction with the reactive polymerizing residue, liquid fractions could be entirely absorbed during exposure at room temperature and the heavy residue got dense and solid. Therefore, stable light fractions were separated immediately after RP.

The residue of APPS radiation processing can be reprocessed for deeper conversion in the more severe radiation conditions or used as a basic lubricant after its chemical stabilization.

Taking into account a strong trend of high-paraffinic oils to polymerization, Al-Sheikhly and Silverman interpreted radiation processes in this type of oils in terms of radiation chemistry of polymers (Al-Sheikhly and Silverman 2008). It was noted that polymerization increased at higher dose rates and temperatures. In particular, high dose rates may suppress radical reactions in polymers because of the more intense radical recombination in spurs. According to this model, it was proposed to use as low values of the main irradiation parameters as possible for a reasonable rate of the process (temperature in the range of 50°C–250°C, dose rates of 0.5–4 kGy/s, and absorbed doses below 3 kGy). Introduction of oxygen (air) and/or ozone into the feedstock before irradiation was proposed for suppressing polymerization.

The temperature range of 50°C–250°C at relatively low dose rates is below a threshold for RTC reactions, and low-rate radiolysis at the doses below 3 kGy cannot provide considerable conversion at high production rates. However, introduction of reactive oxidizers, such as ozone or ionized air, increases the rate of hydrocarbon decomposition and lowers cracking temperature (see Chapter 4).

3.4 RADIATION METHODS FOR HIGH-SULFURIC OIL PROCESSING

The problem of crude oil and oil product desulfurization arises from the trend in supplying heavier and more sulfurous raw material for refining combined with hardening environmental requirements. Today it is solved by using a complex technique that assumes application of catalysts accompanied by hydrogen expense, as well as processing at high temperatures and pressures.

The existing technologies for oil desulfurization can be principally brought to two basic approaches. In the first approach, sulfuric compounds are reduced to H_2S, which is then removed from the processed product and transformed to elemental sulfur. Both of these stages are technologically complicated, comprise multistep catalyst reactions, and cause environmental problems. In this approach, simple technological processes cannot provide a high desulfurization level.

Another approach presumes oxidation of the sulfuric species in oil with subsequent extraction of the oxidized high-molecular sulfurous compounds. The second stage of oxidized species extraction is well advanced and does not come across considerable technological difficulties. The higher the oxidation degree of the sulfurous compounds, the more perfect the appropriate technology.

Both of the two approaches or their combination can be used in radiation methods for oil desulfurization depending on sulfur concentration in the original feedstock.

3.4.1 EARLY STUDIES OF PETROLEUM RADIATION DESULFURIZATION

A patent on oil desulfurization using reactor irradiation was registered by Esso Corporation in the end of the 1950s (Esso 1960). It was found that irradiation of sulfur-containing petroleum oils with high-energy ionizing radiation from atomic piles converts sulfur in the oil from a relatively stable chemical form to a relatively unstable or loosely bound chemical form. Crude oil and oil fractions were irradiated in the presence of metal-oxide catalysts. After irradiation, sulfur species were removed by conventional methods.

Oil samples used in the experiments contained 0.18–1.09 mass% sulfur. Irradiation doses varied in the range from 10^7 to $6 \cdot 10^9$ R of gamma rays, from $7 \cdot 10^{14}$ to 10^{18} slow neutrons/cm^2 and from $7 \cdot 10^{13}$ to 10^{18} fast neutrons/cm^2. Irradiation in the presence of the silica–alumina catalyst was accompanied by partial sulfur transformation to H_2S. As a result of the 10 day oil RP, concentration of total sulfur in oil was reduced by 60%–75%.

Chemical composition of sulfur compounds before and after processing was not identified. In these experiments, a high degree of sulfur conversion required long time RP. An additional difficulty of the method is associated with the residual radioactivity of the processed oil.

In the work (Scapin et al. 2010), gamma radiation was used for sulfur removal from used automotive lubricating oils. The oil samples were irradiated with the doses from 10 to 500 kGy in a Co-60 irradiator. To enhance oil radiolysis, 50–70 vol.% of Milli-Q water and 30 vol.% of hydrogen peroxide were used. The results showed 68% sulfur removal at the absorbed dose of 500 kGy with an addition of 70 vol.% of Milli-Q water.

Most part of the radiation studies on oil desulfurization were based on the effects of radiation-enhanced oxidation of sulfuric compounds.

In the beginning of the 1990s, a research group from Kazakhstan (Nadirov et al. 1991) developed a method for extraction and oxidation of organic sulfurous compounds from hydrocarbon feedstock. A series of experiments on the extraction of the sulfurous concentrate and its subsequent radiation-enhanced oxidation was directed to development of a technology for production of sulfoxides, sulfones, and sulfonic acids from petroleum products.

In these studies, the organic sulfurous concentrate was separated by the method of liquid extraction using organic solvents, such as acetylacetone, good for the extraction of a wide range of sulfurous organic compounds. Ultrasound of the medium frequency (2.6 MHz, 3 W) was used for the more pronounced phase immiscibility. The extracted sulfurous concentrate was placed in the reactor connected with a vessel for gas accumulation and irradiated by gamma rays with the dose in the range of 70–220 kGy at the dose rate of 1.30 Gy/s. After irradiation, the products of oil oxidation were chromatographically separated and studied by the method of IR spectroscopy.

The absorption bands characteristic of sulfoxides were observed up to the doses of 150 kGy. At the higher doses, the absorption bands associated with sulfones,

TABLE 3.16

Dose Dependence of the Degree of Sulfuric Concentrate Oxidation

Sulfur species	Oxidizer	Oxidation conditions	Yield, Mass%		
			Sulfoxides	Sulfones	Sulfo-Acids
Organic	$H_2O_2 + H_2O_4 + HCl$ (0.1 mass%)	30 min, 50°C	35	n/d	n/d
Sulfuric		45 min, 100°C	50	n/d	n/d
Concentrate		45 min, 150°C	69	5	n/d

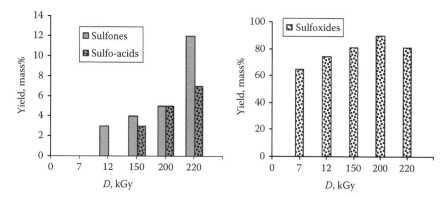

FIGURE 3.28   Dose dependence of sulfone and sulfoxide yields under gamma irradiation.

dibenzo-, and naphthalene benzothiophenes (BTs) appeared in the IR spectra. At the dose of 200 kGy, sulfoxide yields decreased, and deeper oxidation of sulfurous compounds to sulfones and sulfonic acids was observed. Oxidation was carried out under continuous aeration by atmospheric air. Radiation-enhanced oxidation is characterized by a higher yield of the target product compared with that determined in the case of oxidation with hydrogen peroxide without irradiation (Table 3.16 and Figure 3.28).

Radiation pretreatment of the original sample with the subsequent product fractionation resulted in sulfur changes observed in all fractions (Figure 3.29).

As a result of preliminary RP, sulfur contents in the light fractions decreased while sulfur concentration in the residue became higher. Low dose rates of gamma irradiation ($P = 1.10$ Gy/s) required long time processing accompanied by considerable losses of hydrocarbon components in the form of gaseous products that prevented the more complete sulfur extraction.

Application of medium-frequency ultrasound (0.2–2 MHz) was recommended for more pronounced phase immiscibility at the stage of separation of the sulfurous extract from high-viscosity oils and bitumen. Due to ultrasonic processing, the extracted sulfurous concentrate contained less amount of different hydrocarbons and other hetero atoms.

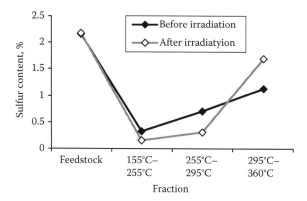

FIGURE 3.29 Sulfur concentration in the separate fractions of high-viscous crude oil before and after irradiation ($P = 1.10$ Gy/s, $D = 200$ kGy). Sulfur concentration was determined by x-ray radiometric analysis.

Even the first experiments in RTC of heavy oil residue (Zhuravlev et al. 1991) have shown a higher desulfurization of light fractions compared with the conventional TC. Most considerable changes were observed in the concentration of the mercaptan sulfur. In the light products of RTC, it was about two times lower compared with the mercaptan contents in the same fractions after thermal processing.

The concentrations of mercaptan sulfur in the light fractions obtained in RTC and TC processes were measured by the method of potentiometric titration (Figure 3.30). A content of mercaptan sulfur in the RTC products was two times lower than that in the same TC fractions.

The essential redistribution of sulfur between the fractions of the bitumen during electron irradiation was observed in the study by Bludenko et al. in 2007 (see

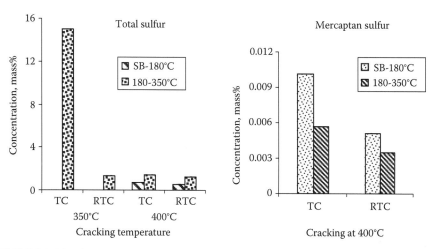

FIGURE 3.30 Concentration of mercaptan sulfur in light TC and RTC fractions. Cracking duration 8.25 h, $T = 400°C$, $D = 1.5 \cdot 10^5$ Gy.

Section 2.4). On the one hand, H_2S and free sulfur were formed. In this experiment, hydrogen sulfide was removed from bitumen by the flow of a bubbling gas. The concentration of sulfur in the end product decreased to 0.5 wt%. On the other hand, the sulfur content of high-molecular-weight compounds increased. The total yield of the sulfur redistribution was about 102 molecules/100 eV indicating a chain mechanism of sulfur transformations. These results were explained in the work (Bludenko et al. 2007) in the frames of the following model.

Thiols and disulfides can often protect other solutes against radiation damage (Woods and Pikaev 1994). In particular, a thiol SH group can regenerate hydrocarbon molecule RH from radical ·R:

$$\cdot R + RSH \rightarrow RH + RS\cdot \tag{3.2}$$

The recombination of RS· radicals leads to the formation of high-molecular-weight products

$$2RS\cdot \rightarrow RSSR \tag{3.3}$$

RS· radicals being added to alkenes can give sulfides by a chain mechanism (the yield is 102–105 molecules/100 eV at room temperature) (Woods and Pikaev 1994):

$$RS\cdot + RHC = CHR' \rightarrow RS - CHR - C\cdot HR' \tag{3.4}$$

$$RS - CHR - C\cdot HR' + RSH \rightarrow RS - CHR - CH_2R' + RS\cdot \tag{3.5}$$

The process has been suggested as a way to remove thiols from petroleum using irradiation to convert the thiols to less volatile sulfides (Nevitt et al. 1959). In turn, electron addition to sulfide (Woods and Pikaev 1994) provides the formation of a heavier sulfur compound:

$$e^- + R_2S \rightarrow R_2S^- \tag{3.6}$$

$$R_2S^- + R_2S \rightarrow RSSR^- + R_2 \tag{3.7}$$

The formation of hydrogen sulfide and free sulfur can be initiated by the H atom and electron:

$$e^- + HSR \rightarrow HS^- + \cdot R \tag{3.8}$$

$$HS^- + H^+ \rightarrow H_2S \tag{3.9}$$

$$\cdot H + HSR \rightarrow H_2 + \cdot SR \tag{3.10}$$

$$\cdot H + HSR \rightarrow H_2S + \cdot R \tag{3.11}$$

Thiols, as well as dissolved sulfur and intermediates from sulfides and thiophenes, participated in reactions (3.8) through (3.11). The accumulated hydrogen sulfide has a low ionization potential and acts as an electron donor in the charge-exchange reactions (Woods and Pikaev 1994):

$$RH^+ + H_2S \rightarrow RH + H_2S^+ \rightarrow RH + H^+ + \cdot SH \tag{3.12}$$

These reactions reduce the total yield of bitumen decomposition. Hydrogen sulfide takes part in the formation of free sulfur, for example,

$$H + H_2S \rightarrow H_2 + HS \cdot \tag{3.13}$$

$$H + HS^{\bullet} \rightarrow H_2 + S \tag{3.14}$$

$$2HS^{\bullet} \rightarrow H_2S + S \tag{3.15}$$

Elemental sulfur is an excellent radical scavenger, and sulfur can be incorporated in organic compounds by irradiation of a solution or dispersion of sulfur in the organic material. Thus, the irradiation of alkanes or alkyl derivatives containing dissolved sulfur yields thiols (RSH), sulfides (R–S–R), and disulfides (R–S–S–R). In these reactions, sulfur acts as a scavenger for radicals produced during radiolysis or chain propagation (Woods and Pikaev 1994).

Further studies (Nadirov et al. 1997b,c, Lunin et al. 1999, Zaikin and Zaikina 2004a,d) have shown a special role of oxygen reactions in sulfur transformations under ionizing irradiation. It was shown that mechanisms of sulfurous-compound transformations caused by saturation of hydrocarbon feedstock with oxygen, ozone, and ozone-containing ionized air and subsequent RTC have much in common with the reactions observed in the conventional ozonolysis of oil feedstock.

In the technology of ozone-initiated cracking developed in the works by Kamyanov and coworkers (Kamyanov et al. 1994, 1997, Kamyanov 2005), at the first stage, oil feedstock is saturated with ozone that leads to deep oxidation of sulfurous compounds. At the second stage, the ozonized feedstock is subjected to TC in the temperature range of 350°C–450°C. Sulfones and sulfoxides formed at the first stage of processing play the role of chain carriers in the cracking reaction.

Low-temperature of the ozonized oil partially solves the problem of oil distillate cleaning from sulfurous compounds. However, rather high concentrations of olefins forming in the process of cracking can cause some decrease in the product stability, especially in the case of their long time storage. At the same time, molecules of product ozonization and oxidation are much more polar than those of the original components in crude oils and can be much easier removed from the straight-run distillates by means of traditional extraction or adsorption methods.

Potentialities of the ozone-initiated desulfurization of diesel fuels were demonstrated on the example of the straight-run distillate 200°C–350°C produced from the high-sulfurous oil. The distillate contained 3.34 mass% sulfur, including 1.80% sulfides, 0.32% sulfoxides, and 1.22% thiophenes. In these experiments, the distillate

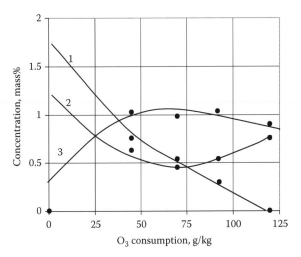

FIGURE 3.31 Effect of the degree of distillate ozonization on the functional composition of the raffinates obtained by the one-stage extraction at the ratio hexane/DMFA/feedstock of 5:2:1: 1—sulfoxides; 2—untitrable sulfur; 3—sulfoxides.

was ozonized at the specific discharge of ozone in the range from 47 to 115 g/kg. The products obtained were dissolved in hexane at the volume ratio of 1:5. The solution was mixed with dimethylformamide (DMFA) at 20°C and the formed immiscible phases were separated. After hexane distillation, the refined product (raffinate) and ozonization product (extract) were obtained by water washing of the DMFA phase.

The changes in the group composition of sulfurous compounds in the case of reaction were controlled by the potentiometric titration analysis of the separated products (Figure 3.31).

The sulfide contents in the raffinates rapidly decreased down to their complete disappearance as the ozonization degree increased. It was accompanied by increase in the share of sulfoxides in total sulfur at the ozone consumption up to 47 g/kg (about 1 mol/mol). At the higher ozone consumption, sulfoxide sulfur stabilized at the level of 1 mass% and decreased when ozone consumption was higher than 97 g/kg (about 2 mol/mol). The same behavior in the changes of sulfoxide sulfur concentration was observed in the extracts, but the absolute values of sulfur concentration were much higher (4.1–4.3 mass% at the ozone expense up to 100 g/kg).

The content of the untitrable sulfur in the raffinates and extracts decreased until ozone consumption reached about 70 g/kg (1.5 mol/mol) and increased at the higher ozone consumption. At the start of the process, the thiophene components of the distillate bound ozone and transformed to oxygen-containing compounds well-soluble in DMFA. These sulfurous compounds entered the extract composition; therefore, their concentration in the raffinates decreased. At the late stages of the process, sulfoxides partially oxidized to sulfones, which were also untitrable.

As a result of the one-stage distillate extraction of the distillate ozonized with the ozone consumption of 96 g/kg (about 2 mol/mol), 54% sulfur was removed from the fuel. The two-stage extraction provided removal of 85% sulfur, and 90% of the initial sulfur was removed after the four-stage extraction. Sulfur concentration in the extracts increased to the maximum of 9.9 mass% that indicated to the presence of up to 70% sulfoxides and sulfones. Sulfoxides predominated in the extracts at the ozone consumption lower than 100 g/kg. Sulfones were the prevailing sulfur species at the higher ozone consumptions.

Thus, ozonolysis of the straight-run diesel fractions provides deep fuel desulfurization without application of catalysts, hydrogen-containing gases, or high temperatures. It also provides concomitant production of sulfoxide and/or sulfone concentrates used in industry as efficient agents for extraction of metals, flotators, and other valuable materials. However, the substantial factors that hinder the full-scale industrial realization of oil ozonolysis are the high consumption of expensive ozone and limited productivity of the industrially produced ozone generators (primarily, for water purification).

These disadvantages of ozone desulfurization can be completely eliminated in radiation technology with the application of ionized air as an oxidizer. Together with ozone, ionized air contains such strong oxidizers as atomic oxygen and excited oxygen molecules. As compared with thermolysis of the ozonized feedstock, RP results in deeper decomposition of sulfurous compounds and can be performed at a lower temperature.

Strict ecological regulations impose severe restrictions on concentrations of sulfur (15–50 ppm) and polycyclic aromatic hydrocarbons (6–11 mass%) in diesel fuel. In conditions of hydro-refining, thiophene sulfur derivatives can be removed with the greatest difficulties while aromatic hydrocarbons practically cannot be converted because of the thermodynamic limitations. Effective simultaneous removal of thiophene sulfur and polycyclic aromatic hydrocarbons requires increase in pressure in the hydro-refining process up to 8 MPa and considerable reduction of the volume rate of the feedstock supply. It brings down the production rate of the hydro-refining facilities, increases capital and operational costs, and lifts the manufacture costs of the commodity fuels.

One of the possible ways of intensification of diesel fuel hydro-refining is preliminary feedstock ozonization. In the end of the 1990s, experiments on the ozonolysis of the straight-run diesel fraction, a feedstock for diesel fuel hydro-refining, were conducted in Moscow State University (Lunin et al. 1999). This fraction was ozonized at room temperature with the specific ozone expense of 6 g/kg; then it was subjected to hydro-refining at the lab facility for industrial hydro-refining simulation.

Sulfur concentration in the original feedstock was 0.902 mass%. After hydro-refining of the straight-run fraction, sulfur concentration dropped down to 0.099 mass%. The subsequent hydro-refining of the ozonized fraction reduced sulfur concentration to 0.032 mass%, that is, a three times lower value.

The same research group attempted combination of oil feedstock ozonolysis and irradiation (Lunin et al. 1999). A series of experiments was conducted on irradiation of commodity crude oil supplied for processing to the Moscow refinery. Crude oils were irradiated with the beam of accelerated electrons at the temperatures close to room temperature. Electron energy was 1.8 MeV; the beam current density was 2–3 µA/cm². Irradiation modes used in these studies could not provide radiation-induced propagation of the sufficiently intense chain reaction of hydrocarbon cracking: experiments resulted in the increased viscosity, density, and temperature of the beginning of oil boiling.

Since dissolved oxygen from atmospheric air and water is always present in crude oils, it was suggested that oxidation processes should take place in radiolysis conditions. This suggestion was confirmed by the data of IR spectroscopy for the original fraction and the product of its RP: increase in the intensity of the absorption band of 1700 cm^{-1} associated with the carbonyl groups was observed for the fraction SB-180°C of the irradiated oil. The changes in oil composition after irradiation were explained by the two competing processes proceeding in the oil system: decomposition of the components of the dispersion medium (in particular, cracking of high-molecular alkanes and dealkylation of alkane chains in hydrocarbons of hybrid structure) and densification due to radical recombination.

Although RP of crude oil did not improve its quality, the results of ozonized oil irradiation have shown that decomposition of ozonides under the electron beam has obvious advantages over thermolysis; a radiation process proceeds at room temperature and the yield of light fractions rises by 10% (Lunin et al. 1999) (Figure 3.32).

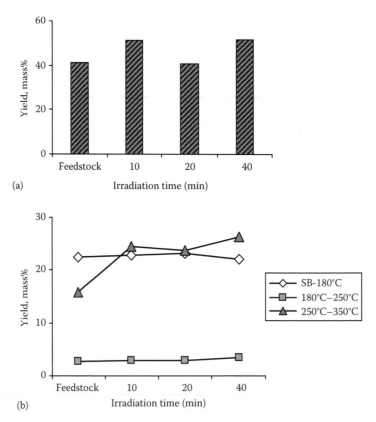

FIGURE 3.32 Effect of irradiation time on the yield of light fractions from the ozonized oil, mass%: (a) total SB-1800°C, (b) narrow fractions SB-180°C, 180°C–250°C, and 250°C–350°C. Specific expense of ozone 20.5 g/L kg original oil.

IR spectra of the separated fractions have shown increase in the intensity of the absorption bands associated with carbonyl, aromatic disulfides, and thiosulfonates. Subsequent lixiviation and water washing of the fuel fractions resulted in decrease in the total sulfur concentration by 20%–70%. Based on these experiments, it was concluded that ozonization of oil feedstock with subsequent irradiation could be used for the deep transformation of sulfurous compounds. Taking into account that ozone attacks metal-binding porphyrin complexes, such technology could be used in parallel for oil demetallization.

More comprehensive studies of oil desulfurization were conducted in the works (Zaikin et al. 1998, Zaikina et al. 2002a,c, 2004, Zaikin 2005). The methods developed in these works were based on radiation-induced oxidative desulfurization.

It was shown that RP can be efficiently applied for demercaptanization and desulfurization of crude oil and petroleum products (Zaikina et al. 2002a,c). The technology is especially simple when initial sulfur concentration does not exceed 2.0 mass%. The proposed method of desulfurization includes two stages. The first stage is RP, and the second one is the standard procedure for extraction of deeply oxidized sulfuric compounds. At the first stage, no desulfurization of the overall product is observed. It results in the strong oxidation of mercaptans and other light sulfuric species to sulfones, sulfur oxides, and acids that do away with their chemical aggressiveness and release their removal. Besides, it causes sulfur redistribution in the overall product leading to the partial desulfurization of its light fractions.

Contact of the product of radiation-thermal processing with reactive ozone-containing air in the radiation-chemical reactor allows combination of RTC with intense oxidation of sulfur compounds. The process is accompanied by sulfur transfer to high-molecular compounds, such as sulfoxides and sulfones. The high level of sulfur oxidation is due to the double activation of oxidation processes both by radiation activation of the feedstock and by activation of the atmospheric air.

One of the channels for mercaptan oxidation and transformation can be illustrated by the following scheme:

$$R \cdot + R_1SH \rightarrow R_1S \cdot + RH$$

$$R_1S \cdot + R_1S \cdot (R_2 \cdot) \rightarrow R_1SSR_1 \ (R_1SR_2)$$

$$R_1SR_2 + O_3 \rightarrow R_1R_2S^+ \ {}^-O\text{-}O\text{-}O^- \longrightarrow R_1\overset{\displaystyle O}{\overset{\|}{S}}R_2 + O_2$$

Sulfoxide

$$R_1\overset{\displaystyle O}{\overset{\|}{S}}R_2 + O_3 \rightarrow R_1R_2\overset{\displaystyle O}{\overset{\|}{S}}{}^+ {}^-O\text{-}O\text{-}O^- \rightarrow R_1\underset{\displaystyle O}{\overset{\displaystyle O}{\underset{\|}{\overset{\|}{S}}}}R_2 + O_2$$

Sulfone

$$R_1S \cdot + H \rightarrow R_1SH$$

Accumulation of hydrogen hampers mercaptan conversion, but the controlled access of ozone-containing air into the reactor allows regulation of the mercaptan conversion.

The application of ozone as an oxidizer for oil mixtures was proposed earlier (Kamyanov et al. 1991, 1994, 1997). However, powerful ozonizers are expensive and do not have sufficient production rate for a large-scale oil processing. Unlike that, the excited air mixture in the irradiation technology is an inexpensive by-product of accelerator operation, comprising monatomic oxygen, excited oxygen molecules, and various oxygen-containing complexes, together with ozone. Many components of this mixture are strong oxidizers, and some of them are stronger than ozone. One of the merits of this technology is the possibility of achieving any desired degree of sulfur oxidation at minimum processing time and maximum process simplicity.

The processing of heavy fuel oil in two irradiation modes for producing light oil fractions from the feedstock and simultaneous sulfur transformation into harmless and easily extracted forms was described in the paper (Zaikin et al. 1998). Feedstock was irradiated by 2 MeV electrons in the temperature range 300°C–400°C at the different values of other operational parameters (dose rate, P; dose, D). Mode 1 ($P = 6$ kGy/s, $D = 30$ kGy) used "severe" irradiation conditions and resulted in high yields of motor fuels. Mode 2 ($P = 2$ kGy/s, $D = 70$ kGy) was "milder." It caused lesser changes in hydrocarbon contents but was more favorable for conversion of sulfur compounds. Eighty percent mercaptan conversion was reached in this "milder" mode and more than 90% of the total sulfur concentrated in the heavy liquid fraction with boiling temperature higher than 350°C.

Figure 3.33 shows dependence of the degree of mercaptan conversion on irradiation dose for a crude oil from Karazhanbas field characterized by high viscosity and high concentrations of sulfur compounds (about 2 mass%) in the same two irradiation modes. Similar relationships are presented (Figure 3.34) for two different kinds of feedstock (Karazhanbas oil and heavy extraction wastes of sulfuric Mangistau oil, so-called lake oil), irradiated in mode 2. Experimental data show that both in

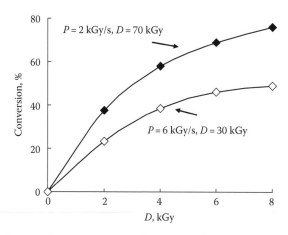

FIGURE 3.33 Degree of mercaptan conversion in high-viscous crude oil versus irradiation dose for different irradiation conditions.

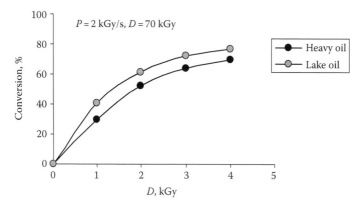

FIGURE 3.34 Degree of mercaptan conversion in mode 2 versus irradiation dose for different types of feedstock.

mode 1 and in mode 2, considerably lower concentrations of mercaptans, disulfides, and sulfides were observed in the light products of radiation processing. In particular, mercaptan concentration decreases more than fivefold, coming down to the low level of 0.02 mass%.

Standard chemical analysis and IR spectroscopy data have shown availability of disulfides, sulfones, sulfoxides, and sulfonic acids in the products of oil RP in mode 2 indicating a higher degree of mercaptan oxidation in these irradiation conditions. RP in mode 1 leads to more considerable destruction of hydrocarbon molecules and appearance of hydrogen in amounts sufficient for the reverse reaction of sulfide formation from disulfides. For this reason, lower disulfide yields were observed in mode 1.

For comparison, "Kazan" technology for oil demercaptanization applied on the Tengiz oil field by Chevron Company (Ormiston et al. 1997) allowed obtaining similar results for light mercaptans only. A decrease in the mercaptan concentrations down to 0.02 wt% was reached due to application of a very complicated catalytic process in an alkaline medium at a heightened pressure of the pumped air (12 atm), and mercaptans transformed only to disulfides after processing, that is, a lower degree of sulfuric compound oxidation could be achieved. Besides, application of this technology for oil demercaptanization involves an environmental problem of waste alkaline solution removal. RP allows obtaining more significant results with much lower energetic and economic expenses.

Deep conversion (cracking) of the oil feedstock in the described modes was accompanied by deep oxidation and partial desulfurization of light fractions as a result of total sulfur redistribution in the overall processed product. So, the sulfur content in the gasoline fraction in fuel oil decreased by 40 times in comparison with its concentration in the original feedstock and even without additional extraction of sulfuric compounds was 430 ppm (Nadirov et al. 1994a). After RTC of highly sulfuric oil from Karazhanbas field (sulfur content in the original oil was 1.7%), sulfur concentration in gasoline produced in a similar manner did not exceed 440 ppm.

If a deep oil destruction is undesirable and the single purpose of its processing is desulfurization, appearance of by-products (gases, cooking rest) could be considered as product yield losses needed to be reduced to a minimum. In this case, other conditions of RP can be applied for intense oxidizing of sulfuric compounds. When the irradiation dose rate or the processing temperature become lower than certain threshold values and the rate of development of the cracking chain reaction becomes comparable with the recombination rate of hydrocarbon radicals generated by radiation, the intensity of hydrocarbon destruction drops down sharply while the intensity of radiation-induced oxidizing remains high. These conditions are optimal for oil desulfurization with the minimum product yield losses (1%–5%).

Distributions of sulfoxides and sulfones forming as a result of oxidation process were similar both in the "severe" and in the "mild" irradiation modes. According to the data of chemical analysis and IR spectroscopy, sulfur did not pass to the gaseous fraction (H_2S was not observed in gases). The sulfur content did not change in coke too or was only slightly higher than that in the original feedstock. So, all the residual sulfur could be found in the liquid products of RP. The higher the boiling temperature of the fraction obtained by RTC, the greater amount of residual sulfur it contained, that is, sulfur was concentrated after processing mainly in heavy liquid fractions.

In the method for cleaning hydrocarbon mixtures from sulfur compounds proposed by Zaikina et al. (1999a), hydrocarbon mixtures were processed by electron beam or gamma irradiation having energy in the range of 0.5–8.0 MeV at temperatures in the range of 0.5–8.0 MeV and at the dose rates of 0.1–10.0 kGy/s with simultaneous radiation stimulation of oxidation reactions. Sulfur compounds, mainly in the forms of high-molecular oxidized compounds (sulfones, sulfoxides, sulfonic acids), were extracted from a heavy residue by conventional methods using such extracting agents as water solutions of sulfuric acid or acetone, ethylene glycol, or monoethyl alcohol.

In one of the tests, crude oil having a sulfur content of 1.6 mass% was irradiated with 2 MeV electrons at the dose rate and doses indicated in Table 3.17. The irradiation temperature was 380°C for run 1 and 280°C for run 2. In run 1, oil was pumped through the reactor at the rate of 10 cm³/s with simultaneous supply of the ionized air at the rate of 3 cm³/s.

Run 1 corresponded to the mode of intense destructive processing of crude oil for the maximal production of motor fuels. As a result of RP, the oxidized sulfur

---

TABLE 3.17

Sulfur Transformation Caused by Crude Oil Radiation Processing

	Irradiation Parameters		Content, %			
Run No.	Dose Rate, kGy/s	Absorbed Dose, kGy	Total Sulfur	Mercaptans	Disulfides	Sulfides
Feedstock			1.6	0.0048		0.1194
1	4	20	1	0.001	0.0136	0.0561
2	1	90	0.9	0.002	0.0068	0.0557

TABLE 3.18

Sulfur Transformation Caused by Fuel Oil Radiation Processing

Run No.		Irradiation Parameters		Content, %			
		Dose Rate, kGy/s	Absorbed Dose, kGy	Total Sulfur	Mercaptans	Disulfides	Sulfides
1	Feedstock			0.66	0.002	0.002	0.1074
	Mode 1	5	40	0.45	0.0002	0.0032	0.0537
2	Feedstock			0.67	0.0037	0.0001	0.0869
	Mode 2	10	10	0.43	0.0019	0.0081	0.255
3	Feedstock			0.96	0.01	0.2	0.6
	Mode 3	1	90	0.8	0.002	0.003	0.04

compounds accumulated in the heavy residue. In run 2, RP caused transformation of the undesirable sulfur compounds in a mode of slight oil destruction that could be used for oil preparation to transportation and further processing.

In another example, fuel oil having sulfur contents of 0.66 mass%, including 0.002 mass% mercaptans, was irradiated by 4 MeV electrons in the modes shown in Table 3.18. Irradiation temperature was 37°C in run 1, 40°C in run 2, and 200°C in run 3. In run 1, fuel oil was pumped through the reactor at the rate of 6.5 cm³/s with simultaneous supply of ionized air at the rate of 40 cm³/s. In run 2, fuel oil was pumped through the reactor at the rate of 53 cm³/s with ionized air supply at the rate of 300 cm³/s. In run 3, fuel oil was pumped through the reactor at the rate of 0.5 cm³/s with ionized air supply at the rate of 3 cm³/s.

Runs 1 and 2 were performed in the conditions of intense fuel oil cracking for the maximal production of motor fuels (mode 1) and base lubricants (mode 2) with simultaneous desulfurization of the commodity products. The oxidized sulfur compounds (sulfoxides) concentrated in the heavy residue. Then sulfoxides were extracted from the heavy residue by a method of liquid extraction using 62% sulfuric acid. Sulfoxides were separated from the extracted phase using 15% soda solution with the subsequent phase settling. Sulfoxides separated in a distinct layer were washed with 10% sodium chloride. The sulfoxide yield was 5.3 mass%. A similar degree of sulfoxide extraction was obtained when using acetone-water solutions, ethylene glycol, or monoethyl alcohol. Run 3 was performed in a mode of slight fuel oil destruction (mode 3) that provides effective desulfurization and increase in the value of the commodity products.

The extracted high-molecular sulfur compounds are valuable commodity products. They are used in mining and metallurgy as extraction agents in the processes of extraction and separation of radioactive and rare metals (uranium, thorium, zirconium, hafnium, niobium, tantalum, rare-earth elements, tellurium, rhenium, gold, palladium, etc.). It was forecasted that a wide application of oil sulfoxides will have a great impact on ore enrichment with many rare and noble metals since oil sulfoxides appear to be much more efficient than tributylphosphate and traditional extraction agents. The oil sulfoxides are also used in hydrometallurgy as the

TABLE 3.19
Feedstock Characteristics

| Sample | Fractional Contents, Mass% | | | Sulfur Concentration, Mass% | |
	Number of C atoms	Boiling Temperature, °C	Density ρ, g/cm³	Total Sulfur	Mercaptan Sulfur
Original oil	Up to C_{40}	Below 550	0.92	3.90	NA
Light gas oil	C_4–C_{22}	34–350	0.78	1.59	0.045
HAGO	C_4–C_{30}	38–450	0.90	3.71	0.054
Heavy residua	C_{22}–C_{42}	350–570	1.00	5.40	—

best extraction agents due to their high efficiency and selectivity in the extraction of gold, palladium, and silver. Applications of oil sulfones and sulfoxides in oil refining as selective solvents for paraffin hydrocarbon purification of unsaturated and organic sulfuric compounds were reported in the paper (Sharipov 1997).

RP of high-sulfuric oil with sulfur contents of 3 mass% and higher becomes more difficult, especially in the cases when feedstock contains high concentrations of dissolved hydrogen sulfide together with mercaptans. One-stage high-temperature RP of such feedstock can provoke oxidation of hydrogen sulfide and its transformation into mercaptans. Some increase in sulfur concentration in light RTC products can result from decomposition of sulfur-containing heteroaromatic cycles of pitches and asphaltenes.

One of the promising methods for control of oxidation–reduction processes in oil feedstock is the application of high-sulfuric oil bubbling by ionized air produced as a result of accelerator operation.

This type of processing was studied using samples of highly sulfuric crude oil from Southern Tatarstan fields (Russia) and products of its primary distillation: light oil distillate (gasoline-diesel fraction C_4–C_{22}), furnace oil (broad gas oil fraction C_4–C_{30}), and fuel oil (heavy residue C_{20}–C_{42} of oil primary distillation). The samples of crude oil and oil products characterized in Table 3.19 considerably differ by their density and concentrations of sulfur and heavy aromatics.

At the next stage, oil samples were irradiated by 2 MeV electrons from the linear electron accelerator ELU-4. Radiation-thermal processing of oil products was performed in the modes of the continuous feed flow and the under-beam distillation.

3.4.2 Sulfur Transformation Due to Oil Bubbling by Ionized Air

A contact with ozone-containing air provokes conversion of practically all components of crude oil or oil products; they participate in the reactions of electrophilic O_3 addition or ozone-induced radical oxidation by molecular oxygen. The rates of chemical reactions that can simultaneously proceed during low-temperature ozonolysis vary in a wide range of values. Sulfide oxidation to sulfoxides at 20°C is noticeable for the high reaction rate of 1500–1900 L/mol s. It proceeds according to the scheme:

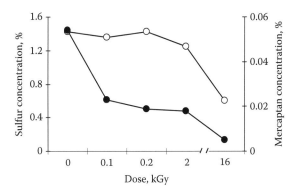

FIGURE 3.35 Changes in concentrations of total sulfur (○) and mercaptan sulfur (●) under irradiation by bremsstrahlung x-rays from 2 MeV electrons ($P = 16$ Gy/min) with continuous bubbling of ionized air (20 mg/s/1 kg of feedstock).

It is known (Kamyanov et al. 1997) that ozonization destroys thiophene cycles. As a result, a part of sulfur eliminates from the liquid phase and another part oxidizes to the corresponding sulfoxide and sulfone groups without destruction of the carbon skeleton. Therefore, prevailing of sulfide (thiacyclane) sulfur can be expected, up to 90% of total sulfur usually being concentrated in the heavy liquid residue of oil processing.

At the stage of oil bubbling with ionized air, a considerable rise in the contents of mercaptans and total sulfur in light fractions was observed. This process was accompanied by sulfur elimination in the liquid phase. As a result, the total sulfur concentration in the overall liquid product decreases by 1.5 times (Figure 3.35).

Continuation of bubbling causes deep oxidation of sulfuric compounds and their transfer to heavy fractions. Oil feedstock processing with ionized air, when combined with gamma irradiation, caused an increase in the concentration of the gasoline fraction by 1.6–2.0 times and a corresponding decrease in the concentration of mercaptan sulfur by 2.5–4.2 times. Decrease in the total sulfur and its predominant concentration in high-molecular compounds improved crude oil quality and prepared it for a more efficient high-temperature processing.

3.4.3 Sulfur Transfer at the Stage of Radiation-Thermal Cracking of Ozonized Oil

At this stage of high-viscosity oil RP, decomposition of polyaromatic structures, including the greater part of sulfur compounds, was accompanied by sulfur release and its transfer to light fractions, partially to the gaseous phase. RTC of crude oil with initial sulfur concentrations up to 4 mass% resulted in the production of the liquid products with sulfur concentrations in the range of 0.5–1.2 mass% that was two to four times lower than the sulfur content in the corresponding products of TC.

A wide fraction of the overall RTC liquid product, including gasoline (up to C_{11}), kerosene (C_{11}–C_{13}), and diesel (C_{14}–C_{21}) components, contained not more than 0.8 mass% sulfur, that is, about 20% of the total sulfur in the feedstock. RTC of high-sulfuric oil carried away on average 25 mass% of sulfur into gaseous phase; in the case of preliminary oil processing by ozone-containing air, the share of sulfur converted to gases increased up to 60 mass%. Sulfur transfer to gases increased at higher RTC temperatures.

Most of all, the observed sulfur release is associated with the least radiation-resistant aliphatic chains serving as links between structural blocs of pitches and asphaltenes (Jewell et al. 1976). These aliphatic chains can be relatively easily broken by ionizing radiation. Detachment of alkyl substituents with double C=C bonds from aromatic rings gives rise to concentration of unsaturated hydrocarbons in light fractions of the liquid RTC product (Figure 3.36). As unsaturated compounds easily react with sulfur atoms, their accumulation hampers sulfur elimination from light fractions.

Thermodynamic calculations (Berg and Habibullin 1986) show that C–S bonds of aromatic nature are three to four times stronger than those in aliphatic compounds: energy necessary to break a C–S–Ar bond is 301.5 J/mol, while that for C–S–Alk bonds is only 227.3 J/mol. To provide sulfur removal, it is important first to weaken sulfur interaction with an arene nucleus. An effective way to release oil desulfurization is deep oxidation of high-molecular sulfur-containing compounds, including asphaltenes. The oxidized high-molecular sulfur compounds, such as ozonides and sulfoxides, can be destroyed thermally or, easier and more efficiently, by RTC.

The coking residue of oil RP was 10–15 mass%; sulfur concentration in the residue was usually 5–6 mass%. Figure 3.37 shows distribution of the total sulfur in the heavy parts of the liquid RTC product, namely, heavy residua with boiling temperature higher than 300°C, higher than 350°C, and higher than 400°C. The heavier the

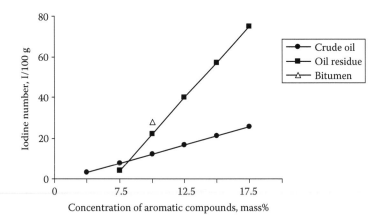

FIGURE 3.36 Dependence of iodine number on concentration of aromatic hydrocarbons in gasoline fraction of the liquid product obtained by RTC of high-sulfuric Tatarstan oil.

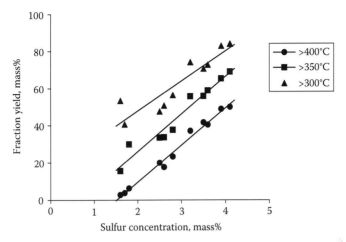

FIGURE 3.37 Total sulfur concentration in heavy residue (T_{boil} >300°C, 350°C, and 400°C) after RTC of high-sulfuric Tatarstan oil versus residue concentration in the overall liquid RTC product (data of gas–liquid chromatography).

residue and the higher its concentration in the overall liquid product after RP, the higher the sulfur concentration in the residue.

The alterations in sulfur contents in the liquid product of low-temperature RP of high-sulfuric Tatarstan oil at 20°C (mode 1) and a product of high-temperature processing of the same feedstock at 400°C (mode 2) are shown in Figure 3.38. In both of the modes, crude oil was bubbled with ionized air at room temperature during 20–30 min at a rate of 20 mg/s/1 kg feedstock in the field of gamma radiation ($P = 0.275$ Gy/s). In mode 2, it was additionally subjected to a high-temperature RTC.

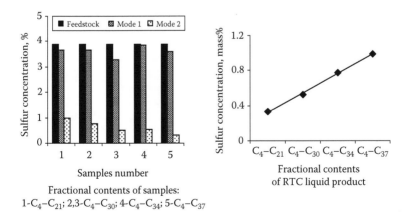

FIGURE 3.38 Concentration of total sulfur in fractions C_4–C_{21}; (1), C_4–C_{30} (2,3), C_4–C_{34} (4), and C_4–C_{37} (5) after RTC of high-sulfuric Tatarstan oil with bubbling by ionized air in the field of x-ray radiation (mode 1) and after RTC of ozonized oil (mode 2).

Processing conditions of crude oil samples differed only in bubbling time and irradiation dose during high-temperature cracking, except sample 3, which was bubbled at a heightened temperature of 60°C. As a result, liquid products of oil processing had different densities and fractional contents. Figure 3.38 shows that increase in the average molecular mass due to a higher concentration of heavy aromatic compounds leads to an increase in sulfur concentration in the RTC liquid product. The heavier the residue and the higher its concentration in the overall liquid product after RP, the higher the sulfur concentration in the residue.

The rate of sulfur accumulation in the heavy residue decreased as residue conversion increased with irradiation dose. At the decreased concentrations of the heavy residue in the total RTC product, sulfur concentration in the heavy residue came to saturation (about 6 mass% in the example shown in Figure 3.39).

Sulfur balance in RTC products of high-sulfuric oil processing shows that 60 mass% of sulfur moved to gaseous phase, provided feedstock was bubbled by ionized air before RTC. Increase in the duration of bubbling up to 2 h and temperature of bubbling up to 60°C (sample 3) caused some greater desulfurization. In the cases when oil was not bubbled by ionized air before high-temperature RP, sulfur concentration in the liquid RTC products was higher than 1.6 mass%, and eliminated sulfur did not exceed 12–18 mass%.

Correlations of vanadium contents and such characteristics of oil products as density, sulfur concentration, and yields of gasoline-gas oil fractions after RTC were observed for both types of feedstock (Figures 3.40 and 3.41). The deeper the radiation-induced oil conversion, the higher the vanadium concentration in the heavy residue.

A pronounced synergetic effect of radiolysis and ozonolysis enhanced by x-ray irradiation was detected when furnace oil (mixture of light fractions and heavy gas oil with initial gasoline concentration of 17.6 mass%) was processed at room temperature (Section 4.2). The observed low-temperature cracking became apparent, first of all, in a considerable increase in the yield of gasoline fraction. Total sulfur concentration in the overall product of furnace oil processing decreased by 23 mass% compared with that in the unprocessed material. Increase in the yields of the gasoline fraction was accompanied by deeper desulfurization of the gasoline fraction (Figure 3.42).

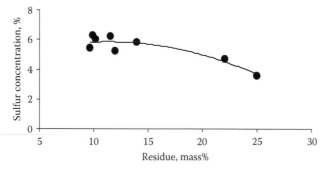

FIGURE 3.39 Total sulfur concentration in the heavy residue after RTC of high-sulfuric Tatarstan oil pretreated with bubbling by ionized air in the field of x-ray radiation.

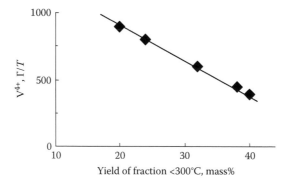

FIGURE 3.40 Dependence of V^{4+} concentration in oil residue on the yields of gasoline-gas oil fraction (start of boiling −300°C) obtained by radiation-thermal cracking of Karazhanbas oil (EPR data).

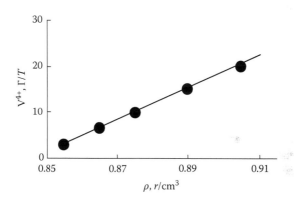

FIGURE 3.41 Dependence of V^{4+} concentration on the density of the fraction boiling above 250°C obtained by radiation-thermal cracking of heavy fuel oil (EPR data).

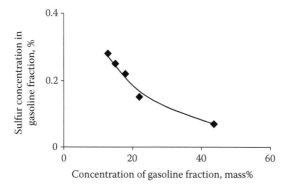

FIGURE 3.42 Total sulfur concentration in gasoline fraction versus concentration of gasoline fraction after furnace oil processing in different conditions.

RP of oil products in the presence of ozone-containing reactive air mixtures results in intensive oxidizing processes that simultaneously cause transformation of sulfur-containing and metal-containing organic compounds. This effect was studied in a series of experiments on RTC of high-viscosity Karazhanbas oil, the most vanadium-rich of all crude oils from Kazakhstan fields, and heavy fuel oil—a product of primary oil processing from Mangystau fields at the Atyrau refinery (Zaikina and Nadirov 2001). This sort of oil was subjected to RTC with accelerated electrons.

Figure 3.43 shows decrease in V^{4+} concentration with irradiation dose. The total vanadium concentration in the coking residue of RTC was 5–10 times higher because vanadium easily transforms to species with a higher oxidation degree not registered by the EPR method. The coking RTC residue contains a high concentration of vanadium in the form of V_2O_5, the latter being of a high commercial value.

The concept of the technology for processing heavy petroleum feedstock developed in Moscow State University by Likhterova, Lunin, and coworkers (Lunin et al. 2002) consists in the chemical pretreatment of the feedstock by reacting with ozone followed by decomposition of ozonolysis products under exposure to ionizing radiation of various types (Section 4.1). Deep oxidation of sulfur compounds transfers them to the heavy residue in the easily removable forms of sulfones, sulfoxides, and sulfuric acids.

This approach to crude oil refining and desulfurization is rather close to that used in the studies described earlier (Zaikin et al. 1998, Zaikina et al. 2002a,c, 2004, Zaikin 2005). The main difference is that, in these studies, preliminary feedstock oxidation was combined with the oxidizing gas bubbling under the electron beam. Another difference is that reactive ozone-containing ionized air produced as a by-product of the accelerator operation was used instead of the expensive ozone. Generally, processing of high-sulfuric crude oil and petroleum products by ionized ozone-containing air provides deep oxidation of sulfur compounds and improves the fractional contents of the products.

A method using x-ray irradiation for deep desulfurization of oil feedstock with original sulfur content of about 50 ppm was proposed by Ayukawa and Ono (2004). A high degree of oil desulfurization is associated with elimination of such hardly removable sulfur species as dibenzothiophenes (DBTs). An example of such refractory sulfur species is 4,6-dialkyl DBT:

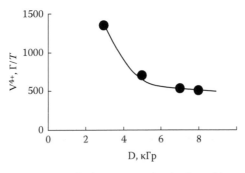

FIGURE 3.43   Dose dependence of V^{4+} concentration in the coking residue (a by-product obtained by RTC of Karazhanbas crude oil).

Thiophene   Benzothiophence   Dibenzothiophene

4, 6-Dialkyl dibenzothiophene

An earlier approach to this problem was based on the application of radioactive rays for activation of sulfur in oil and provision of its contact with a metal powder. It was suggested that thiophenes would relatively easily dissociate under ionizing irradiation and the dissociated sulfur species would react with the metal catalyst. Metal sulfides fell into deposit and could be removed. One of the difficulties in the practical realization of this approach was keeping metal powder dispersed in oil. In the method developed by Ayukawa and Ono (2004), this difficulty was overcome with the aid of the liquid desulfurization catalysts.

The desulfurization device comprised a catalyst liquid–liquid contact part for contacting a metal compound solution as a catalyst and the petroleum product, a beam-irradiating part, and a sulfide-collecting part for separation and collection of the metal sulfide produced by irradiation.

The catalysts proposed in the invention were liquid solutions of a metal compound where a metal was selected from the group consisting of silver, lead, iron, copper, and precious metals. In the particular examples, silver nitrate ($AgNO_3$) dissolved in water and isopropyl alcohol was used. According to the patent, the irradiation unit of the desulfurization plant can be based on an x-ray source. It was supposed that sulfur compounds were decomposed by radiation and sulfur reacted with a metal-containing catalyst so as to be substituted to metal sulfide for sedimentation.

The experiments showed a tendency to reduction in sulfur concentration in the case of the monochrome x-rays of the sulfur-absorption end wavelength (5.0185 Å). Lower wavelengths were undesirable. For example, suppose that x-rays with the wavelength lower than that in silver (3.5 Å) are applied for sulfur activation in the solution of a catalyst like silver nitrate, that is, irradiation energy is higher than that required for desulfurization. In this case, the silver nitrate would be decomposed before the sulfur decomposition. As a result, formation of silver or silver oxide instead of silver sulfide would be observed. Therefore, it was important to utilize the absorption end wavelength of sulfur by cutting the x-ray wavelength at the absorption end for silver or other metal used in the process.

The method developed by Ayukawa and Ono involves nonionizing irradiation: ionization effects are minimal for x-ray energies of about 1–3.5 keV and play no role in the process. However, the idea of this patent to use liquid catalysts instead of solid particles seems to be productive. It radically improves conditions of catalyst contact with sulfur-containing compounds and at the same time eliminates the problems of liquid dynamics associated with feedstock transmission through the reaction zone. Together with radiation activation of the catalytic reactions, this approach can be used in the methods based on application of high-energy ionizing irradiation.

Basfar and Khaled (2012) proposed a method for sulfur removal from crude oil using gamma irradiation. This method comprises oxidizing a sulfur compound present in a crude oil or diesel fuel by adding at least one of the oxidizing agents, an extracting salt, and a solvent to create a chemically reactive crude oil or diesel fuel for the subsequent desulfurization. Such a chemically modified oil was subjected to gamma irradiation. The oxidizing agent was added to oil prior irradiation while an extraction salt and a solvent were introduced after irradiation.

The recommended solvents are acetic acid, sodium chloride aqueous solution, acetone, acetonitrile, methanol, hexane, heptane, toluene, and petroleum ether. The absorbed dose was in the range of 50–200 kGy. The following processing steps were applied for crude oil and straight-run diesel desulfurization, each process conducted separately: A—irradiation only, B—irradiation with extraction (salt solution, C—irradiation with extraction (acetonitrile), D—irradiation with extraction (methanol), E—irradiation with oxidizer (10% H_2O_2), F—irradiation with oxidizer (10% H_2O_2) and extraction (salt solution), G—irradiation with oxidizer (10% H_2O_2) and extraction (acetonitrile), H—irradiation with oxidizer (10% H_2O_2) and extraction (methanol), I—irradiation with oxidizer (10% H_2O_2) + 2% CH_3COOH, J—irradiation with oxidizer (10% H_2O_2) and extraction (salt solution), K—irradiation with oxidizer (10% H_2O_2) + 2% CH_3COOH and extraction (acetonitrile), L—irradiation with oxidizer (10% H_2O_2) + 2% CH_3COOH and extraction (methanol).

The efficiency of these modes of radiation desulfurization is illustrated in Figures 3.44 and 3.45. The best of the modes proposed for oil radiation desulfurization

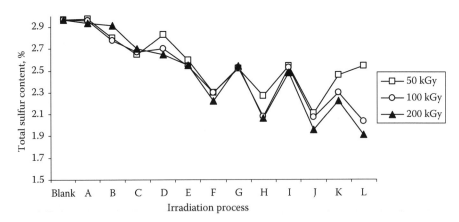

FIGURE 3.44 Total sulfur removal from Arabian heavy crude oil in various irradiation processes.

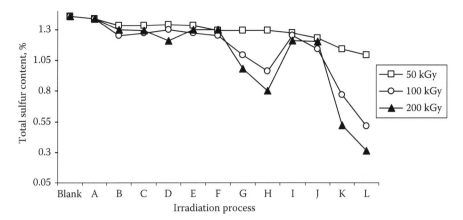

FIGURE 3.45 Total sulfur removal from straight-run diesel fuel in various irradiation processes.

by this method resulted in reducing the sulfur content by 36% in Arabian heavy crude oil and by 78% in straight-run diesel fuel with the absorbed irradiation dose of 200 kGy.

A series of experiments were conducted on radiation-enhanced oxidative desulfurization of the hardly removable sulfur species in the model petroleum-simulating solvents (Qu et al. 2006, 2007, Zhao et al. 2004). The removal of sulfur compounds in petroleum was based on the radiochemical reactions induced by gamma irradiation. DBT dissolved in dodecane was employed as simulated petroleum. Irradiation was conducted in the presence of a catalyst: zirconium oxide impregnated on alumina (Zr/Al_2O_3).

The dose of gamma irradiation was 179.1 kGy at the dose rate of 3.85 kGy/h; all the tests were performed with aeration at 298 ± 2 K. The air flow rate was about 30 mL/min. The loading contents of every metal oxide on Al_2O_3 were all about 10%, and all the catalysts were calcined at 673 K for 2 h. The ratios of the employed catalyst amount to the simulated oil volume, C/O, were all about 0.1 g/mL. The main product of DBT conversion in this process was dibenzothiophene sulfone ($C_{12}H_8SO_2$).

DBT conversion of 98.9% and an increase of over 80% compared to that without catalyst were achieved as a result of the 2 h special catalyst preparation and radiation-catalytic processing during 46.5 h (Figure 3.46).

The experiments have shown that the sulfur compounds are more degradable when present in low concentrations (Figure 3.47). In the example shown in Figure 3.47, DBT was dissolved in dodecane and irradiated at 298 K in static conditions. The radiation dose rate was 3.95 kGy/h.

Dose dependences of DBT conversion at the different dose rates of gamma irradiation are shown in Figure 3.48. DBT was dissolved in dodecane and irradiated at 298 K a static condition.

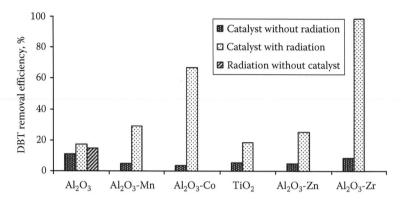

FIGURE 3.46   Removal efficiency of DBT with various catalysts in different conditions.

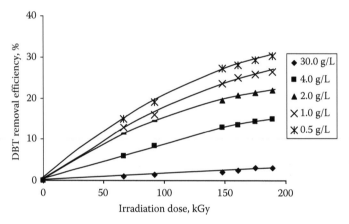

FIGURE 3.47   Removal efficiency of DBT with a different initial DBT concentration at various radiation doses.

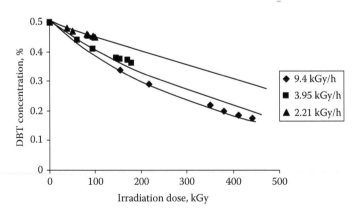

FIGURE 3.48   DBT concentration in the simulated oil versus the irradiated dose at various radiation dose rates.

Figure 3.48 shows that a higher radiation dose rate corresponds to higher removal efficiency at the same irradiation dose. This tendency differs from that observed for the radiation depletion of dodecanethiol or dibutyl sulfide (Yan et al. 2004), the removal efficiency of which was higher at a lower dose rate. It was explained (Qu et al. 2006) by the specificity of DBT's molecular structure. DBT is made up of a conjugate π bond, in which the π electrons are nonlocalized. When irradiated by gamma rays, the energy absorbed by these molecules quickly distributes in the whole molecule. Therefore, the probability of energy focusing in a single bond is low. It inverts a π-electron from a high-excitation state to a low-excitation state, and the probability of the reaction decreases. The higher the dose rate, the higher the reaction probability.

The experiments with aeration during irradiation of the simulated oil have shown that oxygen was not helpful in the depletion of DBT in the reaction system with dodecane and DBT only. However, addition of hydrogen peroxide accelerated radiation degradation of BT and DBT. Hydrogen peroxide demonstrated an apparent synergistic behavior with radiation for BT and DBT degradation. Radiation-induced conversion of DBT was more considerable in the presence of acetic acid and hydrogen peroxide than that of hydrogen peroxide alone (Figure 3.49). The content of H_2O_2 in the aqueous solution was about 30 wt%. The volume ratio of the simulated oil, H_2O_2 solution, and acetic acid was 5:1:1. The oil and aqueous phases were blended by bubbling with air (30 mL/min). The radiation dose rate was 3.95 kGy/h at 298 K.

Cobalt oxide has also shown considerable promotion to DBT removal under irradiation; in its presence, efficiency of DBT removal under gamma irradiation increased by over 40% (Figure 3.50). DBT was dissolved in dodecane and irradiated at 298 K. The dose rate was 3.95 kGy/h. The flow rate of nitrogen or air was 30 mL/min. The content of the catalyst in the simulated oil was about 0.1 g/mL. In all runs, the main detected products of DBT degradation were sulfones.

Removal of dodecanethiol from simulated petroleum by gamma irradiation was studied in the work (Tian et al. 2003). *n*-Dodecanethiol (DCT) dissolved in

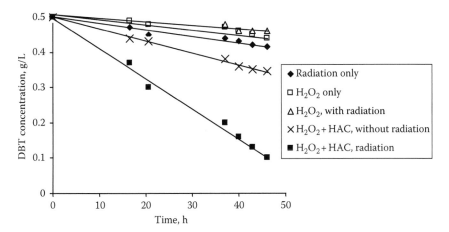

FIGURE 3.49 DBT concentration in the simulated oil versus the reaction time in the presence of H_2O_2 and acetic acid.

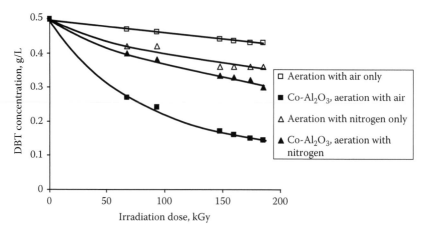

FIGURE 3.50 DBT concentration in the simulated oil versus the irradiated dose in the presence of the catalyst Co-Al$_2$O$_3$.

n-dodecane was used as a simulated petroleum product. The results showed that the degradation efficiency of DCT was proportional to the irradiation dose, and a lower dose rate seemed to correspond to higher degradation efficiency at the same dose. In the presence of oxygen, the dose needed for degradation efficiency of 90% fell from 700 to 250 kGy. The end product of DCT degradation was sulfate in the presence of oxygen, and hydrogen sulfide in the absence of oxygen.

Asphaltene and DBT sulfur conversion in aromatic solvents under high dose-rate electron irradiation was studied in a series of our experiments. Radiation-induced processes were studied in the model solutions of neat asphaltenes and asphaltenes with added DBT in toluene. Asphaltene concentration in all the original compositions varied from 5 to 10 mass%. In the samples with added DBT, its concentration was about 3 mass%.

All the samples were irradiated in the same conditions to the total dose of 300 kGy at the dose rate of 80 kGy/s and at the temperature of about 50–70°C. Gas evolution in the irradiated samples did not exceed 2 mass%.

The fractional contents of the products of RP for the fractions C$_4$–C$_{16}$ obtaining using liquid-gas chromatography are shown in Figure 3.51.

Comparison of graphs (a) and (b) in Fig.3.51 shows increase in the concentrations of fraction C$_{12}$ after radiation processing. In supposition that asphaltenes and DBT react independently and the observed increase in C$_{12}$ is unreacted DBT (C$_{12}$H$_8$S), the DBT conversion will make 92%–94%. It approximately corresponds to the detected DBT deposition from the liquid phase (86–90 mass%) presumably in a polymerized form.

In another experiment, 15.1 g asphaltenes were dissolved in 41 g (50 mL) of kerosene and 8.9 g (6 mL) of chloroform (sample 1S). This solution was subjected to RP in static conditions at ambient temperature; the time-averaged dose rate was 6.0 kGy/s, and the absorbed dose was 780 kGy. Asphaltene conversion under these conditions was discussed in Section 2.3 (Figure 2.42).

In the same experiment, disulfide sulfur was introduced in the form of 2.5 g of 4,4-diaminobiphenil-2,2 disulfo-acid (C$_{12}$H$_{12}$N$_2$O$_9$S$_2$) into the solution described

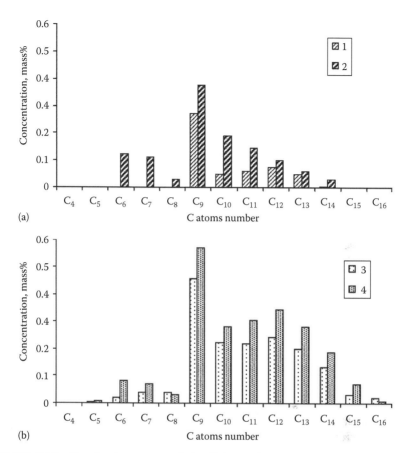

FIGURE 3.51 Fractional contents of the light product after electron irradiation of (a) asphaltene solutions in toluene and (b) asphaltene and DBT solutions in toluene: (1) 5 mass% asphaltenes, (2) 10 mass% asphaltenes, (3) 5 mass% asphaltenes, 3 mass% DBT, (4) 10 mass% asphaltenes, 3 mass% DBT.

earlier (sample 1S). This mixture (sample 2S) was subjected to RP at 70°C in the same conditions. Total sulfur was determined in the original solution and in the liquid product of RP (sample R2S).

Then, thiophene sulfur was introduced in the form of 5.6 g of 1% methyl thiophene solution in toluene into sample 2S. The solution obtained was identified as sample 3S. The mixture was subjected to RP at 70°C in the same conditions as those for samples 1S and 2S. Again, total sulfur was determined in the original solution and in the liquid product of RP (sample R3S).

Table 3.20 shows that RP releases both disulfide and thiophene sulfur.

The radiation-chemical yield, G, for decomposition of the heavy residue (fraction boiling out above 450°C) made 12.1 molecules/100 eV. The G-values for thiophene and disulfide sulfur decomposition were 278 S atoms/100 eV and 61.7 (S atoms)/100 eV, respectively. The high G-values for decomposition of asphaltenes and sulfur compounds indicated to radiation-induced chain reactions.

TABLE 3.20

Sulfur Transformation in Asphaltene Solutions under Electron Irradiation

	Sulfur, Mass%					
		Product of Radiation Processing			Disulfide	Thiophene
	Original			Total Sulfur	Sulfur	Sulfur
Sample	Solution	Gaseous	Liquid	Conversion, %	Conversion, %	Conversion, %
(R)1S	2.88	0.96	1.92	33.3		
(R)2S	3.38	1.04	2.34	30.8	16.0	
(R)3S	3.81	1.35	2.46	35.4		72.1

3.4.4 RADIATION METHODS FOR DESULFURIZATION OF COAL AND HEAVY OIL RESIDUA

One of the basic approaches to reducing sulfur compounds in the coal combustion products is sulfur removal from coal prior to combustion. Application of radiation methods for coal desulfurization contributes to higher selectivity and lower energy consumption of the desulfurization process.

A method for coal desulfurization using electron beams was proposed by Ray and Feldman (1983).

Removable forms of sulfur species include elemental sulfur (melting point 118.2°C), which may be removed by melting or by solvent extraction, gaseous compounds that can be swept away in a stream of air, water-insoluble sulfur compounds that can be filtered, and water-soluble sulfur compounds that can be washed away. In the process developed by Ray and Feldman (1983), coal was reacted with the electron beam in the presence of water and, particularly, in the presence of free radicals generated from the water by the action of the electron beam. In addition to the presence of water, a significant aspect of the method was that the coal was pulverized to a powder of approximately −60 to +200 mesh and slurred in water. It made the distribution of sulfur compounds much more uniform and increased efficiency of energy transfer from the electron beam. Irradiation dose was at least 15.8 kGy.

The following mechanism of coal conversion under ionizing irradiation was proposed. Hydrogen atoms, hydroxyl radicals, molecular hydrogen, and hydrogen peroxide are produced as a result of water radiolysis:

$$H_2O + h\nu \rightarrow H_2O^+ + e$$

$$H_2O + H_2O \rightarrow H_3O^+ + OH\cdot$$

$$e + H_2O \rightarrow OH\cdot + H + e' \hspace{2cm} (3.16)$$

$$2OH \rightarrow H_2O_2$$

$$2H_2 \rightarrow H_2$$

A reaction with inorganic sulfur compounds in the form of pyrite can be written as

$$FeS_2 + 2H \rightarrow FeS + H_2S \tag{3.17}$$

Although FeS is insoluble, it is subsequently converted to soluble $FeSO_4$:

$$2FeS_2 + 16H_2O_2 \rightarrow Fe_2(SO_4)_3 + H_2SO_4 + 16H_2O \tag{3.18}$$

$$FeS + H_2SO_4 \rightarrow FeSO_4 + H_2S \tag{3.19}$$

$Fe_2(SO_4)_3$, H_2SO_4, and $FeSO_4$ are all water-soluble.

The hydrogen sulfide produced in reaction (3.19) can be converted to elemental sulfur:

$$H_2S + H_2O_2 \rightarrow S + 2H_2O \tag{3.20}$$

Other reactions occurring are

$$Fe_2(SO_4)_3 + H_2S \rightarrow 2FeSO_4 + H_2SO_4 + S \tag{3.21}$$

$$2FeS_2 + 32OH \cdot \rightarrow Fe_2(SO_4)_3 + H_2SO_4 + 16H_2O \tag{3.22}$$

$$Fe_2(SO_4)_3 + FeS \rightarrow 3FeSO_4 + S \tag{3.23}$$

The reactions with organic sulfur in coal were summarized as follows:

$$R_2S + 2H \rightarrow R=R + H_2S \tag{3.24}$$
$$\text{organic sulfur compound in coal}$$

$$H_2S + H_2O_2 \rightarrow 2H_2O + S \tag{3.25}$$

$$R_2S + 2OH \cdot \rightarrow R_2O + H_2O + S \tag{3.26}$$

$$R_2S + H_2O_2 \rightarrow R_2SO + H_2O \tag{3.27}$$
$$\text{Sulfoxide}$$

$$R_2S + 2H_2O_2 \rightarrow R_2SO_2 + 2H_2O \tag{3.28}$$
$$\text{Sulfone}$$

$$R_2S + H_2 \rightarrow R=R + H_2S \tag{3.29}$$

$$R=R + H_2 \rightarrow 2RH \tag{3.30}$$

Generally, R_2O, R_2SO, and R_2SO_2 are water-soluble and removable by washing. Finally, as a further benefit, coal dechlorination results as a consequence of the subject desulfurization:

$$RCl + OH \cdot \rightarrow RO \cdot + HCl \tag{3.31}$$

$$RO \cdot + H \rightarrow ROH \text{ (alcohol)} \tag{3.32}$$

$$RO \cdot + OH \cdot \rightarrow ROOH \text{ (acid)} \tag{3.33}$$

A series of studies was conducted on radiation desulfurization of brown coal with recommendations of similar method application to heavy oil residua.

In the work by Mustafaev (1996), desulfurization of brown coal with the sulfur content of 2.7% was carried out using the electron accelerator ELU-4 ($E = 3.5$ MeV, $P \leq 1$ kW).

Coal was irradiated in flows of argon, hydrogen, methane, steam, and oil products. Sulfur concentration in coal was controlled by measurements of the residual sulfur in the coal composition and identification of sulfur-containing gaseous products (H_2S, COS). The process temperature was varied in the range of 300°C–450°C; the flow rate of gases was $v \leq 10$ L/h.

In the most favorable conditions, more than 75% sulfur was extracted in the form of liquid and gaseous products. The highest degree of desulfurization (82%) was achieved in the methane atmosphere. The high selectivity of radiation-thermal coal desulfurization was explained by a heightened electron density in the neighborhood of sulfur atoms in the organic mass of coal.

Radiation-thermal technology for coal and heavy oil residua processing by accelerated electron beams based on the data on high-temperature radiolysis of nitrogen and sulfur-containing compounds (Solovetskii et al. 1995) was proposed by Lunin and Solovetskiy (1996).

Practically complete decomposition of nitrogen and sulfurous compounds was reported as a result of brown coal RP. Application of electron irradiation provided reduction in the time of feedstock preparation for deep coal conversion with simultaneous decrease in the process temperature threshold. The degree of oil desulfurization in "severe" irradiation conditions reached 75%–85%.

3.5 RADIATION-THERMAL PROCESSING FOR REGENERATION AND UTILIZATION OF DESULFURIZATION CATALYSTS AND SORBENTS

Radiation-thermal processing of deactivated catalysts and sorbents for oil and gas technologies can be applied for regeneration of desulfurization catalysts for the repeated use, catalyst utilization with extraction of useful components, and neutralization of toxic waste products of catalytic production. The technology developed in the works (Solovetskii 1989, Adigamov et al. 1990, Ajiev et al. 1991, Lunin et al. 1993,

Solovetskii et al. 1995, 1998) is based on the application of electron accelerators. The methods developed allow perfect purification of catalysts from sulfur and coke during 2–20 min, the structure of the material being preserved. In the ascertained processing modes, valuable catalyst components were separated from coke and transferred to environmentally safe forms.

Long-term exploitation of oil and gas processing catalysts leads to the formation of carbon- and sulfur-containing structures of coke and dense products on the catalyst surface. They block active catalyst sites and reduce the catalytic activity. RP leads to practically complete destruction and removal of C- and S-containing structural groups formed in the bulk and on the surface of catalytic systems. The temperature limit for catalyst regeneration is considerably lower than that in the case of conventional thermal processing. The degree of radiation regeneration was 99.999% for a number of deactivated catalysts (Solovetskii 1989).

3.5.1   REMOVAL OF CARBON PRECIPITATIONS BY RADIATION PROCESSING OF COKED CATALYSTS

Aluminum oxide is the most widely spread industrial catalyst carrier. Comparison of carbon decrease in traditional burning and radiation-thermal processing (Solovetskii 1989) shows considerable intensification of desorption under ionizing irradiation (Figure 3.52).

Time necessary for the same reduction in carbon-containing compounds is two to three times lower in the case of RP compared with conventional thermal processing or ozone treatment in similar temperature conditions. Electron-beam processing provides practically complete carbon removal within 10–15 min (Solovetskii 1989).

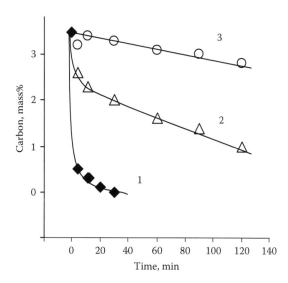

FIGURE 3.52   Kinetics of carbon removal from a catalyst on the base of Al_2O_3 in the cases of (1) radiation processing, (2) ozone treatment, and (3) traditional regeneration at 673 K.

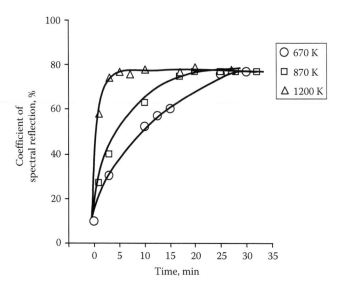

FIGURE 3.53 Dependence of spectral reflection coefficient from Al_2O_3 surface on temperature and time of radiation processing. 1–670 K, 2–870 K, 3–1200 K.

Electron spectra of diffusion reflection from the surface of Al_2O_3 after RP demonstrate perfect purification of the catalyst carrier from coke (Figure 3.53).

Rather detailed information on radiation-induced coke decomposition and desorption is available for a wide spectrum of catalytic systems used in oil processing, such as sorbents on the base of Al_2O_3: Pt/Al_2O_3 and $Co-Mo/Al_2O_3$, catalysts of the Claus process (production of elemental sulfur by processing of exhaust gases containing H_2S and SO_2), and iron oxide catalysts for direct reduction of hydrogen sulfide to sulfur (Adigamov et al. 1990). Practically for all catalysts, the residual carbon content did not exceed 0.01 mass% after irradiation in the regeneration mode and 0.001 mass% in the mode of radiation-thermal utilization.

Together with coke sediments, catalyst poisoning by sulfuric compounds makes one of the main contributions to catalyst deactivation (Adigamov et al. 1990). RP accelerates radiolysis and desorption of sulfur-containing compounds (Figure 3.54).

The process is most intense at the initial stages of irradiation. At the temperature of 873 K, concentration of elemental sulfur in the catalyst grain drops by 80%–90% and then slowly decreases. The observed changes in the intensity of radiation-induced sulfur removal are determined by different forms of sulfuric compounds in the catalyst grains. Elemental sulfur can be easily evaporated from the pores in the surface layers. Removal of the residual sulfur is associated with decomposition of sulfide and, later on, sulfate structures.

Irradiation was carried out using specially designed reactors with the controlled atmosphere of the inert gas that did not allow oxygen accumulation under irradiation (Solovetskii 1989). In these conditions, it was impossible to avoid considerable sulfate production. If deactivated contacts originally contain sulfates, their removal,

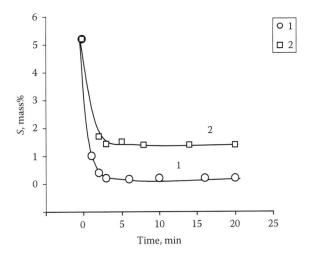

FIGURE 3.54 Kinetics of sulfur desorption from the catalyst grains for Claus process: 1—electron beam irradiation; 2—traditional calcination. $T = 673$ K.

in contrast to the traditional technology, can be carried out by irradiation at temperatures higher than those characteristic of sulfate radiolysis.

The main advantage of RP is that it allows purification of the catalyst surface from sulfur in a short period of time and obtaining oxide systems with the preserved matrix characteristics good for further application that is impossible in the traditional method conditions of thermal burning.

3.5.2 Effect of Radiation Processing on Basic Technological Characteristics of Catalysts

3.5.2.1 Retention of the Specific Surface of Catalytic Systems

The most important characteristic of catalysts' quality is their ability to retain a high original specific surface during the entire period of their operation. Radiation regeneration of catalysts may significantly affect the catalytic reactivity due to the changes in the specific surface of the carrier, provided irradiation time is long enough and irradiation temperature is higher than that characteristic of phase transitions in the matrix material.

For example, decrease in the specific surface of the catalyst on a titanium oxide base after long-time high-temperature electron irradiation is a result of anataz–rutyl phase transfer. Specific surface of the high-temperature TiO_2 modification (rutyl) is markedly lower and lies in the range of 20–25 m²/g. This mode of RP was recommended only for neutralization of the worked-out catalysts in the recycling process (Solovetskii 1989).

In the case of polluted catalysts and sorbents, especially when the deactivation mechanism is based on pore blockage, short-term high-temperature radiation-thermal treatment as a part of regeneration process leads to radiolysis of coke and dense surface products, opening of pores, and even to some increase in the specific surface (Solovetskii 1989).

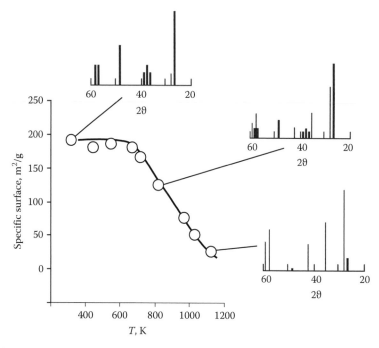

FIGURE 3.55 Dependence of specific surface on irradiation temperature and diffraction patterns of TiO$_2$ carriers. Irradiation time 3–20 min.

As a rule, RP in the time and temperature ranges characteristic of regeneration modes guarantees complete retention of the original specific surface (Solovetskii 1989). It was found that rather strong but short-time overheating of catalyst granules by an electron beam (e.g., 1–2 min at 1373 K) does not cause noticeable changes in the carrier-specific surface (Figure 3.55). The diffraction pattern of the "dead" surface incredible for the proper modes of RP is shown for comparison.

3.5.2.2 Retention of the Reactive Component in Regeneration Modes of Catalyst Radiation Processing

Radiation-thermal catalyst processing leads to profound alterations in the phase composition, crystal size, and specific surface of the reactive components in the cases when irradiation dose is high enough and irradiation temperature is substantially higher than the temperature of cluster sintering (Lunin et al. 1993). High reliability of irradiation temperature control allows perfect preservation of the active component of the used catalyst.

Sintering of the reactive components of used Pt, Pt–Re, and Pd reforming catalysts can be easily prevented in the technology of catalyst radiation regeneration (Lunin et al. 1993).

The curve characterizing Pt crystals sintering in a "severe" radiation mode is shown in Figure 3.56 for comparison. This mode of RP is never used for catalyst

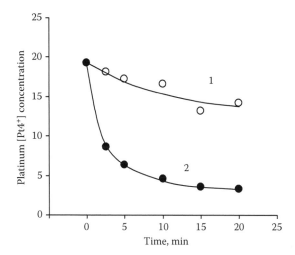

FIGURE 3.56 Retention of the reactive component (platinum) in radiation technology of (1) 0.5% Pt/Al_2O_3 catalyst regeneration and (2) its sintering in lab conditions.

regeneration. It is useful only for coke burning in the process of precious-metal separation from the worked-out catalysts.

3.5.2.3 Pore Structure in Radiation-Processed Catalysts

Changes in the pore structure of carriers, sorbents, and catalysts indicate to direct correlation with the changes in their specific surface (Ajiev et al. 1991).

Bidispersial structure typical for pure Al_2O_3 resulting from locally ordered packing of the original particles of about 5 nm in size and original pores of 4–7 nm does not change their distribution after radiation (Figure 3.57). Only insignificant reduction in the amount of pores was observed.

Similar effects of ionizing irradiation on catalyst porosity were observed for Al_2O_3-based reforming catalysts produced in France (Figure 3.58), and a similar under-beam behavior was observed for the pores 50–1000 nm in size (Ajiev et al. 1991).

3.5.2.4 Effect of Radiation-Thermal Processing on Mechanical Characteristics of Industrial Catalysts

Mechanical strength is the most important consumer property for high-temperature industrial catalysts. In conditions of thermal oxidation or steam regeneration of deactivated catalysts, overheating and recrystallization of carriers are usually inevitable. Crystal growth and agglomeration accompanied by local overheating lead to decrease in the mechanical strength of catalyst granules for many oxide systems.

Reactions of hydrocarbon and carbon contaminants with hydrogen and water steam from the gas phase lead to the appearance of gaseous products in the intra-porous matrix and result in differential pressure. In this case, a great amount of microcracks forming in this process can transform into failure cracks.

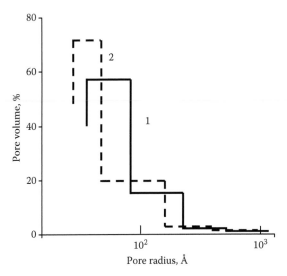

FIGURE 3.57 Pore distribution (secondary porosity) in KP-104 (Pt/Al$_2$O$_3$ reforming catalyst, Russia) (1) before and (2) after irradiation.

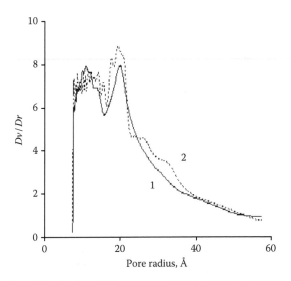

FIGURE 3.58 Pore distribution (primary porosity) in CSR-2 (Al$_2$O$_3$-based reforming catalyst) (1) before and (2) after irradiation.

For the majority of catalysts based on aluminum and titanium oxides, the original strength of granules and extrudates is in the range from 4 to 20 kG, decreasing during operation and regeneration by 1.5–2.5 times (Lunin et al. 1993).

Figure 3.59 shows the time dependence of the mechanical strength of the CSR-2 spherical catalyst granules (Catalysts and Chemicals Industries Co. Ltd., Japan),

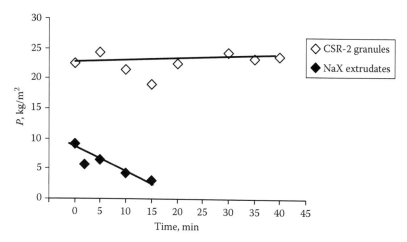

FIGURE 3.59 (a) Mechanical characteristics of CSR-2 catalyst granules and (b) extrudates of NaX zeolite. Electron energy, $E = 1.2$ MeV.

deactivated as a result of three years of operation at Mubarek gas-processing plants and regenerated by radiation-thermal processing and conventional ignition in comparable thermal conditions (Lunin et al. 1993).

Radiation regeneration in the modes used did not noticeably affect the strength of the worked-out catalyst. No changes in the structure and composition of CSR-2 matrix and, therefore, no deterioration in strength properties of the industrial samples were observed after irradiation.

3.5.3 CATALYST UTILIZATION

The complete change of deactivated catalysts or sorbents from the converter at the end of their service life requires their preliminary neutralization. In this case, the aim of RP is the maximally deep radiolysis of organic contaminants and stabilization of inorganic impurities in the matrix. The product obtained must meet international sanitary standards used for material certification for the out-of-door storage. Probably, the most prospective direction of catalyst and sorbent utilization after their radiation neutralization may be their application in the industry of construction materials as refractory additives or pigments.

The necessary stage of used catalyst utilization is preliminary granule grinding to the particle size down to 1000–5000 nm. Application of preliminary radiation-thermal processing can make this most energy-consuming industrial stage considerably less expensive.

RP at the temperatures two to three times higher than catalyst operation temperature results in strong disordering of the granule material. Usually, the main cause of decrease in the strength of catalysts and sorbents is phase transitions accompanied by alterations of mole volume. For example, deterioration in mechanical properties of the CSR-2 catalyst is caused by production of the alpha-phase of aluminum oxide (corundum) (Solovetskii 1989).

The most effective way to drastically reduce the strength of a porous material is introduction of easily vaporizable liquids into the granule volume. Presence of the gaseous radiolysis products reduces strength characteristics of the catalyst granules subjected to RP up to 50% (Solovetskii 1989). For example, NaX zeolite has smaller pores (0.8 nm) compared with those in the CSR-2 catalyst (10 nm and higher); hence, desorption of gaseous products becomes difficult. Matrix decomposition in conditions of intense water evaporation from moist channels of the catalyst granules leads to formation of the powder fraction 0.2–0.4 mm in size. Longer RP at higher temperatures leads to the material fusion and its total sintering into a glass-like mass.

Thus optimal conditions of radiation-thermal processing provide complete retention of the mechanical strength of catalyst and sorbent granules in the regeneration mode and effective neutralization of catalysts and sorbents with reduction of their strength properties by more than 50% in the neutralization mode.

3.5.3.1 Application of Radiation-Thermal Processing for Molybdenum Extraction from Catalysts of Hydrodesulfurization Processes in the Process of Solid Waste Utilization

The mass of the catalysts used in the petroleum industry for hydrodesulfurization is second in tonnage after the mass of cracking zeolites. Replacement of the catalyst systems caused by their irreversible deactivation is accompanied by significant losses of valuable metals, such as Ni, Co, and Mo. Water extraction and alkaline treatment permit separation of Ni (Co) and Mo compounds with the degree of molybdenum extraction of 90 mass%. Application of radiation-thermal processing for molybdenum extraction allows practically complete molybdenum extraction.

The results of comparative experiments on traditional thermal calcination of catalyst systems and their radiation-thermal processing (Adigamov et al. 1990) are shown in Figure 3.60.

The temperature threshold for the start of MoO_3 sublimation is well pronounced in the experimental curves. In the case of thermal burning, MoO_3 sublimation is primarily a result of sulfides decomposition. MoO_3 sublimation caused by radiation-thermal processing in the same time and temperature conditions proceeds more intensely and leads to more complete decomposition of all other Mo compounds.

A similar effect was observed in the case of radiation-thermal processing of deactivated the Co-Mo/Al_2O_3 catalyst. Selective decomposition of the Mo-containing phase with retention of the Ni-containing phase under the electron beam is characteristic of different components of hydrodesulfurization catalysts. In specially designed coolers, MoO_3 sublimated in the process of high-temperature radiolysis can be easily extracted from the gas phase in the form of small light-yellow crystals of 99.99% purity.

3.5.3.2 Catalytic Activity of Oxide Contacts after Radiation-Thermal Processing

Experimental data on the surface conditions, composition, and matrix structure of the catalysts regenerated by radiation-thermal processing indicate to their high catalytic reactivity. The catalytic reactivity of irradiated systems is illustrated in Figure 3.61 by the data for three catalyst samples of the Claus process: a neat sample,

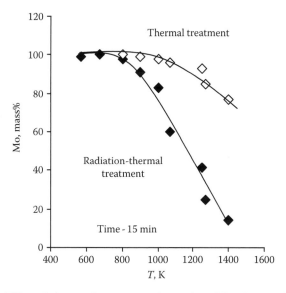

FIGURE 3.60 Effect of electron-beam processing and traditional thermal heating on the decrease in molybdenum mass due to MoO_3 sublimation from $Ni–Mo/Al_2O_3$ catalyst.

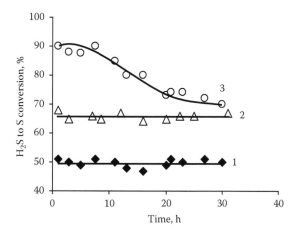

FIGURE 3.61 Conversion of hydrogen sulfide to sulfur at (1) deactivated, (2) regenerated, and (3) neat CSR-2 catalyst.

a sample deactivated after 1 year of operation in industrial conditions, and a sample after radiation-thermal regeneration (Solovetskii et al. 1998).

Catalyst CSR-2 regenerated by radiation-thermal processing demonstrated reactivity comparable to the stationary reactivity of the neat catalyst. Decrease in the sulfur yield of the neat catalyst in the beginning of its operation is associated with the production and accumulation of sulfates due to the presence of oxygen admixture in the reaction mixture. Conversion of H_2S to S was only 41–44 vol.% on the deactivated contact and reached 65–67 vol.% on the regenerated contact.

The industrial tests of Claus catalysts regenerated by radiation-thermal processing were performed with the batch of worked-out CSR-2 catalysts in the amount of 2.0 tons (Solovetskii et al. 1998). In the period of time from 1991 to 1994, the contact mass regenerated at the pilot line on catalyst RP of Ecort, Ltd. had been working with operation characteristics close to initial ones at the Claus facility of a gas-processing plant in Otradny, Russia.

Thus 100% recovery of the stationary catalyst reactivity can be reached in optimal conditions of radiation-thermal regeneration of reversibly deactivated catalysts. Radiation technology for used catalyst and sorbent regeneration provides reduction of process duration by 10–100 times, maximal preservation of the carrier structure, and a strict control of the modes of high-temperature radiation-thermal annealing of harmful admixtures.

Radiation technology for utilization of an irreversibly deactivated catalyst and sorbent provides radiation heating of the processed material up to 1300–1400 K during 1–2 min, decomposition of toxic organic admixtures and transference of a catalyst into an ecologically safe condition, and complete decomposition of toxic organic admixtures and catalyst transfer to environmentally safe conditions.

For example, purification of catalysts and sorbents from sulfur and carbon-containing compounds of S and C is carried out in atmospheric air and the process duration is 1–15 min depending on characteristics of the deactivated material.

4 Complex Radiation-Thermal Treatment and Radiation Ozonolysis of Petroleum Feedstock

4.1 APPLICATION OF THE IONIZED AIR AND OZONE-CONTAINING AIR MIXTURES FOR ENHANCED THERMAL AND RADIATION-THERMAL CRACKING

Changes in the composition and properties of crude oils and oil fractions caused by ozonolysis and decomposition of their primary conversion products due to subsequent thermal treatment are described in the works by Kamyanov et al. (1997). The ability of ozone to selectively interact with saturated sulfides, arene derivatives of thiophene, pyrrole, and furan (with the exception of dibenzo-derivatives and polycycloaromatic hydrocarbons) was used in the development of the methods for oil distillate cleaning from sulfur and polyarene compounds.

The ozonides of oil polyarenes, as well as other organic peroxides, undergo homolytic decomposition at temperatures above 110°C–120°C (Kamyanov et al. 1997). Sulfoxides formed as a result of interactions of oil sulfides with ozone have sill lower thermal resistance and decompose even at temperatures below 100°C. Both ozonides and sulfoxides can play the role of initiators in radical chain reactions. The initiating properties of these compounds can be used for the intensification of thermo-destruction of the heavy components in crude oils and bitumen. In particular, it provides increase in the yields of distillate fractions and reduction in the yields of heavy residues in the distillation process.

For the maximal accumulation of ozonides and for preventing their decomposition, the ozonation process was carried out in the flow reactor at a short-time (less than 3 min) contact of feedstock with the ozone-containing mixture. The temperature in the reactor was kept at the level of 20°C when low-viscous West Siberian oils were ozonized and was increased to 70°C–80°C in the case of high-viscous oils and bitumen treatment. In all the cases, the estimations were made for the amount of ozone being rapidly (by electrophilic mechanism) bound with the feedstock components.

One of the test subjects was a commodity mixture of West Siberian crude oils having a density of 0.865 g/cm^3 and containing 1.28 mass% sulfur, 11.5% pitches, 1.24% asphaltenes, and 6.6% paraffins. Concentrations of the fractions boiling below 200°C, 350°C, and 490°C were 24.0, 54.0, and 77.8 mass%, respectively (Kamyanov et al. 1997). Ozone consumption was 26 g/kg. The ozonized oil was subjected to

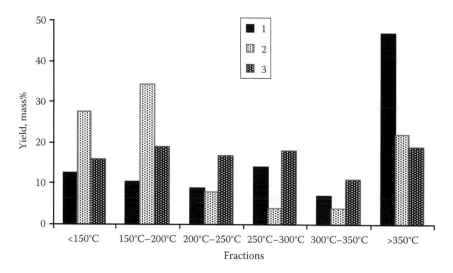

FIGURE 4.1   Distillation curves of West Siberian crude oil before and after ozonation. 1—Crude oil; 2—Ozonized oil; 3—Ozonized oil after thermal processing.

thermolysis at the temperature of 350°C during 1 h. The changes in oil fractional contents after ozonation and thermal processing are shown in Figure 4.1.

The analyses of the products treated indicated decomposition of big polyarene nuclei with parallel decomposition of the sulfuric compounds and the presence of 10–16 mass% olefins, the direct products of C–C bond breaks in the saturated fragments of oil components in the distillates of ozonized and thermolized oil. Thus, the thermolysis of ozonized oil was accompanied with intense initiated cracking at temperatures lower by about 150°C than those of the conventional thermal cracking (TC) at practically atmospheric pressure instead of 45–50 atm.

The effects of initiating thermo-destruction reactions of oil components by radicals formed due to sulfoxide decomposition at relatively low temperatures become more pronounced when ozonized oil is rich in sulfur compounds. This phenomenon was studied on the example of the blended crude oil having a density of 0.9266 g-cm³ at 20°C and viscosity of 384 cSt at 20°C and 34.4 cSt at 50°C (Kamyanov et al. 1997). Sulfur concentration in oil was 4.6 mass%. It contained 17.4 mass% pitches and asphaltenes, 5.1 mass% solid paraffins, 13.5 mass% gasoline fraction, and about 33% light fuel distillates.

The experiments have shown that this type of oil is capable of a rapid electrophilic addition up to 58 g ozone per 1 kg of feedstock; however, the optimal ozone expense in oil processing was about 12 g/kg. The ozonized oil was subjected to thermal processing at the temperature of 350°C during 20–60 min. As a result of initiated cracking of ozonized oil, the yield of the gasoline distillate SB-350°C increased by 2.6 mass%, the yield of diesel fraction 200°C–350°C increased by 8.2 mass%, and the total yield of motor fuels increased by 10.8%.

The total increase in the yield of a wide distillate SB-400°C was 14.5 mass% and 79.14 mass% of fraction was distilled at the temperature of 480°C (by 20.2 mass%

or one-third higher than that distilled from crude oil). As a result of oil processing, sulfur concentration decreased in all distillates: in gasoline fraction (SB-180°C), it decreased from 0.51 to 0.07 mass%, while in diesel fraction 180°C–350°C, it decreased from 3.57 to 0.43 mass%.

TC provides a higher conversion of heavy oil components and proceeds at lower temperatures if radical chain reactions are initiated by introduction of special additives into the feedstock composition. In turn, decrease in cracking temperature leads to lower gas and coke formation. Ozonation is one of the methods for generation of such additives (initiators of chain reactions) in the feedstock.

A series of experiments was conducted to study the opportunities of increasing the yields of fuel fractions by ozonide-initiated cracking of natural bitumen (Kamyanov et al. 1997). The organic fraction of bitumen from Mortuk field, Kazakhstan, was used as feedstock (Mortuk bitumen processing by means of RTC is described in Section 2.4). Bitumen density was 0.94 g/cm^3 at 20°C, its average molecular mass was 500 g/mol, and concentration of fractions boiling below 350°C was 24.2 mass%.

Because of the high bitumen viscosity (234 cSt at 80°C; 63 cSt at 100°C), ozonation was applied to bitumen solution in toluene, kerosene, or kerosene mixed with xylene. The process was conducted in a flow reactor at 500°C. Ozone consumption was 34 g per 1 kg bitumen. The solvent from the ozonized toluene solution was distilled away in vacuum at a temperature not higher than 70°C. The remaining material was heated up to 350°C and kept at this temperature until termination of cracking gas evolution.

The data obtained demonstrated substantial advantages of ozone-initiated cracking over high-temperature TC. Ozonolysis application allowed avoiding considerable mass losses for gas and coke formation totally reaching 14% in the case of TC. The yield of gaseous products in initiated cracking did not exceed 3.5 mass% without considerable coke formation (only coke traces were detected). A similar method was proposed for the intensification of brown coal liquefaction (Kamyanov et al. 1997).

A new promising approach to the processing of high-viscous and high-sulfuric oil is application of synergetic action of the two types of initiated cracking: RTC and cracking initiated by ozonolysis (Nadirov et al. 1997b; Zaikin and Zaikina 2002, 2004a,d; Zaikin 2005). This approach is also based on the ability of ozonides and sulfoxides to initiate radical chain reactions that can be effectively used for intensified thermal destruction of heavy, high-sulfuric crude oil, bitumen, and different types of oil wastes. An additional advantage of such combination is an opportunity to utilize ionized ozone-containing air, a by-product of radiation facility operation, instead of expensive ozone. This combination preserves and amplifies advantages of both types of initiated cracking and makes oil processing more efficient and economic.

The temperature of the ozone-initiated cracking is lower than that of the conventional TC by 150–200 K and close to the temperature of radiation-thermal cracking (RTC). To take the advantages of both the processes, the two types of initiated cracking can be combined in a more efficient and economic method for heavy oil processing (Nadirov et al. 1997b). Since the end of the 1990s, two main approaches to the combined application of RTC and ozonolysis were developed. In the first approach (Likhterova et al.), the ozonization technique earlier developed in the works by Kamyanov et al. (1997) was used for feedstock preparation to radiation-thermal processing. In the other approach (Zaikina and Zaikin), such a strong oxidizer as ionized ozone-containing air, a by-product of an electron accelerator operation, was used instead of expensive ozone.

In the study by Lunin et al. (1999), ozonized and nonozonized oil was irradiated with accelerated electrons at room temperature (see Section 3.4.1). Although radiation processing of the original crude oil did not improve its quality, irradiation of the preozonized oil resulted in the increase in light fractions by 10%.

Likhterova et al. (2005) represented a comparative analysis of the data obtained by radiolysis of the straight-run and ozonized fuel oil. The straight-run fuel oil was subjected to ozonation (4.4 g O_3/kg) with subsequent irradiation using the electron accelerator LUE-5 (average beam current of 0.5 mA and electron energy of 6.5 MeV). Irradiation was performed at the temperature of 16°C–17°C and atmospheric pressure in the dose range of 100–2000 kGy. Oil samples were packed in plastic bags with openings for the exit of the radiolysis gaseous products. The irradiated samples were subjected to visbreaking and atmospheric–vacuum distillation. Visbreaking was conducted at temperatures of 360°C–380°C during 40–60 min.

It was shown that preozonation leads to considerable changes in the colloid structure of oil feedstock and affects concentration of vanadyl porphyrins in the ozonated products. However, the yields of decomposition products obtained by visbreaking were, on average, 5 mass% lower in the case of preozonated fuel oil.

In the work by Lunin et al. (2012), it was noted that industrial introduction of ozone-consuming technologies is held back because of the considerable consumption of electric energy for ozone production (20 kWh per 1 kg ozone). Therefore, the specific ozone consumption per mass unit of the target product is crucial for the process industrial realization.

The concept of using ozone-containing ionized air both for oil preparation to irradiation and for under-beam radiation processing was set forth in the paper (Nadirov et al. 1997b). In the method for cleaning hydrocarbon mixtures from sulfur compounds proposed by Zaikina et al. (1999a), stimulation of oxidation reactions with ionized air resulted in conversion (up to 40%–45%) of heavy residue ($T_b > 450°C$) (see the following examples).

In a modified and more complicated method proposed by Likhterova et al. (1998), feedstock was processed with an air–ozone mixture obtained from air irradiated by fast electrons at a temperature of 80°C–100°C. The feedstock was mixed with water and subjected to irradiation by electrons having an energy of 0.5–2.0 MeV. After irradiation, gas and water were separated. The dissolved products of ozonolysis and radiolysis were removed from water. Water was cleaned from contaminating impurities under combined action of air–ozone mixture and electron irradiation and then was used again for mixing with ozonized feedstock. After the separation of gases and liquid hydrocarbons, water was alkalized with 2% NaOH and processed with HCl to obtain pH = 2–3. The precipitated sediment was separated by filtration and dried. A de-emulsifying agent produced in this manner was used for oil dewatering and desalting at the stage of its primary processing. The gaseous products were used as fuel for feedstock heating.

The main parameters of oil radiation processing (temperature, dose, dose rate, and detailed conditions of feedstock processing with ionized air) were not disclosed by the authors. However, it was noted that application of the method allowed increase in the yields of light fractions by 7.3%–15% compared with conventional methods.

In more details, application of ionized air for the improved yields and properties of the products of heavy oil radiation processing is described in the papers (Zaikin and Zaikina 2004a,d). In a series of experiments, oil samples were irradiated by 2 MeV electrons and bremsstrahlung x-rays from a linear electron accelerator ELU-4.

The basic processing modes included (1) feedstock pretreatment by bubbling with ionized air at 20°C–70°C in the field of x-ray radiation and (2) subsequent RTC (electron irradiation) at a heightened temperature with simultaneous feedstock bubbling by ozone-containing air.

Oil feedstock was bubbled at room temperature by atmospheric air at the rate of 60 mg/kg-s in the zone of maximum air ionization and a high concentration of reactive ozone-containing mixtures during 10–40 min. After bubbling, feedstock was subjected to RTC at the minimum values of dose rate and temperature required for the start of intense cracking reactions in the ranges of 1.2–1.5 kGy/s and 380°C–400°C, respectively.

Experiments conducted in these conditions have shown that successive application of radiation ozonolysis and RTC is associated with the following characteristic effects: (1) decrease in the dose necessary for the maximum yield of liquid RTC products and decrease in the temperature of cracking beginning on average by 40°C, (2) decrease in the yield of the liquid RTC product on average by 8%–10%, (3) increase in the concentration of gasoline in the overall liquid product by 10%, and (4) favorable alterations in the hydrocarbon contents of gasoline fraction (its enrichment with isomers and aromatics, increase in the octane numbers). Decrease in the RTC temperature due to the oil pretreatment with ozonized air is illustrated in Figure 4.2.

The points where temperature comes to a maximum correspond to the beginning of an intense RTC reaction. The experiments have shown that application of oil treatment with ozonized air allows reduction in the cracking temperature by

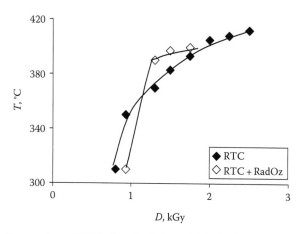

FIGURE 4.2   Temperature of RTC of crude oil ($\rho_{20}=0.94$ g/cm^3, $M=382$) versus irradiation dose. RadOz—radiation ozonolysis.

20°C–40°C. The maximal reduction by 50°–70° was observed for high-sulfuric Tatarstan crude oil.

The effects observed were caused by alterations in the feedstock chemical composition under the action of ionized air, predominantly its most active components, such as atomic hydrogen and ozone. Therefore, the considered two-stage process could be compared with TC of ozonized feedstock.

Kamyanov et al. (1997) marked out the following basic reactions that accompany oil feedstock ozonization:

I. Electrophilic 1-addition of ozone to sulfide atoms with subsequent spontaneous oxygen detachment and formation of sulfoxides and sulfones;

II. Electrophilic 1,3-addition of ozone to 9,10 bonds in regularly condensed polycycloaromatic nuclei
In the experiments on combined radiation-thermal treatment and radiation ozonolysis, the following chain reactions are important:

III. Radical chain oxidation of aromatic compounds accompanied by attachment of C–C β-bonds in substituents and

IV. Ozone-initiated radical chain oxidation of saturated hydrocarbons and molecular fragments

$$RO_2^{\cdot} + RH \rightarrow ROOH + R^{\cdot}$$

with subsequent oxidation of a peroxide compound through the two possible channels:

This type of chain reaction proceeding through the second channel was observed at the temperature of 40°C–60°C (Razumovskiy and Zaikov 1974).

Reactions (I) and (II) are most important for ozonolysis of high-sulfuric oil and bitumen; to a great extent, they define contents of ozonolysis products in these substances. In the absence of radiation, radical chain reactions of oxidation (III) and (IV) proceed very slowly at room temperature. However, their rates rapidly increases as temperature rises, and contributions of these chain reactions, especially reaction (III), to the overall conversion product may become considerable at heightened temperatures.

The role of subsequent thermal or radiation-thermal processing is destruction of ozonides and other ozonolysis products accompanied by deep hydrocarbon conversion.

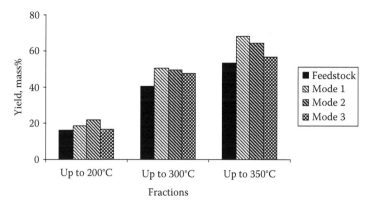

FIGURE 4.3 Radiation-thermal cracking of high-sulfuric oil in different conditions of radiation processing: mode 1, RTC of crude oil without preliminary bubbling ($T=420°C$; $P=1.3$ kGy/s); mode 2, RTC of oil after bubbling by ionized air during 30 min ($T=380°C$, $P=1.3$ kGy/s); mode 3, TC after oil ozonolysis and thermal processing at 350°C. (From Kamyanov, V.F. et al., *Ozonolysis of Petroleum Feedstock*, Pasko, Tomsk, Russia, 1997, 271pp.)

Figure 4.3 shows that the total yield of the liquid fraction boiling below 350°C is much higher when feedstock processing by ionized air is applied with subsequent RTC than in the case of TC with preliminary conventional ozonolysis (Kamyanov 1997). Application of preliminary oil bubbling by ionized air raises the gasoline yield, but at the same time, it makes the total yield of RTC liquid product less. This is a result of cracking reaction proceeding at the lowered temperature when RTC rate becomes lower and oxidation–reduction reactions, such as (I), (II), (III), and (IV), essentially contribute to chemical conversion.

Due to direct initiation of cracking chain reaction, RTC proceeds at temperatures 200°C–250°C lower than the temperature characteristic of the thermal process. It allows saving 40%–60% of energy consumed for feedstock heating. Ozonolysis provides additional decrease in the cracking temperature by 20°C–40°C and, therefore, additional energy saving.

Figure 4.4 shows that most noticeable changes in hydrocarbon contents of the gasoline fraction obtained by RTC from oil pretreated with bubbling by ionized are considerable increase in the isoalkane concentration and decrease in the concentration of olefins. Both of these effects can be explained by decrease in temperature and dose rate of electron irradiation that creates more favorable conditions for isomerization. A lower rate of RTC leads to lower yields of olefins.

The observed increase in the amount of bi- and monoaromatic hydrocarbons can be associated with more intense decomposition of polyarene nuclei of pitches and asphaltenes by ionizing irradiation after feedstock bubbling with ionized ozone-containing air. The increase in gasoline octane numbers is the effect of increase in concentrations of isomers and aromatic compounds. Similar but less pronounced changes in hydrocarbon contents of RTC gasoline fractions were observed as a result of oil bubbling with nonionized air.

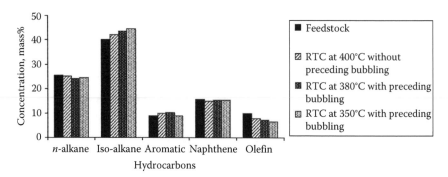

FIGURE 4.4 Hydrocarbon contents of the gasoline fraction (boiling out at 180°C) distillated from the liquid RTC product in different processing conditions.

Still more considerable changes caused by radiation ozonolysis were observed in heavy-oil residue fractions boiling at temperature higher than 350°C, which contained considerable amount of pitches and asphaltenes. Application of radiation ozonolysis reduced the concentration of high-molecular aromatic compounds due to disintegration of condensed polyarene nuclei in molecules of pitches and asphaltenes that lead to considerable decrease in their average molecular mass, partial elimination of sulfur and nitrogen, and enrichment of the residue with oxygen. At the same time, cracking was facilitated by opening a part of aromatic rings.

Splitting of pyrrole and thiophene cycles reduces sulfur concentration and simultaneously gives rise to oxygen concentration. Data of IR spectroscopy confirm that oil bubbling by ionized air causes decomposition of aromatic nuclei and oxidation of sulfuric compounds. The consequence of polyaromatic structure decomposition is enrichment of light fractions by bi- and monoaromatic compounds. Concentration of aromatic compounds in RTC gasoline fraction depends on the time of preliminary bubbling with ionized air. In the time interval of 2 h, this dependence is almost linear.

The advantages of oil pretreatment with ionized air before RTC can be summarized as follows: (1) radiation ozonolysis lowers the irradiation dose and the temperature necessary for the maximum yield of liquid RTC products, (2) improves hydrocarbon contents of the liquid RTC product (higher gasoline yields), and (3) enhances desulfurization and upgrades gasoline quality.

Examples of the favorable action of the preceding radiation ozonolysis on the contents and quality of RTC products (Zaikin and Zaikina 2004a, Zaikin 2005) are given below. Figure 4.5a shows molecular mass distributions of paraffin hydrocarbons in the liquid product of RTC of highly viscous Tatarstan oil for different doses of 2 MeV electron irradiation. Similar dependences in Figure 4.5b are given for the case of feedstock pretreatment with ionized air in the field of background x-rays and subsequent RTC in similar conditions.

Comparison shows that molecular mass distributions in Figure 4.5b are more uniform for all irradiation doses and have maxima shifted to lower values of molecular

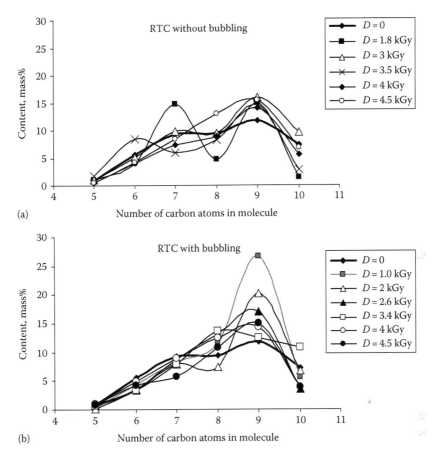

FIGURE 4.5 Molecular mass distribution of paraffins after radiation-thermal cracking in different conditions: Electron dose rate $P=0.5$ kGy/s. (a) RTC at 370°C–400°C; (b) pretreatment by bubbling with ionized air for 30 min at the rate of 20 mg/s-kg at 20°C and RTC.

mass. Therefore, application of feedstock prebubbling with ionized air provides a higher degree of feedstock conversion and improved hydrocarbon contents of the gasoline fraction at lower irradiation doses.

Similar changes in contents, quality characteristics, and optimal irradiation conditions after feedstock bubbling by ionized air were observed for all types of hydrocarbon mixtures studied, including the heaviest feedstock—bitumen (Table 4.1).

Data of Table 4.1 show that application of bubbling with ionized air prior to RTC raises the yield of the liquid RTC product not higher than by about 5%. At the same time, it noticeably lowers RTC temperature and the electron dose needed for the given conversion. The essential effect of radiation ozonolysis is a considerably higher desulfurization of the light products. Increase in the temperature of the ionized air bubbling affects rather RTC conditions than the yields and quality of the product.

TABLE 4.1

Bitumen High-Temperature Radiation Processing Using Preliminary Radiation Ozonolysis (RadOz); Dose Rate $P = 1.3$ kGy/s

Processing Conditions	Yield of Liquid Product, Mass%	Desulfurization, %	Process Temperature, °C	Dose, kGy
RTC	84	27	410	3.5
RadOz at 20°C + RTC	88	53	400	3.2
RadOz at 70°C + RTC	89	52	370	2.8
Bubbling by nonionized air + RTC	85	28	410	3.4

RadOz, radiation ozonolysis.

4.2 APPLICATION OF THE IONIZED AIR IN LOW-TEMPERATURE RADIATION PROCESSING

Effect of oil treatment with ionized air on its conversion at relatively low-dose rates of gamma (or bremsstrahlung x-ray) irradiation was studied for various types of petroleum feedstock.

4.2.1 GASOLINE

The straight-run natural gasoline was irradiated with bremsstrahlung x-rays from accelerated electrons having an energy of 2 MeV (a tungsten target was used for electron deceleration) at nominal room temperature. Increase in temperature during irradiation did not exceed 30°C. Irradiation was conducted at the dose rate of 16.5 Gy/min. The absorbed dose was 0.5 kGy. During irradiation, the feedstock was bubbled with ionized air at the rate of 20 mg/s per 1 kg of feedstock.

The original gasoline had a density of 0.738 g/cm³ and viscosity of 0.91 cSt at 20°C. The total sulfur concentration was 0.43 mass% including 0.034 mass% mercaptan sulfur.

As a result of gasoline radiation processing, concentration of iso-alkanes increased by 2.2 mass% (Figure 4.6).

Much higher level of isomerization was observed when heavy aromatic fractions (heavy residue or bitumen) were added to the original gasoline (Figure 4.7).

The total concentration of aromatic compounds did not considerably change in irradiated gasoline. However, essential changes were observed in the hydrocarbon contents of the aromatic fraction of gasoline. Increase in benzene and toluene concentrations was most considerable (Figure 4.8).

4.2.2 DIESEL FUEL

Processing of diesel fuel with ionized air during 15 min in the field of bremsstrahlung x-ray irradiation ($P = 16.7$ Gy/min) at the rate of ionized-air bubbling of 20 mg/s

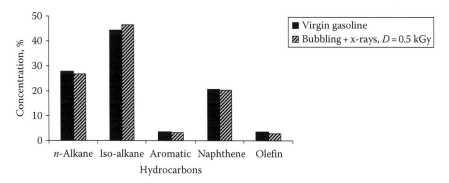

FIGURE 4.6 Hydrocarbon group contents of gasoline after x-ray irradiation with ionized air bubbling.

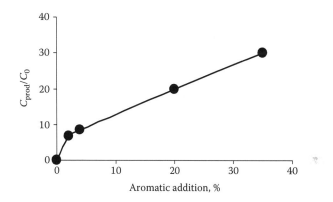

FIGURE 4.7 Effect of heavy aromatic addition on iso-alkane concentration in irradiated gasoline.

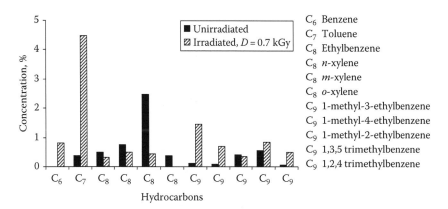

FIGURE 4.8 Hydrocarbon contents of the aromatic fraction in irradiated gasoline (7.5% bitumen was added to original feedstock).

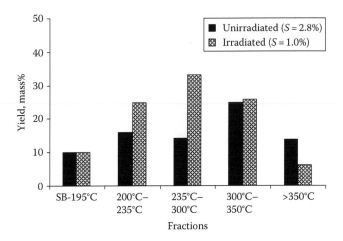

FIGURE 4.9 Fractional contents of diesel fraction bubbled with ionized air in the field of bremsstrahlung x-rays.

per 1 kg of feedstock did not result in considerable changes in the feedstock fractional contents. However, similar processing during 30 min considerably changed the fractional content of diesel fuel (Figure 4.9) and caused decrease in its density from 0.860 to 0.812 g/cm³.

4.2.3 LIGHT GAS OIL FRACTION (SB-250°C)

Fraction SB-250°C was bubbled with ionized air in the field of bremsstrahlung x-rays from 2 MeV electrons at the rate of 20 mg/s per 1 kg of feedstock. The dose rate of x-ray irradiation was 1 kGy/h; the maximal dose was 2 kGy. A considerable increase in the concentration of gasoline fraction (by 13 mass%) was observed after 2 h of oil processing in this mode (Figure 4.10). In the optimal process mode, fractional contents of

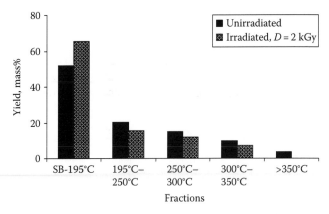

FIGURE 4.10 Fractional contents of light gas oil (SB-250°C) bubbled with ionized air in the field of bremsstrahlung x-rays.

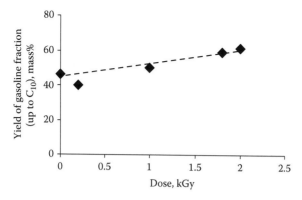

FIGURE 4.11 Dose dependence of the yield of gasoline fraction in gas oil bubbled with ionized air in the field of bremsstrahlung x-rays.

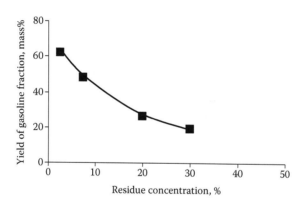

FIGURE 4.12 Correlation between the yield of gasoline fraction and changes in concentration of tail heavy residue in gas oil bubbled with ionized air in the field of bremsstrahlung x-rays.

gas oil fraction changed with irradiation dose as shown in Figure 4.11. Total sulfur concentration decreased from 2 to 0.7 mass% due to sulfur transfer to the gaseous phase.

Increase in the yield of the gasoline fraction (Figure 4.10) was accompanied by a higher conversion of the tail residue ($T_b > 350°C$) (Figure 4.12).

Hydrocarbon distribution in the gas oil fraction (Figure 4.13) demonstrates higher concentrations of light hydrocarbons in the processed products.

4.2.4 Gas Oil (SB-300°C)

A heavier gas oil fraction had a density of 0.785 g/cm³ and viscosity of 1.7 cSt at 20°C, total sulfur concentration of 1.42 mass% including 0.054 mass% mercaptan sulfur. Gasoline fraction (SB-180°C) was 30 mass% of gas oil composition and had a density of 0.738 g/cm³ and viscosity of 0.91 cSt at 20°C and total sulfur concentration of 0.43 mass% including 0.034 mass% mercaptan sulfur.

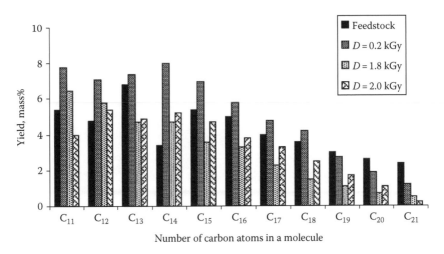

FIGURE 4.13 Molecular mass distribution in gas oil fraction after radiation ozonolysis.

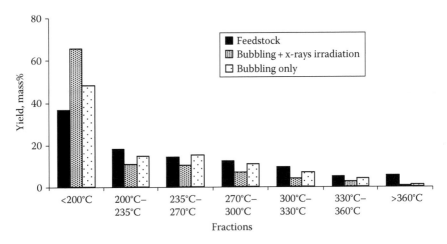

FIGURE 4.14 Fraction distribution in gas oil (SB-300°C) processed with ionized air in the field of bremsstrahlung x-rays.

The dose rate of x-ray irradiation was 16 Gy/min. Ionized air was continuously bubbled during irradiation at the rate of 20 mg/s per 1 kg of feedstock. The combined action of ionized air and x-ray irradiation resulted in a very considerable increase in the concentration of the gasoline fraction (Figure 4.14).

Comparison of the results of gas oil processing in two modes, "ionized air bubbling + x-ray irradiation" and "air bubbling only," indicates synergetic action of ozone-containing air and x-ray radiation.

In another series of experiments, gas oil with a lower concentration of gasoline fraction (23 mass%) was used as feedstock. Oil was processed using different combinations of x-ray and electron irradiation and oil treatment with ozonized air:

Mode 1—X-ray irradiation combined with feedstock bubbling by the gases formed in RTC of another sample (fuel oil) placed under the beam of 2 Mev electrons (dose rate of x-ray irradiation—6 Gy/min, process duration—14 min)

Mode 2—electron irradiation at the temperature of 330°C in the presence of iron bore chips

Mode 3—feedstock bubbling with ionized air (20 mg/s per 1 kg of feedstock) in the field of x-ray radiation during 1 h

Mode 1 is schematically illustrated in Figure 4.15.

Changes in the fractional content of gas oil in different process modes are shown in Figure 4.16.

Each of the irradiation modes used for gas oil processing considerably increased feedstock conversion. The highest yield of light fractions was observed in the mode of gas oil bubbling with ionized air in the field of x-ray irradiation although it required a long-time processing.

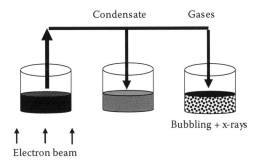

FIGURE 4.15 X-ray irradiation combined with feedstock bubbling by the gases formed in RTC of fuel oil.

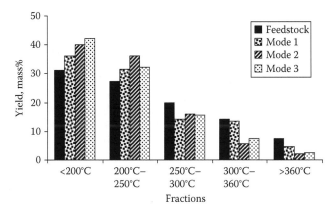

FIGURE 4.16 Fraction contents of gas oil (SB-300°C) processed in different modes of gas bubbling and x-ray irradiation.

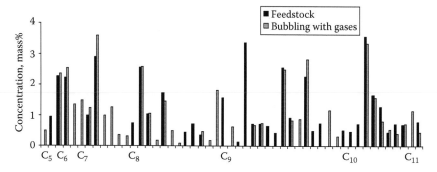

FIGURE 4.17 Hydrocarbon distribution in the gasoline fraction of gas oil (SB-300°C) bubbled with hydrocarbon gases under x-ray irradiation.

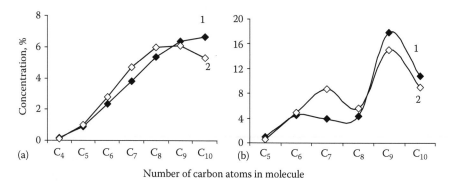

FIGURE 4.18 Distribution of *n*-alkanes and iso-alkanes in the gasoline fraction of gas oil (SB-300°C): 1—feedstock; 2—bubbling with hydrocarbon gases in the presence of x-ray background.

Figure 4.17 demonstrates that bubbling of hydrocarbon gases under x-ray irradiation leads to decrease in the average molecular mass of all components of the gasoline fraction including *n*-alkanes and iso-alkanes (Figure 4.18).

4.2.5 Furnace Oil

The synergetic effect of ionizing radiation and ozonolysis on oil feedstock conversion near room temperature to a great extent depends both on processing conditions and, especially, on the original contents of the feedstock processed (Zaikin and Zaikina 2004a,c,d). In the temperature range of 20°C–40°C, strong synergetic effects showing themselves in high radiation-chemical yields of light fractions characteristic chain cracking reactions were observed for furnace oil. The fractional contents of two sorts of furnace oil used in the experiments (Zaikin and Zaikina 2004a,d) are shown in Table 4.2. They were characterised by (a) high concentration (~30 mass%) of heavy aromatic compounds (availability of the fraction boiling out at temperatures

TABLE 4.2

Fractional Contents of Furnace Oil

Boiling Temperature, °C	Fraction Concentration, Mass%	
	Furnace Oil from Southern Tatarstan	Furnace Oil from Northern Kazakhstan
SB-200	17.6	23.1
200–300	32.6	58.3
300–350	20.0	17.1
>350	29.8	1.5

higher than 250°C) and (b) availability of the gasoline fraction boiling at temperatures below 200°C (not less than 10%).

Irradiation of this type of feedstock by 2 MeV electrons or by bremsstrahlung x-rays in the condition of continuous bubbling with ionized ozone-containing air leads to considerable increase in the yields of light fractions and profound alterations in composition of the heavy residue.

Furnace oil produced from high-viscosity Tatarstan crude oil had a density of 0.904 g/cm³. The hydrocarbon composition of this type of feedstock allowed observation of the pronounced synergetic effects due to the combined action of radiolysis and radiation ozonolysis. The feedstock was subjected to bubbling with ionized air at the rate of 40 mg/s per 1 kg of feedstock in the absence and in the presence of x-ray background ($P = 16.5$ Gy/min; $D = 0.5$ kGy). The changes in its fractional contents due to radiation processing are shown in Figure 4.19. Comparison of the two modes shows that x-ray irradiation noticeably reinforces the action of ionized air.

Oil conversion considerably increased as process duration was increased to 2 h. Figure 4.20 shows that x-ray irradiation of furnace oil accompanied by continuous

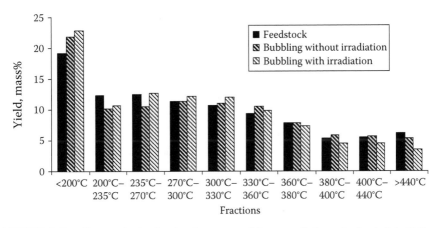

FIGURE 4.19 Changes in the fractional contents of furnace oil due to ionized air bubbling and x-ray irradiation during 30 min.

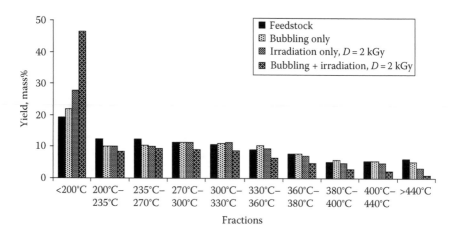

FIGURE 4.20 Changes in the fractional contents of furnace oil due to ionized air bubbling and x-ray irradiation during 2 h ($P = 16.7$ Gy/min).

feedstock bubbling by ionized air results in the increase in gasoline concentration by 28.4 mass%. The result of combined action of the two factors (x-ray irradiation and continuous ozonolysis) was higher than the sum of the effects of their separate action by 1.7 times, that is, the synergetic effect was 70%.

In this case, the radiation-chemical yield of gasoline was 13,800 molecules/100 eV, which is characteristic of chain cracking reactions. It was suggested (Zaikin and Zaikina 2004d) that reactions responsible for hydrocarbon cracking initiation were radiation-induced chain reactions of aromatic compounds with oxidation accompanied by the attachment of C–C β-bonds to substituents and ozone-initiated radical chain oxidation of saturated hydrocarbons and molecular fragments with subsequent oxidation of peroxides through the possible channels (Kamyanov et al. 1997). Under the action of ionizing radiation, these reactions can proceed with rather high rates at near room temperatures in conditions of continuous ozone supply to the feedstock.

An additional result of furnace oil radiation processing was decrease in the total sulfur concentration from 3 mass% in the original feedstock to 0.7 mass% in the irradiated product.

The changes in the hydrocarbon content of the gasoline fraction distilled from the product of furnace oil radiation processing (Figure 4.21) show that furnace oil bubbling with ionized air combined with x-ray irradiation suppresses alkane cyclization and intensifies its isomerization. Increase in the olefin concentration testifies to the cracking of alkanes.

As a result of radiation processing, the total concentration of aromatic compounds in the gasoline fraction decreased (Figure 4.21) with a pronounced increase in ethylbenzene concentration in the aromatic composition (Figure 4.22). Total sulfur concentration decreased from 3 mass% in the original feedstock to 0.7 mass% in the product of radiation processing in mode 2.

Furnace oil processing under conditions similar to mode 2 but at a lower dose rate of 10 Gy/min leads to decrease in the yield of light fractions (Figure 4.23).

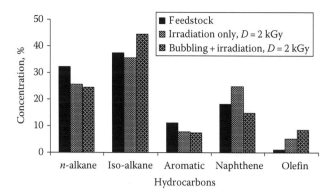

FIGURE 4.21 Hydrocarbon contents of the gasoline fraction of furnace oil after x-ray irradiation and bubbling ionized air.

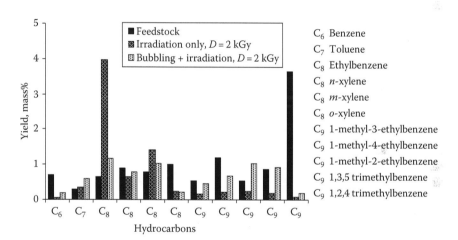

FIGURE 4.22 Distribution of aromatic hydrocarbons in the gasoline fraction of furnace oil after radiation processing. Hydrocarbons are named in the order of their appearance in the graph (left to right).

This was accompanied with a decrease in the concentrations of isomers and olefins in the gasoline fraction that indicates lower intensity of hydrocarbon decomposition and isomerization (Figure 4.24).

Under bubbling with ionized air, the yield of light fraction ($T_b < 200°C$) reached maximum at a dose of x-ray irradiation of 0.1 kGy (Figure 4.25). At higher doses, light fraction yields decreased because of the exhaustion of the potential source of light fractions in the aliphatic part of a branched polyaromatic system accompanied by polymerization processes.

Electron irradiation of furnace at a temperature of 200°C without feedstock pretreatment with ionized air leads to lower yields of light fractions compared with those provided by a long-time combined action of ionized air and x-ray

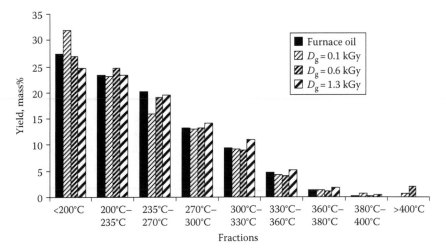

FIGURE 4.23 Changes in the fractional contents of furnace oil (Tatarstan) due to ionized air bubbling and x-ray irradiation ($P = 10$ Gy/min).

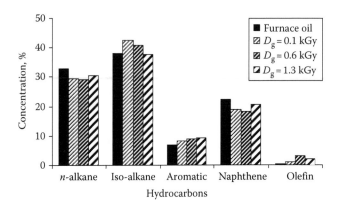

FIGURE 4.24 Hydrocarbon distribution in the gasoline fraction of furnace oil (Tatarstan) irradiated by x-rays ($P = 10$ Gy/min) with ionized air bubbling.

irradiation (Figure 4.26). However, irradiation time was much lower, and the process rate was much higher in the case of electron irradiation.

A series of experiments was conducted on radiation processing of another sort of furnace oil produced at Shymkent refinery, Kazakhstan, and characterised by the high contents of high-molecular paraffins. The feedstock was bubbled with ionized air at room temperature at the rate of 40 mg/s 1 kg of feedstock. The dose rate of bremsstrahlung x-ray irradiation from 2 MeV electrons was 16.7 Gy/min. The changes in the fractional contents of furnace oil due to radiation processing are shown in Figure 4.27.

The highest conversion of heavy residue and maximal increase in the yield of the light fraction ($T_b < 200°C$) by 21.2 mass% was observed after 40 min of ionized

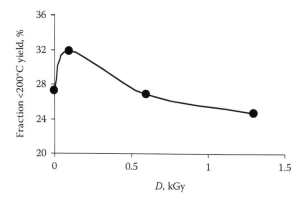

FIGURE 4.25 Dependence of the light fraction yield ($T_b < 200°C$) on x-ray dose ($P = 10$ Gy/min) in conditions of furnace oil bubbling with ionized air.

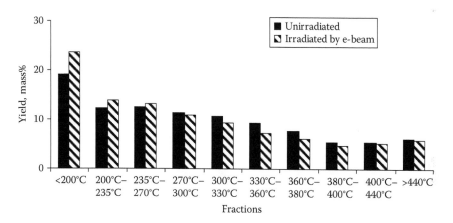

FIGURE 4.26 Fractional contents of furnace oil irradiated by 2 MeV electrons at 200°C ($P = 1$ kGy/s, $D = 0.5$ kGy).

air bubbling in the field of bremsstrahlung x-rays. Decrease in the yields of light fractions with further increase in the process duration was connected with intense radiation-induced polymerization at higher irradiation doses.

The hydrocarbon content of the gasoline fraction in the product of furnace oil radiation processing was characteristic of increased concentrations of iso-alkanes and aromatic compounds and lower concentrations of n-alkanes (Figure 4.28). Low olefin yields testified to considerable polymerization of alkanes.

Molecular mass distributions of n-alkanes (Figure 4.29a), iso-alkanes (Figure 4.29b), and aromatic compounds (Figure 4.29c) in the gasoline fraction of furnace oil bubbled with ionized air under x-ray irradiation indicate increase in the average molecular mass of all basic gasoline components. X-ray irradiation of these types of feedstock in similar conditions without ionized air bubbling did not cause considerable changes in its fractional contents.

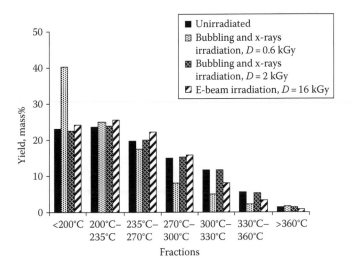

FIGURE 4.27 Changes in fractional contents of furnace oil (Shymkent refinery) due to ionized air bubbling and x-ray irradiation at ambient temperature and RTC at 360°C.

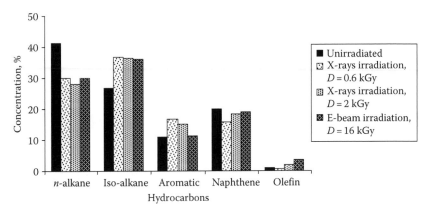

FIGURE 4.28 Changes in hydrocarbon contents of the gasoline fraction of furnace oil (Shymkent refinery) due to ionized air bubbling and x-ray irradiation at ambient temperature ($P = 16.7$ Gy/min).

RTC of furnace oil irradiated with 2 meV electrons at a temperature of 360°C ($P = 2.5$ kGy/s, $D = 16$ kGy) resulted in lower yields of light fractions compared to those obtained by feedstock air bubbling and x-ray irradiation. As a result of RTC, the yield of light fractions ($T_b < 270$°C) increased by 11 mass% (Figure 4.27). Together with the increased concentrations of iso-alkanes and aromatic compounds, the hydrocarbon composition of the RTC gasoline fraction was characterised by higher olefin concentrations (Figure 4.29). It indicates more intense cracking reactions compared with those proceeding in the feedstock bubbled by ionized air in the field of bremsstrahlung x-rays.

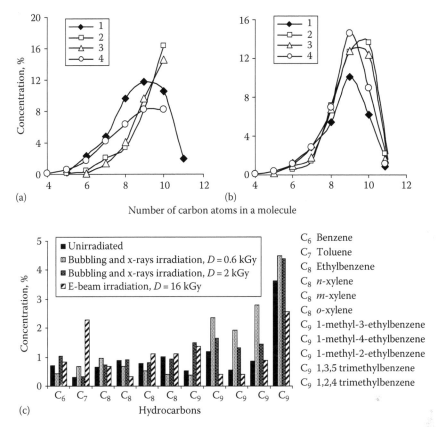

(a)

(b)

Number of carbon atoms in a molecule

(c)

Hydrocarbons

C₆ Benzene
C₇ Toluene
C₈ Ethylbenzene
C₈ n-xylene
C₈ m-xylene
C₈ o-xylene
C₉ 1-methyl-3-ethylbenzene
C₉ 1-methyl-4-ethylbenzene
C₉ 1-methyl-2-ethylbenzene
C₉ 1,3,5 trimethylbenzene
C₉ 1,2,4 trimethylbenzene

■ Unirradiated
▦ Bubbling and x-rays irradiation, D = 0.6 kGy
▨ Bubbling and x-rays irradiation, D = 2 kGy
▧ E-beam irradiation, D = 16 kGy

FIGURE 4.29 Molecular mass distributions of n-alkanes (a), iso-alkanes (b), and aromatic compounds (c) in the gasoline fraction of furnace oil due to ionized air bubbling and x-ray irradiation at ambient temperature during 40 min (1) and 2.2 h (2) and RTC at 360°C (3).

X-ray irradiation and ionized air bubbling lead to a preferential increase in the concentrations of C_8–C_9 hydrocarbons in the light fractions of the products. Distributions of n-alkanes and iso-alkanes in gasoline fractions of furnace oil subjected to this type of radiation processing have also maximums at C_8–C_9, a maximum of iso-alkanes being always higher than that of n-alkanes. These facts indicate that such molecular fragments of the medium molecular mass can be most easily detached from the aromatic and paraffinic structures of heavier fractions. Apparently, alkanes C_8–C_9 can change their electron structure and have a pronounced tendency to isomerization. Bubbling of ionized air to the feedstock intensifies these processes. Feedstock treatment with nonionized air has no considerable effect on the product yields and hydrocarbon contents.

Dealkylation is a very substantial mechanism that affects formation of light fractions. Moreover, alkyl substituents detached from aromatic rings are the main source of potential isomers. In a heavier feedstock, relatively light alkyl substituents in aromatic rings can be easier torn away from heavy aromatic structures. Therefore, the higher the concentration of branched heavy aromatics is, the higher is the yield of light fractions and the higher is the maximum isomer concentration in the gasoline fractions (Figures 4.24 and 4.28). This conclusion is important for the interpretation of the mechanism of radiation-chemical conversion in hydrocarbon mixtures under combined action of ionizing irradiation and radiation ozonolysis.

4.2.6 HEAVY GAS OIL OF ATMOSPHERIC DISTILLATION

A residue of heavy-oil atmospheric distillation having a density of 0.904 g/cm³ (25° API) was another type of feedstock used in this series of experiments. The feedstock was subjected to RTC at a temperature of 390°C and a dose rate of 2.8 kGy/s. Fractional contents of the liquid RTC product are shown in Figure 4.30. The yields of the liquid condensates were 94 and 77 mass% for the doses of 9 and 33 kGy, respectively.

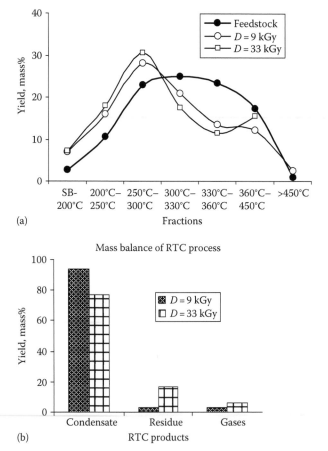

FIGURE 4.30 Distillation curve and mass balance for RTC of heavy gas oil ($T = 390°C$, $P = 2.8$ kGy/s).

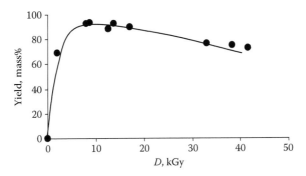

FIGURE 4.31   Dose dependence of the yield of the liquid product obtained by RTC of heavy oil residue ($P = 2.8$–3.2 kGy/s, $T = 390°C$).

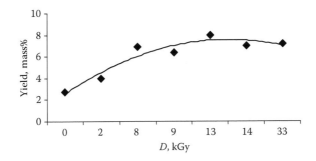

FIGURE 4.32   Dose dependence of the yield of gasoline fraction obtained by RTC of heavy gas oil ($T = 390°C$, $P = 2.8$–3.2 kGy/s).

The dose dependence of the yield of liquid condensate is shown in Figure 4.31.

In these RTC conditions, yields of the gasoline fraction came to saturation and practically did not change at the absorbed doses above 8–9 kGy (Figure 4.32).

In the temperature range of 375°C–400°C, temperature dependence of the yields of gaseous products was practically linear (Figure 4.33).

Irradiation by 2 MeV electrons at the dose of 14 kGy resulted in the changes in the fractional content of heavy residue shown in Figure 4.34. The yield of light fractions ($T_b < 270°C$) increased by 14.5 mass%.

As a result of radiation processing, increase in the concentrations of aromatic compounds and olefins was observed in the gasoline fraction of the liquid RTC product (Figure 4.35). The aromatic part of the gasoline fraction was characterised by considerably increased concentrations of toluene, ethyl benzene, and methyl benzene (Figure 4.36).

In the other series of experiments, the samples of heavy gas oil were subjected to additional processing according to the following scheme. The samples were aerated during 15 min to transform them to the foamed state. Then, the beam of 2 MeV electrons was turned on. One of the samples was placed directly under the electron beam and another one was placed behind a tungsten screen used for transformation of electron radiation to bremsstrahlung x-rays of nearly the same energy.

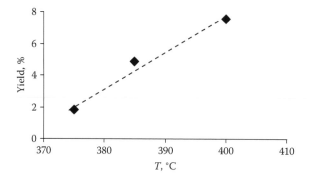

FIGURE 4.33 Temperature dependence of the gas yield in RTC of heavy gas oil ($D = 8–10$ kGy/s, $P = 3.2$ kGy/s).

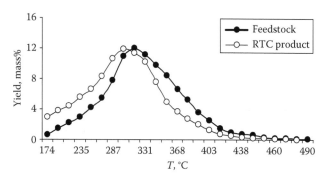

FIGURE 4.34 Fractional content of liquid product obtained by RTC of heavy oil residue ($T = 390°C$, $P = 3.2$ kGy/s). The total yield of the liquid product was 92.2 mass%.

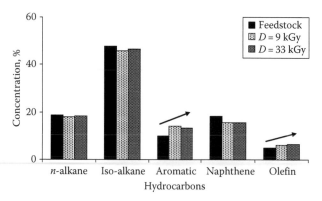

FIGURE 4.35 Hydrocarbon group contents of the gasoline fraction obtained by RTC of heavy oil residue at 390°C.

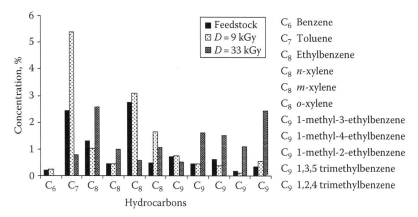

FIGURE 4.36 Distribution of aromatic hydrocarbons in the gasoline fraction obtained by RTC of heavy oil residue at 390°C.

The first sample was irradiated by accelerated electrons at the dose rate of 3.2 kGy/s. Electron irradiation dose was 0.5 kGy. Increase in the sample temperature due to e-beam heating did not exceed 200°C. The other sample was irradiated with bremsstrahlung x-rays at the ambient temperature and a dose rate of 16.7 Gy/min for a longer time; the absorbed dose was 0.45 kGy. Both of the samples were irradiated in the foamed state under continuous bubbling with ionized air at the rate of 40 mg/s per 1 kg of feedstock.

The changes in the fractional content of heavy gas oil as a result of such processing are shown in Figure 4.37. The yield of light fraction ($T_b < 270°C$) increased by 24.5 mass% due to x-ray irradiation at ambient temperature under these conditions. For comparison, RTC at 390°C in the experiment described earlier resulted in an increase in the yield of light fractions by 14.5 mass%.

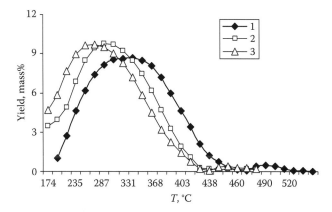

FIGURE 4.37 Fractional contents of heavy gas oil and liquid product of its radiation processing in the foamed state. 1—feedstock; 2—bubbling with ionized air + electron irradiation ($P = 3.2$ kGy/s, $T < 200°C$, $D = 0.5$ kGy), 3—bubbling with ionized air in the field bremsstrahlung x-rays at ambient temperature ($P = 16.7$ Gy/min, $D = 0.45$ kGy).

In the case of short-time electron irradiation of heavy gas oil in the foamed state under continuous bubbling with ionized air, increase in the yield of fraction $T_b < 270°C$ of 15.2% was still higher than that obtained by RTC at 390°C although the process temperature was below 200°C.

The gasoline fraction of heavy gas oil irradiated in the foamed state was characterised by higher concentrations of aromatic compounds (Figure 4.38) compared with those in RTC gasoline (Figure 4.35).

Compared with x-rays, electron irradiation provided higher concentrations of lighter aromatic compounds $C_6–C_7$ (Figure 4.39). As distinct from RTC mode (Figure 4.35), a pronounced increase in iso-alkane concentration was observed when heavy gas oil was irradiated in the foam state. Another characteristic feature of gasoline fraction obtained in these conditions was a decreased olefin concentration indicating dimerization and polymerization reactions.

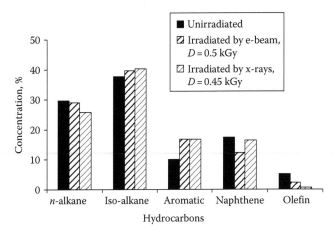

FIGURE 4.38 Hydrocarbon distribution in the gasoline fraction of heavy gas oil irradiated in the foamed state.

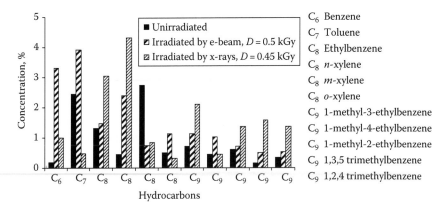

FIGURE 4.39 Distribution of aromatic hydrocarbons in the gasoline fraction obtained by radiation processing of heavy gas oil in the foamed state.

4.3 MECHANISM OF COMBINED ACTION OF LOW-TEMPERATURE IONIZING IRRADIATION AND RADIATION OZONOLYSIS

A marked effect of the simultaneous action of radiation ozonolysis and ionizing irradiation is most pronounced in hydrocarbon mixtures of a certain chemical composition. Experiments using the heaviest types of oil feedstock with the highest concentrations of heavy branched aromatics (bitumen and heavy residua of their vacuum distillation) help in the separation of the roles of ionizing radiation alone and low-temperature radiation ozonolysis of hydrocarbon mixtures.

The results of bitumen low-temperature radiation processing (LTRP) are represented in Figure 4.40, where the fraction yields after bitumen bubbling with ionized air are represented. The samples were kept in a container protected against x-ray radiation so that integral irradiation dose did not exceed 0.2 kGy; preliminary heating at 60°C–80°C was necessary for lower bitumen viscosity.

Figure 4.40 demonstrates that, in contrast to lighter oil feedstock, radiation ozonolysis of bitumen without additional x-ray irradiation provides considerable yields of light fractions. Subsequent electron irradiation of the ozonized samples at 400°C practically does not change bitumen hydrocarbon contents; increase in temperature of preliminary air bubbling does not affect RTC yields, too. It shows that the "work" on bitumen molecule destruction that requires radiation processing at 400°C can be done by bitumen ionized air bubbling at near room temperature.

Analysis of fractional and hydrocarbon contents of LTRP products has shown that light fractions of the product are always enriched with light aromatic compounds, their concentrations increasing as molecular mass of the original aromatic fractions becomes higher. Therefore, in the heavier feedstock, relatively light alkyl substituents in aromatic rings can be easier torn away from heavy aromatic structures.

Comparison of this observation with the regularities listed earlier in the LTRP of hydrocarbon mixtures leads to the conclusion that the main role of oil mixture saturation with ionized air is destabilization of heavy aromatic structures. In turn, it facilitates detachment of alkyl substituents. Simultaneous gamma or electron

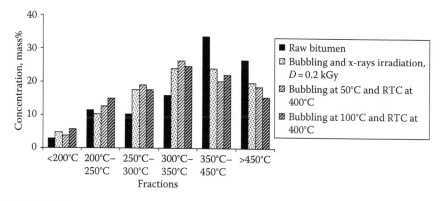

FIGURE 4.40 Fraction yields after low-temperature and high-temperature bitumen radiation processing.

irradiation provides high rates of destructive chain reactions in which gasoline fraction can be presumably considered as a source of light alkyl radicals that appear as a result of gasoline radiolysis. Evidently, radiation-induced cracking of ozonized oil mixtures with destabilized aromatic structure requires much lower concentrations of light radicals compared with high-temperature RTC. Therefore, activation energy for chain continuation for these types of reactions is much lower than that characteristic of RTC; it strongly depends on the average molecular mass of aromatic fraction and tends to zero in highly branched heavy aromatic structures.

LTRP based on the combination of x-ray or electron irradiation and radiation ozonolysis can be applied to practically any type of oil feedstock mixed with suitable inexpensive oil products, such as low-grade crude oil, bitumen, wastes of oil extraction, and low-quality products of oil primary processing. Optimal concentrations of additives and optimal processing conditions for the specific type of hydrocarbon mixtures can be determined in lab conditions.

5 High Dose-Rate Radiation Processing of Petroleum Feedstock in a Wide Temperature Range

Practically, all experiments described in earlier chapters, as well as the methods for oil radiation processing based on these experiments, were conducted at the time-averaged dose rate of ionizing irradiation, which usually did not exceed a value of 5 kGy/s. Such dose rates are much smaller than the highest dose rates (100 kGy/s and higher) provided by modern electron accelerators serially produced by industry for various technological applications. An important role of the dose rate of ionizing radiation in radiation cracking of petroleum was shown in Chapter 1.

A few works were published on the effect of dose rate on radiolysis of light hydrocarbons (Nevitt and Wilson 1961, Barker 1968, Burns and Reed 1968, Gabsatarova and Kabakchi 1969, Massaut et al. 1991). Experiments on cyclohexane irradiation at heightened dose rates using cyclohexene for hydrogen scavenging or deuterium labeling gave no evidence for dose rate effects occurring in radiation spurs (Nevitt and Wilson 1961). It corresponds to the conclusion of the work (Burns et al. 1966) that a major part of radicals in liquid hydrocarbons recombines out of spurs. A very pronounced dose rate effect in gamma radiolysis of C_6–C_{17} hydrocarbons was observed in the work (Massaut et al. 1991). However, behavior of the chain reactions of hydrocarbon cracking at the high dose rates was unstudied till recently.

One of the first attempts of oil processing with pulse electron irradiation with a very high dose rate in a single pulse was undertaken in the works (Gaisin et al. 2003, Remnev et al. 2005). A high-current pulse electron accelerator used in these studies had the following characteristics: electron energy, 350–550 keV; beam current, 6.5 kA; pulse length, 60 ns; energy in a pulse, 200 J.

Irradiation of crude oil and n-pentane was carried out in an open container in a plasma–chemical reactor where volatile components were partially dissolved in the feedstock processed. The absorbed dose was varied by changes in the number of pulses applied to the same target. Oil samples were irradiated by five shots with a time interval of 5 s between two shots. As a result, a dose of 25 kGy was taken. To take a dose of 100 kGy, a sample was irradiated with 20 shots, 5 shots in a series with a 5 min break between two series.

The feedstock was originally kept at room temperature. The beam action was accompanied by explosive gas evolution, oil heating by 5°C–10°C per pulse, and increase in the weight of irradiated feedstock by 2.5% during five successive pulses.

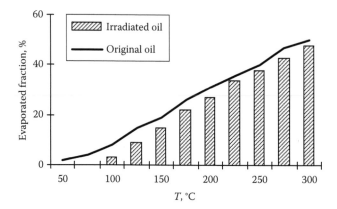

FIGURE 5.1 Changes in oil fractional contents after pulse electron irradiation. (Plotted using the data from Remnev, G.E. et al., A study of the processes of hydrocarbon feedstock decomposition under a pulse electron beam, *Proceedings of the 2nd International Conference on Non-Traditional Methods for Oil Exploration, Preparation and Refining*, Almaty, Kazakhstan, 11pp, 2005.)

Figure 5.1 shows that radiation processing in this mode leads to increase in the concentration of heavy fractions.

Similar effects were observed in the experiments on *n*-heptane irradiation in similar conditions. Table 5.1 shows that the concentration of heavy hydrocarbons in the liquid product of oil radiation processing was higher than that in the original *n*-heptane. The maximal total yield of unsaturated compounds at the absorbed dose of 45 kGy was 7.7 mass%.

The results obtained in the works (Gaisin et al. 2003, Remnev et al. 2005) can be considered from the positions of the radiation cracking theory set out in Chapter 1.

Considerable gas evolution during irradiation of crude oil and *n*-heptane, the high values of radiation-chemical yields of products, and the high yields of unsaturated products testify to radiation-initiated cracking reactions. In these experimental conditions, the cracking rate can be estimated using Equation 1.61, which can be written in the form

$$W \approx k_p R C^* \tag{5.1}$$

where C^* is the concentration of radical states responsible for the reaction chain propagation $k_p \approx 10^{12}$ s^{-1}.

In the state of dynamic equilibrium, concentration of radicals recombining in the first-order reactions is determined by Equation 1.38

$$R = \sqrt{\frac{GP}{k_{t2}}}$$

and the lifetimes of such radicals can be found from

$$GP = \frac{R}{\tau_R} \tag{5.2}$$

TABLE 5.1

Radiation-Chemical Yields of Products Obtained by *n*-Heptane Irradiation with the Electron Beam of Nanosecond Pulse Length

Product		G, Molecules/100 eV
2,5-Dimethyldecane	$C_{12}H_{26}$	803
Dodecane	$C_{12}H_{26}$	804
6-Methyloctadecane	$C_{19}H_{40}$	478
6-Methylundecane	$C_{14}H_{30}$	641
4,5-Dipropyloctane	$C_{13}H_{28}$	701
2,6-Dimethylundecane	$C_{14}H_{30}$	648
6-Methylridecane	$C_{14}H_{30}$	649
3-Methylridecane	$C_{14}H_{30}$	624
2,5-Dimethyldodecane	$C_{14}H_{30}$	604
Tetradecane	$C_{14}H_{30}$	644

Source: Based on the data from Remnev, G.E. et al., A study of the processes of hydrocarbon feedstock decomposition under a pulse electron beam, *Proceedings of the 2nd International Conference on Non-Traditional Methods for Oil Exploration, Preparation and Refining*, Almaty, Kazakhstan, 11pp, 2005.

It implies that

$$\tau_R = \frac{1}{\sqrt{k_{t2}GP}} \tag{5.3}$$

For *n*-heptane, we shall assume that $G = 5$ radicals/100 eV $= 5 \cdot 10^{-8}$ kg/J. In the experiment described earlier, the dose rate in a pulse was $P = 5000$ Gy/60 ns $= 8.3 \cdot 10^{10}$ Gy/s. Assuming that $k_{t2} = 10^9$ s^{-1}, we shall find from Equation 3.21 that $\tau_R = 4.9 \cdot 10^{-7}$ s, which exceeds the pulse length by eight times. After a time τ_R, a chain reaction becomes impossible. According to the estimations of Chapter 1, the lifetime of the radical states responsible for chain continuation, $\tau = 1/k_{t1} \approx 2 \cdot 10^{-4}$ s^{-1}, which is much longer than the duration of an electron radiation pulse in the case considered. Therefore, dynamic equilibrium between radiation generation and recombination of radicals cannot be reached during the time of pulse action. With the assumption that no radical recombination occurs during the time τ_i, concentration of radical pairs responsible for chain reaction propagation can be estimated as

$$R = C^* = GP\tau_i \tag{5.4}$$

The calculation yields the value $R \approx C^* \approx 2.5 \cdot 10^{-4}$. Let us suggest that the average cracking rate before reaction termination is $W_i/2$, where W_i is a reaction rate during the time of pulse action. The chain reaction stops because of the complete recombination of chain carriers as time τ_R passes after pulse termination. Therefore, the average cracking rate between two pulses can be written in the form

$$\overline{W} \approx \frac{W_i}{2} \frac{\tau_R}{\tau_R + t_0} = k_p (GP\tau_i)^2 \tag{5.5}$$

Calculation using Equation 5.5 yields $\overline{W} \approx 3 \cdot 10^{-3}$ s^{-1}.

In the simplest cracking kinetics, not complicated with the consideration of polymerization and chemical adsorption of light fractions, concentration of the original component changes with time according to the law

$$C = 1 - e^{-W_0 t} \tag{5.6}$$

where W_0 is the conversion rate at $t = 0$.

According to Equation 5.6, conversion of about 6% will be achieved at the initial cracking rate of $3 \cdot 10^{-3}$ s^{-1} during 20 s (five pulses), which approximately corresponds to the result of *n*-heptane pulse irradiation.

In the experiment discussed, the lifetime of radicals is much longer than pulse duration. After termination of the pulse action, the rate of radical generation is equal to zero, and radical recombination becomes a prevailing reaction. As a result of radical recombination, hydrocarbons of higher molecular mass compared with that in the original feedstock appear in the product composition.

The first experiments demonstrating an opportunity of obtaining high yield of light fractions by initiation and propagation of chain cracking reactions in high-viscous oil at lowered temperatures, down to room temperature, were conducted in 2005–2006 and gave rise to the development of the new method for oil upgrading and refining (PetroBeam process) (Zaikin and Zaikina 2012).

5.1 CLASSIFICATION OF RADIATION CHAIN CRACKING REACTIONS IN HYDROCARBONS

For the proper formulation of the experimental conditions of self-sustaining cracking reactions, three temperature intervals characteristic of the specific mechanisms of radiation cracking were considered (Zaikin and Zaikina 2008a, 2012): (1) radiation-thermal cracking (RTC), or high-temperature radiation cracking proceeding in the temperature range of 350°C–450°C; (2) low-temperature radiation cracking (LTRC) in the range of 200°C–350°C; and (3) cold radiation cracking (CRC) proceeding at temperatures below 200°C (Figure 5.2).

In the case of RTC, chain cracking reactions are predominantly initiated by ionizing irradiation while chain propagation is predominantly thermally activated. LTRC is characteristic of radiation initiation of chain cracking reaction with the combined contribution of radiation and thermal components to chain propagation. In the case of CRC, chain cracking reactions are both initiated and propagated by ionizing radiation. However, all types of radiation cracking, including CRC, LTRC, and RTC, are endothermic chain reactions. Therefore, the maintenance of the reaction rates at certain level requires supply of the reaction heat.

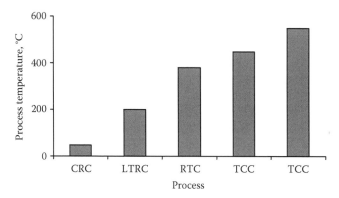

FIGURE 5.2 Characteristic temperatures of cracking processes. CRC, cold radiation cracking; LTRC, low-temperature radiation cracking; RTC, radiation-thermal cracking; TCC, thermocatalytic cracking; TC, thermal cracking.

Dependence of the hydrocarbon molecule cracking reaction rate, W, on the radiation dose rate, P, is different for RTC and CRC. In the case of RTC at heightened temperatures and/or low-dose rates, $W \sim P^{1/2}$; in the case of CRC at low temperatures, $W \sim P^{3/2}$. In this dependence, radiation generation of radical states responsible for chain propagation at heightened dose rates is taken into account. An increase in the radiation dose rate provokes a significant increase in the reaction rate observed in CRC at any temperature. The same relates to LTRC and RTC characterized by still higher reaction rates.

In the whole temperature range of radiation cracking (20°C–450°C), the reaction rate and the limiting dose of radiation can be regulated by the variation of the dose rates in the range of greater than about 5 kGy/s and through additional treatment with processes such as preheating the petroleum feedstock to temperatures less than 150°C, mechanical, acoustic, or electromagnetic processing, to structurally and/or chemically modify the petroleum feedstock. The temperatures of less than about 350°C are not sufficient for thermal activation of the chain propagation reaction; however, when combined with irradiation, thermal enhancement of chain propagation may contribute to an increase in the reaction rate.

In the CRC process, both chain carriers and excited molecules are produced by the interaction of the ionizing radiation at a predetermined dose rate with the petroleum feedstock at temperatures below or equal to about 200°C. The propagation of the chain reaction is provided by interaction of chain carriers with the unstable molecular states (radical pairs) generated by radiation. At sufficiently high dose rates, concentrations of chain carriers and excited molecules generated by irradiation are sufficient for the high rate of chain reaction. Since no, or minimal, thermal heating is required, treatment of the petroleum feedstock can be carried out at temperatures unusually low for hydrocarbon molecule cracking reactions.

By varying the parameters of the radiation self-sustaining cracking (such as temperature, total absorbed dose, dose rate, type of petroleum feedstock, the use of agents, and/or additional feedstock processing), the rate and yield of the radiation cracking chain reaction, as well as the production of desired commodity petroleum products, the final viscosity of the treated petroleum feedstock, the degree of conversion of the petroleum feedstock, and the stimulation of alternate chemical reactions (such as polymerization, polycondensation, isomerization, oxidation, reduction, and chemisorption) may be controlled by the user. The petroleum feedstock may be irradiated either in a continuous or in a pulsed mode. As shown in Chapter 1, continuous irradiation is usually preferable.

The typical irradiation doses were about 20 kGy. However, in each of the temperature range considered, the total absorbed dose of irradiation was selected below the limiting dose, as defined by the stability of the treated petroleum feedstock, the commodity petroleum products desired to be produced, or the desired characteristics of the treated petroleum feedstock. The limiting dose of radiation can be impacted by other parameters of the reaction, such that the limiting dose of radiation for a particular feedstock can be different if other parameters of the reaction are varied. In addition, for each of HTRC, LTRC, and CRC, additional agents might be added before and/or during processing and/or the petroleum feedstock may be treated with a secondary process before and/or during processing.

Generally, the rates of irradiation-induced reactions of chain initiation and chain propagation increase as the dose rate, P, increases. Therefore, the dose necessary for the given degree of petroleum feedstock conversion depends on the dose rate. In the case of RTC at sufficiently high temperatures, this dose is approximately proportional to $P^{1/2}$, while in the case of CRC at sufficiently low temperatures, it is approximately proportional to $P^{-1/2}$. The stronger dependence $D(P)$ for the CRC provides a high-rate processing of petroleum feedstock at low temperatures and at the high dose rates of electron irradiation. The recommended dose rate of ionizing irradiation is, at least, 5 kGy/s in the temperature range of 200°C–350°C and, at least, 10 kGy/s below 200°C.

The heat balance of CRC process is shown in Figure 5.3 in comparison with that for thermocatalytic cracking. CRC provides the most economic process by allowing the highest degree of energy saving through the elimination of energy expenses for the petroleum feedstock heating. Application of RTC and LTRC assumes

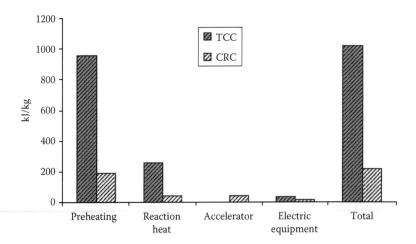

FIGURE 5.3   Heat balance for CRC and TCC bitumen upgrading process. CRC, cold radiation cracking; TCC, thermocatalytic cracking.

preliminary petroleum feedstock heating to temperatures up to 350°C–450°C and 200°C–350°C, respectively, which is associated with additional energy consumption compared with CRC.

However, in the case of LTRC, and to a lesser extent RTC, energy expense for petroleum feedstock heating is much lower than that characteristic of conventional thermo-catalytic cracking or RTC due to the increased and controllable yields of commodity petroleum products produced. At the same time, due to additional thermal excitation of hydrocarbon molecules, the HTRC and LTRC reaction and, therefore, production rates are higher compared with those in CRC at the same dose rate of electron irradiation. Moreover, RTC and LTRC maintain temperature as an additional parameter for initiation and control of thermally activated reactions with low activation energies at relatively low temperatures; the latter can be useful for the provision of the desired properties of products obtained from the special types of petroleum feedstock.

Generally, the technology proposed by Zaikin and Zaikina (2012) is a method for upgrading and deep destructive processing of hydrocarbons in a wide temperature range from room temperature to 400°C, where the known phenomenon of RTC enters as a particular case. The high rates of the self-sustainable cracking reactions of hydrocarbons are attained even at lowered temperatures due to application of the heightened dose rates of electron irradiation and favorable conditions for radiation-induced excitation of hydrocarbon molecular states responsible for chain propagation.

In this method, unique combinations of temperature, absorbed dose of radiation, and dose rate of ionizing irradiation are used to initiate and/or maintain the chain cracking reactions. Irradiation can be carried out at temperatures from 20°C to 450°C and from about atmospheric pressure to 3 atm. While the reaction vessel in which radiation processing occurs is not pressurized, gas evolution generated during such processes can increase the pressure in the reaction vessel to greater than atmospheric pressure. In the majority of applications, heavy oil feedstock is preheated in the feed tank to the reaction temperature, usually in the range of 70°C–150°C, and passes to the reactor where it is subjected to electron irradiation. The gaseous products are separated; usually the amount of gases does not exceed 2–4 mass%. The gaseous products can be recycled back to the reactor and partially or completely utilized for the improvement of the process conditions and the product quality. The vapors of light fractions are condensed and mixed with the total liquid product of oil processing.

An essential unit of the PetroBeam facility is an electron accelerator used as a source of radiation for basic initiation and propagation of different reactions in hydrocarbons. Electron accelerators for technological purposes are presently available on the world market. The accelerator producers provide powerful and highly reliable electron accelerators for numerous applications in different scientific and industrial fields of (e.g., sterilization of medical items, food conservation, sewage water purification, and polymer production). With more than a 40 year history of industrial use, the safety record of electron accelerators is stellar. Application of electron beams of energy up to 10 MeV excludes possibility of any residual radioactivity that makes both processing and products quite safe.

A general scheme for the RTC, LTRC, and CRC process is given in Figure 5.4.

In the radiation-chemical reactor, the petroleum feedstock is exposed to the electrons having energy in the range of 2–10 MeV at the time-averaged dose rate of at

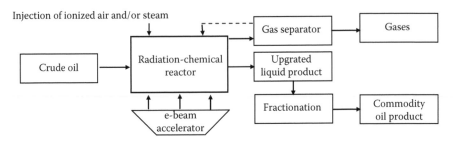

FIGURE 5.4 Layout of the radiation cracking process.

least 5 kGy/s and the absorbed radiation dose rate sufficient to provide the required degree of petroleum feedstock treatment.

To prevent the heating of metal parts of the radiation-chemical reactor vessel, water and/or liquid nitrogen cooling may be used. When more homogeneous irradiation and higher reaction rates are desired, the petroleum feedstock may be injected to the reaction camera in a dispersed form through atomizers or water vapor (such as steam) and/or ionized ozone-containing air may be injected into the reactor vessel. The ionized air used for injection may be obtained as a by-product of the electron accelerator operation. The water vapor and/or ionized air may be pumped into the reactor vessel during irradiation of the petroleum feedstock or may be bubbled into the petroleum feedstock before introduction into the reactor vessel.

The limiting dose or irradiation and the reaction rate may be varied through the use of the feedstock modification as a result of additional processing. Furthermore, the limiting dose or irradiation and the reaction rate may be varied through altering the time-averaged irradiation dose rate and the flow condition parameters.

The petroleum feedstock may be irradiated in a static or a non-static state. However, the process efficiency is much higher when oil feedstock is processed in a flow with high shear rates (see Section 1.7). The flow depth (1–4 cm) is limited by electron penetration to oil at the given particle energy.

RTC is accompanied by considerable gas evolution (10–15 mass%). Just as in the case of CRC, the gases produced may be partially or completely recycled and used for upgrading the products of the process. Oil treatment in flow conditions provides a radiation-chemical yield of light fractions ($<C_{14}$) of not less than 100 molecules per 100 eV consumed by the feedstock. The total product of LTRC is upgraded oil with lower average molecular weight, lower density, viscosity, higher quality, and higher concentrations of light fractions.

Similar to other radiation technologies, the production rate of a radiation facility, Q (kg/s), designated, can be evaluated using formula (3.1)

$$Q = \alpha \frac{N}{D}$$

where
    N is the electron beam power (kW)
    D is the dose (kJ/kg)
    α is the coefficient that takes into account beam power losses

For the given characteristics of an electron accelerator, the production rate of a radiation facility depends only on the dose required for the process.

Examples of radiation cracking (PetroBeam process) applications to different types of high-viscous oil in a wide temperature range are given in the following text.

5.2 HIGH-TEMPERATURE RADIATION-THERMAL CRACKING (350°C–450°C)

The main difference of the approach described in this section from RTC technique discussed in Chapter 3 is that it takes into account radiation contribution to propagation of the chain cracking reactions at heightened dose rates and the effect of feedstock structure on the efficiency of radiation processing.

The next two examples demonstrate the effect of the basic process parameters (temperature, irradiation dose, and especially, dose rate) and feedstock pretreatment on the results of radiation-thermal processing of high-viscous oil feedstock at relatively high temperatures (RTC).

In one of the runs, fuel oil characterized in Table 5.2 was processed in a mode of RTC with the following parameters: pulse irradiation mode (pulse width of 5 μs and pulse frequency of 200 s^{-1}) using electrons with an energy of 2 MeV in flow conditions at a temperature of 410°C and time-averaged dose rate of 2 kGy/s for a total absorbed electron dose of 3 kGy. The total yield of liquid product (fraction boiling out below 450°C) was 76 mass%, and the yield of motor fuels (fraction with boiling point up to 350°C) was 45 mass%. However, the liquid commodity petroleum products produced were unstable and demonstrated a strong tendency toward coking. After a 10 day storage postprocessing, the concentration of the fraction with BP < 350°C (motor fuels) decreased by 10 mass%.

In this example, for the given type of petroleum feedstock utilized (fuel oil) and the RTC processing conditions employed, the limiting dose of irradiation as defined by the stability of the commodity petroleum products was lower than 3 kGy.

To increase the limiting dose of irradiation and to increase the yields of desirable commodity petroleum products (in this case, motor fuels such as gasoline), the same

TABLE 5.2
Characteristics of Fuel Oil and Products of Its Radiation Processing

Properties	Feedstock	RTC Product
Density ρ20, g/cm^3	1.003	0.87
Gravity, °API	7	31.5
Sulfur, wt%	>5.0	1.0
Pour point, °C	27	—
Coking ability, %	12.4	—
Kinematic viscosity at 80°C, mm^2/s,	71.1	2.6

fuel oil petroleum feedstock was pretreated by bubbling with ionized air produced as a by-product of the electron accelerator operation for 7 min at a temperature of 180°C before being subject to RTC processing. The ionized air aids in the destruction of the radiation-resistant cluster structure present in the fuel oil petroleum feedstock, reducing the tendency toward coking and increasing the stability of the produced commodity petroleum products. It allowed the limiting dose of radiation to be increased. At the same time, the ionized air increased desulfurization of the light fractions and caused oxidation reactions that facilitated destruction of high-molecular compounds. As a result, the temperature required for RTC could be lowered.

To further increase the limiting dose of irradiation and to increase the yields of motor fuels, the fuel oil petroleum feedstock was irradiated with an increased electron dose rate. In this example, the fuel oil was processed using RTC with the following parameters: pulse irradiation mode (pulse width of 5 μs and pulse frequency of 200 s⁻¹) using electrons with an energy of 2 MeV in flow conditions at a temperature of 380°C and time-averaged dose rate of 6 kGy/s for a total absorbed electron dose of 3.5 kGy.

The total yield of liquid product under these conditions was 86 mass%, the yield of gases was 8.6 mass%, and the yield of the coking residue was 5.4 mass%. The yield of motor fuels ($T_b < 350°C$) was 52 mass% (Figure 5.5).

With these conditions, the products obtained were stable. The fractional contents of the treated petroleum feedstock in 1 year after RTC processing did not show considerable changes (Figure 5.6). In this example, the limiting dose of irradiation increased over 3.5 kGy due to bubbling of ionized air into the petroleum feedstock prior to RTC and application of the heightened dose rate of irradiation.

An additional result of the RTC processing was decrease in the total sulfur content in the liquid product. The sulfur content was reduced by 1 mass%, which is three times lower than the sulfur concentration in the liquid product of a similar distillation cut of the original fuel oil.

In the next series of experiments, a high-viscosity oil and fuel oil characterized in Table 5.2 were processed in a mode of RTC with the following parameters: pulse

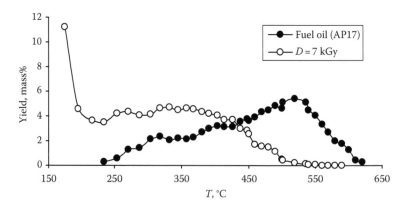

FIGURE 5.5   Fractional contents of fuel oil and the liquid product of its processing in the RTC mode.

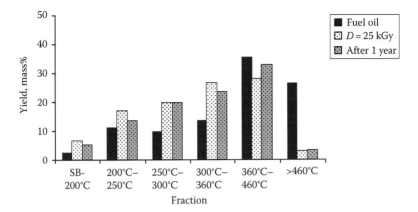

FIGURE 5.6 Changes in the fractional contents of the liquid product of fuel oil radiation processing after 1 year storage.

irradiation mode (pulse width of 5 μs and pulse frequency of 200 s^{-1}) using electrons with an energy of 2 MeV in flow conditions at a temperature of 430°C, the time-averaged dose rate of 1 kGy/s, and the total absorbed electron dose of 7 kGy.

In this example, a desired commodity petroleum product was the basic material for lubricant production characterized by longer hydrocarbon chains and higher molecular mass compared with motor fuels.

As distinct from the requirements for the optimal production of commodity petroleum products such as motor fuels, an important role in RTC processing for the production of lubricants is performed by radiation-induced polymerization, which reduces the mono-olefin content in the lubricant-containing fraction and attenuates its oxidation (see Section 2.4). The heavy polymer deposit forming during HTRC processing is the result, in part, of the high adsorption capacity of such compounds. The intense olefin polymerization combined with radiation-induced adsorption causes efficient release of the lubricant-containing fraction from pitches, asphaltenes, mechanical impurities, if available, and further easy extraction of purified lubricants. The combination of high rates of destruction and olefin polymerization is provided by RTC processing at temperatures higher than the temperature characteristic of the start of intense RTC.

This experiment shows that variation of irradiation parameters, such as temperature, dose rate, and total dose and petroleum feedstock allows control of a required length of the hydrocarbon chain and provides different types of products obtained from the same feedstock.

5.3 LOW-TEMPERATURE RADIATION CRACKING (200°C–350°C)

For many technological operations at the refineries, such as processing of heavy residues after crude oil atmospheric and vacuum distillation, LTRC provides optimal combinations of moderate process temperature and moderate irradiation conditions.

The next examples illustrate application of heavy oil radiation processing in the temperature range of 200°C–350°C (LTRC). A heavy crude oil (viscosity $\nu_{20} = 2200$ cCt, density $\rho_{20} = 0.94$–0.95 g/cm³, about 2 mass% sulfur, and 100–120 μg/g vanadium) was processed in a mode of LTRC with the following parameters: pulse irradiation mode (pulse width of 5 μs and pulse frequency of 200 s⁻¹) using electrons with an energy of 2 MeV in static conditions at temperatures of 160°C–250°C and a time-averaged dose rate of 5–10 kGy/s for a total absorbed electron dose of 500–1800 kGy. Oil was irradiated in the sealed aluminum or aluminized plastic envelopes cooled by a nitrogen vapor. Probes of the gas product were taken from the envelope after irradiation. Liquid and gaseous products of oil radiation processing were analyzed using gas–liquid chromatography.

Figure 5.7 provides comparison of pure thermal and radiation thermal action on the composition of heavy crude oil. Oil heating to 280°C without irradiation deteriorates its fractional contents. Oil heating by the electron beam during the 1 minute simultaneous irradiation ($P = 10$ kGy/s, $D = 600$ kGy) leads to considerable increase in the yields of light fractions ($T_b < 350$°C).

Fractional contents of the liquid product after radiation processing of heavy oil from Karazhanbas field, Kazakhstan, in static conditions at a temperature 280°C, the dose rate of 5 kGy/s, and absorbed dose of 900 kGy are shown in Figure 5.8. Radiation processing resulted in considerable reduction of the heavy residue and high yields of light fractions. The yield of the gaseous product was 5.3 mass%.

Figure 5.9 shows that nearly the same result was obtained by irradiation of the same sort of oil at a lower temperature of 250°C and lower dose of 550 kGy but a higher dose rate of 9 kGy/s. As a result of radiation processing, oil density measured at 20°C decreased from 0.936 to 0.894 g/cm³.

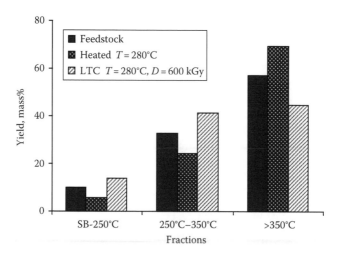

FIGURE 5.7 Effect of thermal heating ($T = 280$°C) and radiation-thermal processing ($T = 280$°C, $P = 10$ kGy/s, $D = 600$ kGy) on fractional contents of heavy crude oil.

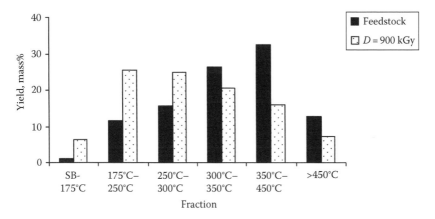

FIGURE 5.8 Fractional contents of heavy crude oil and the liquid product of its processing in the LTRC mode at 280°C by the number of carbon atoms: $P=5$ kGy/s; $D=900$ kGy.

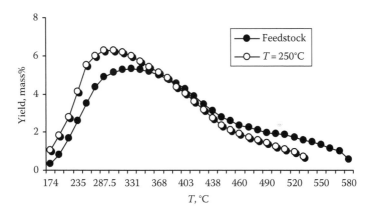

FIGURE 5.9 Fractional contents of heavy crude oil and the liquid product of its processing in the LTRC mode at 250°C by boiling temperatures: $P=9$ kGy/s; $D=550$ kGy.

The yields of gaseous products increased from 2 to 4 mass% as temperature increased from 160°C to 250°C. The hydrocarbon compositions of the gaseous products are shown in Figure 5.10. Increase in the process temperature leads to higher yields of olefins, propane, and butane isomers and heavier hydrocarbon gases. It is accompanied by reduction of the hydrogen concentration in the overall gaseous product (Figure 5.11).

More intense cracking at higher temperatures leads to higher yields of light fractions. At the same time, the hydrogen concentration in the gaseous products decreases. This correlation is shown in Figure 5.12.

A high absorbed dose needed for considerable conversion of heavy crude oil is characteristic of static irradiations without using oil pretreatment. However,

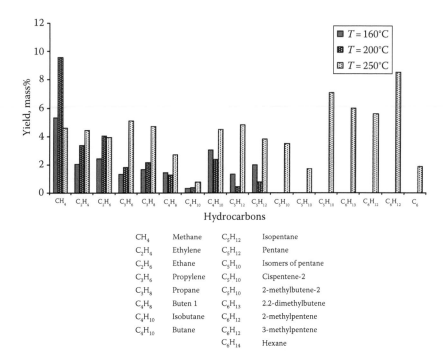

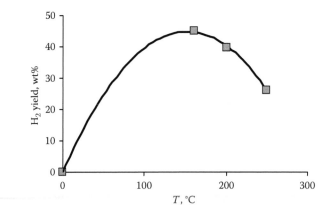

FIGURE 5.10 Hydrocarbon contents of the gaseous products obtained by radiation cracking of high-viscous oil: $P = 9$ kGy/s, $D = 550$ kGy.

FIGURE 5.11 Hydrogen concentration in the gaseous product obtained by radiation cracking of high-viscous oil: $P = 9$ kGy/s, $D = 550$ kGy.

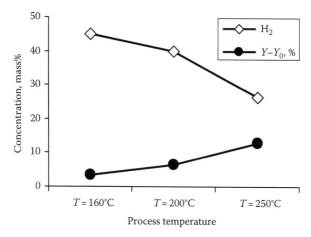

FIGURE 5.12 Correlation between the yield of liquid light fraction (SB-350°C), $Y-Y_0$ and hydrogen concentration in the gaseous product obtained by radiation cracking of high-viscous oil: $P=9$ kGy/s, $D=550$ kGy.

irradiation dose can be considerably lowered using oil pretreatment with ionized air and flow conditions characterized by high shear rates.

5.3.1 COLD RADIATION CRACKING (PETROBEAM PROCESS)

CRC (PetroBeam process) is the technology of the highest energy savings (Figure 5.2), especially advantageous for high-viscous oil and bitumen upgrading near the sites of their exploration.

The next examples relate to the modes of CRC radiation processing (<200°C). A high-viscous crude oil (viscosity $\nu_{20}=496$ cCt, density $\rho_{20}=0.92$ g/cm³, sulfur concentration, 1.4 mass%) was processed with the following parameters: pulse irradiation mode (pulse width of 3 μs and pulse frequency of 60 s⁻¹) using electrons with an energy of 7 MeV in the mode of feedstock distillation under the electron beam and bubbling of ionized air into the petroleum feedstock during radiation processing inside the reactor vessel at a temperature of 170°C and a time-averaged dose rate of 2.7 kGy/s for a total absorbed electron dose of 300 kGy.

The CRC processing resulted in an increased yield of lighter hydrocarbon fractions and a decreased yield of heavier hydrocarbon fractions (Figure 5.13).

Another sort of heavy crude oil (viscosity $\nu_{20}=2200$ cCt, density $\rho_{20}=0.95$ g/cm³, 2 mass% sulfur, and 100–120 μg/g vanadium) was processed in the sealed bags of aluminized plastic at the temperatures of 120°C–150°C at the dose rate was 10–20 kGy/s. Figure 5.14 shows that a considerable increase in the yields of diesel fractions was observed in these experimental conditions.

The effective conversion of the heavy fraction boiling above 360°C was a function of irradiation time, which is represented in Figure 5.15. In these experiments, the dose rate was 20–24 kGy/s and the process temperature was 130°C–150°C.

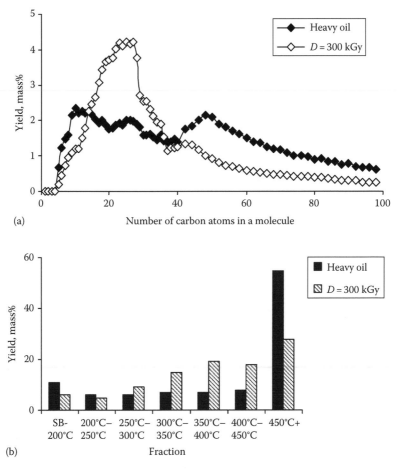

FIGURE 5.13 Fractional contents of heavy crude oil and the liquid product of its processing in the CRC mode at 170°C with ionized air bubbling by (a) the number of carbon atoms and (b) the boiling temperature.

The conversion reached a maximum at the irradiation time of about 180s (or the dose of about 360 kGy) and decreased at higher doses.

The same sort of oil was processed at a still lower temperature of 50°C but at a higher dose rate of 36–40 kGy/s (Figure 5.16). The total absorbed dose was 320 kGy. The chromatographic data in Figure 5.16 show that CRC process caused considerable changes in fractional contents of the untreated versus treated petroleum feedstock. Notably, after CRC processing, the concentration of heavy fractions (represented by fraction having molecules with over 27 carbon atoms and boiling points greater than about 400°C) decreased, and the average molecular mass of the component in the various fraction contents became considerably lower indicating to products with smaller hydrocarbon chains.

The effects of CRC processing were the decrease in the heavy residue content and increase in the concentration of light fractions. The degree of petroleum feedstock

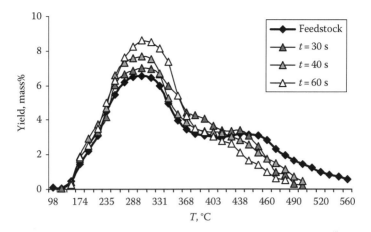

FIGURE 5.14   Fractional contents of the liquid product of heavy oil processing in static conditions at 130°C: $P = 15$ kGy/s.

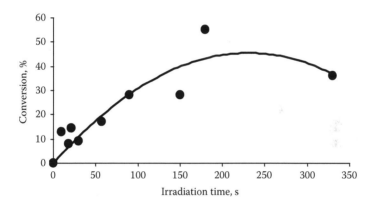

FIGURE 5.15   Dependence of heavy residue ($T_b < 360°C$) conversion on time of heavy oil irradiation.

conversion was conventionally defined by the relative changes in the concentration of the heavy residua boiling out at temperatures higher than 450°C. In this example, the conversion reached 47% after 9 seconds of radiation processing; the rate of conversion was 5.2% per second.

In another series of experiments, the petroleum feedstock was bitumen. Density of the bitumen samples was in the range of 0.97–1.00 g/cm³; molecular mass was 400–500 g/mol; kinematic viscosity at 50°C was in the range of 170–180 cSt; sulfur concentration was 1.6–1.8 mass%. The bitumen was processed in a CRC mode with the following parameters: pulse irradiation mode (pulse width of 5 μs and pulse frequency of 200 s⁻¹) using electrons with an energy of 2 MeV in static conditions at a temperature of 50°C, a time-averaged dose rate of 20–37 kGy/s, and a total absorbed dose of radiation of 360 kGy.

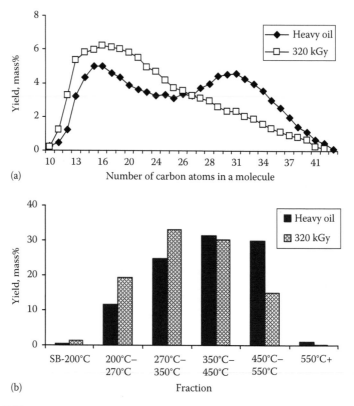

(a)

(b)

FIGURE 5.16 Fractional contents of heavy crude oil and the liquid product of its processing in the CRC mode at 50°C by (a) the number of carbon atoms and (b) the boiling temperature.

In the next example, two types of high-viscous crude oils were used: Sample 1 had viscosity $\nu_{20} = 2200$ cCt, density $\rho_{20} = 0.95$ g/cm³, and contained 2 mass% sulfur; Sample 2 had viscosity $\nu_{20} = 496$ cCt, density $\rho_{20} = 0.92$ g/cm³, and sulfur concentration of 1.4 mass%. Sample 1 was processed in a CRC mode with the following parameters: continuous irradiation mode using electrons with an energy of 2 MeV in static conditions at a temperature of 50°C and a time-averaged dose rate of 80 kGy/s. Sample 2 was processed in the same conditions but using the time-averaged dose rate of 120 kGy/s. The total absorbed dose of radiation was 100 kGy in Sample 1 and 50 kGy in Sample 2. The results are displayed in Figure 5.17a for Sample 1 and Figure 5.17b for Sample 2.

Figure 5.17 shows that although the degree of bitumen feedstock conversion is somewhat lower than that observed after the processing of the petroleum feedstocks comprising lighter hydrocarbon chains, CRC processing significantly altered the hydrocarbon chain length in the fractional contents of the bitumen feedstock. The exposure to the electron beam led to an increase in the amount of the lower boiling fractions.

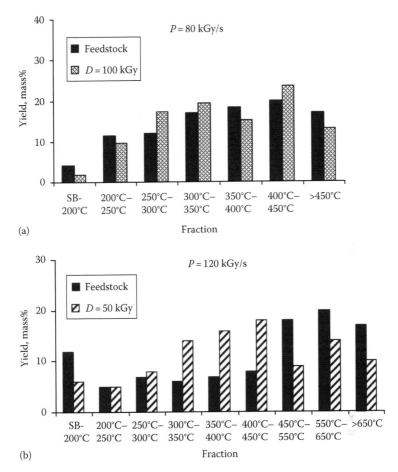

FIGURE 5.17 Fractional contents of two types of heavy crude oil and liquid products of its processing in the CRC mode at 50°C: (a) sample 1 and (b) sample 2.

Concentration of the total sulfur in the fractions that compose motor fuels (fractions boiling out at temperatures less than 350°C) decreased more than twofold after CRC processing compared with sulfur concentration in the products of primary thermal distillation of the original bitumen petroleum feedstock. The degree of the feedstock conversion increased proportionally to the time of exposure, reaching 45% conversion after 18 seconds of radiation processing; the average rate of conversion was 2.5% per second.

Comparison of Figure 5.17a and b shows that nearly the same degree of oil conversion (about 50% by mass) can be attained at the dose rate of 80 kGy/s and the total dose of 100 kGy or at the dose rate of 120 kGy/s and the total dose of 50 kGy.

According to the dependence of the cracking reaction on the dose rate characteristic for the process at a temperature below 200°C, the ratio of these two doses must be (120 kGy/s/80 kGy/s)$^{3/2}$, which is approximately equal to 1.8. Therefore,

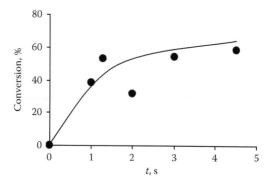

FIGURE 5.18 Conversion of heavy oil versus irradiation time.

the experimentally observed dose ratio is in accordance with the concepts stated in Chapter 1.

Figure 5.18 shows the degree of petroleum feedstock conversion as a function of irradiation time for Sample 1. The feedstock conversion increased proportionally to the time of exposure reaching about 50% conversion after 3 seconds of radiation processing; the average rate of conversion was about 17% per second.

Similar results were obtained for Sample 2.

Increase in the yields of the light fraction ($T_b < 350°C$) and decrease in the concentration of heavy residue ($T_b > 450°C$) in the product of radiation processing in a single series of experimental runs are given in Figure 5.19 for Sample 1 as functions of irradiation dose. Heavy crude oil was irradiated at the dose rate of 80 kGy/s. Electron energy was 1 MeV. The process temperature did not exceed 100°C.

A higher yield of light fractions at much lower irradiation doses can be achieved by oil processing in flow conditions. Fuel oil ($\rho_{20} = 0$, 975 g/cm³ [13.5 API], $\mu_{100} = 9$ cSt, $S_{total} = 2.9$ mass%, pour point = 28°C, coking ability = 14.2%) was subjected to the following treatment. The feedstock was preheated to 150°C (heating

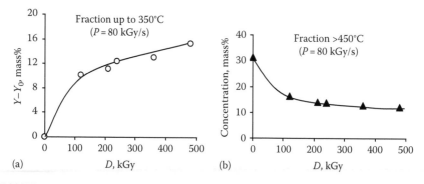

FIGURE 5.19 Dose dependence of the yields of (a) light fraction ($T_b < 350°C$) and (b) heavy residue ($T_b > 450°C$) in the product of heavy oil radiation processing at the dose rate of 80 kGy/s. $Y - Y_0$ is difference between fraction concentration in the product of radiation processing and in the feedstock.

was not maintained during CRC, which was carried out at 50°C) and irradiated in a CRC mode in flow conditions (with the flow rate of 60.1 kg/h in a layer 2 mm thick) with the following parameters: pulse irradiation (pulse width of 5 μs and pulse frequency of 200 s^{-1}) using electrons with an energy of 2 MeV at the time-averaged dose rate of 6 kGy/s. The feedstock was continuously bubbled with ionized air supplied into the reactor during radiation processing. The total absorbed dose was 1.6 kGy.

In this example, the limiting dose of irradiation as defined by the stability of the petroleum commodity products and the rate of cracking reaction was regulated by feedstock preheating and continuous supply of ionized air into the reactor.

As a result of the CRC processing, the degree of the feedstock conversion reached 53% after irradiation with a dose of 1.6 kGy (Figure 5.20). The same result could be obtained in static conditions at a total absorbed dose about 60 times higher and a dose rate about 15–20 times higher compared with irradiation parameters used in this example.

A similar mode was used for CRC processing of high-viscous oil (viscosity $\nu_{20} = 2200$ cCt, density $\rho_{20} = 0.95$ g/cm³, 2 mass% sulfur). The feedstock was preheated to 110°C and irradiated in CRC mode in flow conditions (with the average linear flow rate of 20 cm/s in a layer 2 mm thick) at the time-averaged dose rate of 6 kGy/s. Feedstock preheating was necessary for lower oil viscosity and the higher rate of its passage under the electron beam in a thin layer. The total absorbed dose of radiation depends on the time of exposure of the petroleum feedstock to the radiation. The total absorbed dose of radiation was 10–30 kGy.

The petroleum feedstock was not heated during irradiation. The temperature of the liquid product accumulated in the receiving tank after processing was 30°C–40°C. The products were analyzed in 3–10 h after processing.

Figure 5.21 shows that the degree of heavy residue conversion was about 48% at the dose of 10 kGy and slowly changed with the dose reaching 52% at the dose of 60 kGy. More significant changes with the variation of irradiation dose were observed in the concentration of light fractions boiling below 250°C; it reached the

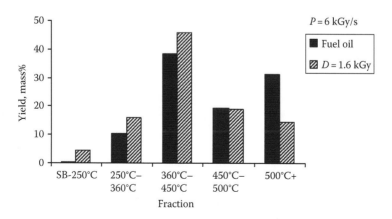

FIGURE 5.20 Fractional contents of fuel oil and liquid product of its processing in flow conditions at 50°C.

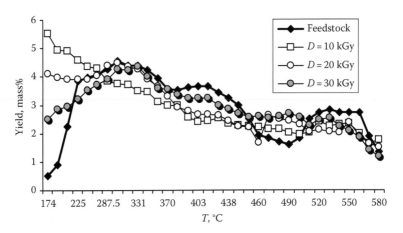

FIGURE 5.21 Fractional contents of heavy oil and liquid products of its processing in flow conditions (feedstock preheating to 150°C).

highest values at the doses below 10–15 kGy. At the same time, concentration of the high-boiling fractions slowly and monotonically decreased with the dose. It shows that the major yield of light fraction was obtained from the middle boiling part of feedstock (350°C–450°C).

The commodity petroleum products obtained by irradiation with total absorbed doses higher than 10–15 kGy were unstable; their hydrocarbon contents changed in a time-dependent manner with higher total absorbed doses of irradiation. The liquid CRC commodity petroleum products obtained by irradiation with the total absorbed dose of 10 kGy at 6 kGy/s demonstrated sufficiently high stability. Figure 5.22a shows that its hydrocarbon content has not noticeably changed after 30 days of exposure. However, in the case of the sample irradiated with the dose of 14 kGy (Figure 5.22b), increase in its molecular mass by 10%–15% was observed already in 12 days after irradiation. In this example, the total absorbed dose of 10–12 kGy limited the product stability. The light fractions obtained by irradiation with a total dose of 10 kGy had the highest concentration in the overall liquid product and the highest stability.

As discussed in Sections 1.7 and 2.4, highly paraffinic petroleum oils are especially difficult for high-temperature radiation processing because of their extremely high tendency for polymerization. The next examples show that these difficulties can be overcome by using CRC processing modes. The feedstock used in CRC processing was high-paraffin crude oil (density $\rho_{20}=0.864$ g/cm³ (32 API), $\mu_{50}=18.8$ mm²/s, $S_{tot}\leq1.0$ mass%, pour point +10°C, asphaltenes and resins 18%, paraffins 20%, and coking ability = 3.5%). The petroleum feedstock was preheated to 35°C and irradiated in CRC mode in flow conditions (with the flow rate of 30 kg/h in a layer 2 mm thick) using 2 MeV electrons at the time-averaged dose rate of 5.2 kGy/s.

As a result of radiation processing, the pour point of the upgraded product dropped down to −20°C. Figure 5.23 illustrates changes in the fractional contents of the products of high-paraffin oil CRC processing under flow conditions for different irradiation doses. It shows that a highest conversion degree and higher yields of light fractions were observed after CRC processing with a total absorbed dose of 8.5 kGy.

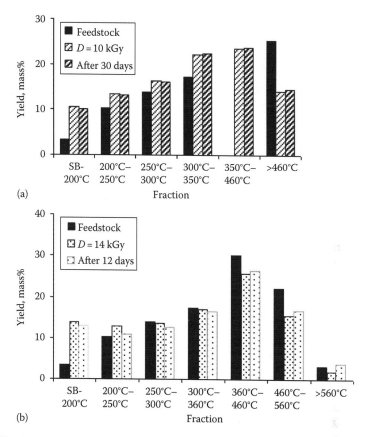

(a)

(b)

FIGURE 5.22 Changes in the fractional contents of heavy crude oil (a) irradiated with a dose of 10 kGy after 30-day exposure and (b) irradiated with a dose of 14 kGy after 12-day exposure.

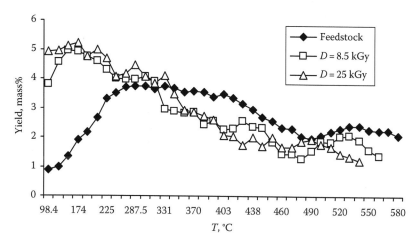

FIGURE 5.23 Fractional contents of high-paraffinic crude oil and liquid product of its CRC processing in flow conditions <100°C.

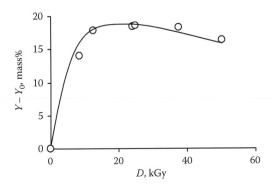

FIGURE 5.24 Dose dependence of the yield of fraction ($T_b < 360°C$) after radiation processing of high-paraffinic crude oil. $Y - Y_0$ is difference between concentrations of fraction ($T_b < 360°C$) in the product of radiation processing and in the feedstock.

The dose dependence of the light fraction yield after radiation processing of high-paraffinic oil is shown in Figure 5.24. Increase in the dose over 10 kGy not only reduces the yields of light fractions but also degrades stability of the liquid petroleum commodity products due to accumulation of the reactive polymerizing residue.

Heating high-paraffinic oil to high temperatures characteristic of RTC provokes thermal activation of intense polymerization that reduces yields of light fractions and makes them unstable. Therefore, CRC processing at heightened dose rates is most effective and advantageous for high-paraffin oil upgrading or deep processing.

A similar mode of radiation processing was applied to high-paraffin fuel oil, a product of high-paraffin crude oil primary distillation (density ρ_{20}, 0.925 g/cm³ [21 API], sulfur content <1 mass%, pour point +29°C, coking ability 6.8%, and kinematic viscosity 16.8 cSt at 80°C). This type of petroleum feedstock is especially difficult for traditional methods of oil processing, due to the presence of high-molecular paraffins that results in a very high pour point (+29°C).

The feedstock was preheated to 60°C and irradiated in CRC mode in flow conditions (with the flow rate of 30 kg/h in a layer of 2 mm thickness) with 2 MeV electrons at the dose rate of 5.2 kGy/s. The irradiation dose was 24 kGy. For comparison, CRC processing was also accomplished using the earlier stated parameters in static mode at the time-averaged dose rate of 20 kGy/s. The irradiation dose was 300 kGy.

The pour point of the upgraded product was −10°C compared with that of +29°C in the original feedstock. Comparison of the efficiencies of CRC processing in flow and static conditions (Figure 5.25) shows that flow conditions provide a considerably higher effect compared with static conditions even at much lower total doses and dose rates of electron irradiation.

Figures 5.26 and 5.27 show the results of low-temperature radiation processing of heavy oil feedstock at the PetroBeam pilot line in optimal irradiation and flow conditions.

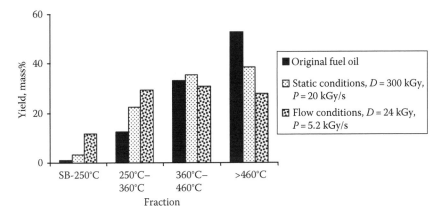

FIGURE 5.25 Fractional contents of high-paraffinic fuel oil and liquid product of its CRC processed in static and flow conditions flow conditions at 60°C.

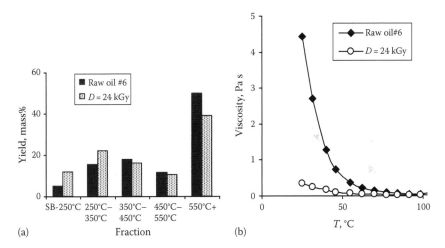

FIGURE 5.26 (a) Fractional contents and (b) temperature dependence of viscosity for oil (#6) and product of its radiation processing at 120°C.

The heavy fuel oil (oil #6) was processed at the PetroBeam pilot line. Conversion of 35% for the heavy residue boiling out above 450°C was reached at the electron irradiation dose of 24 kGy. The changes in the fractional contents of the feedstock were accompanied by the drop in the product viscosity by 86% (Figure 5.26).

A still heavier feedstock, bitumen, was characterized by a density of 0.9781 g/cm^3 (API gravity 13.20) and a kinematic viscosity of 2912 cSt. The C/H ratio was 8.7. Radiation processing of this type of feedstock also resulted in considerable improvements in the fractional contents and rheological properties of the upgraded product (API gravity 22°) (Figure 5.27).

The tests of the technology for heavy oil and bitumen radiation processing at the PetroBeam pilot line have demonstrated a high efficiency of low-temperature

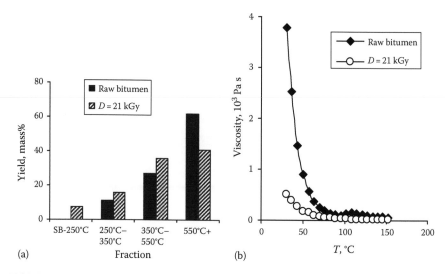

FIGURE 5.27 (a) Fractional contents and (b) temperature dependence of viscosity for bitumen and the product of its radiation processing at 145°C.

radiation processing at heightened dose rates of electron irradiation for different types of high-viscous oil feedstock. Application of this type of oil radiation processing allows combination of a high production rate, a high degree of oil conversion, and the highest energy savings compared with any conventional methods of hydrocarbon processing.

References

Adigamov, B.Y., Lunin, V.V., Miroshnichenco, I.I., Panteleev, D.M., Solovetskii, Y.I., Taletsky, Y.V., and Fesenko, O.G. 1990. Radiation-thermal treatment of hydrodesulphurization carbides Co-Mo catalyst. *J. Kinet. Catal.*, 31: 666.

Ajiev, A.Y., Adigamov, B.Y., Lunin, V.V., Miroshnichenko, I.I., Panteleev, D.M., Riabchenko, P.V., Sadovnichaya, M.V., Solovetsii, Y.I., and Taletsky, Y.V. 1991. Phase transformations of Fe-containing catalyst structures during the process of direct hydrogen sulphide oxidation into sulphur as a result of radiation-thermal treatment. *J. Kinet. Catal.*, 32: 434.

Alfi, M., Da Silva, P., Barrufet, M., and Moreira, R. 2012a. Utilization of charged particles as an efficient way to improve rheological properties of heavy asphaltic petroleum fluids. *SPE Latin America and the Caribbean Petroleum Engineering Conference*, Mexico City, Mexico, Vol. 2, pp. 1379–1387.

Alfi, M., Da Silva, P.F., Barrufet, M.A., and Moreira, R.G. 2012b. Electron induced chain reactions of heavy petroleum fluids—Effective parameters. *Proc. SPE Heavy Oil Conference*, Calgary, Alberta, Canada, Vol. 2, pp. 1230–1239.

Allara, D.L. 1980. A compilation of kinetic parameters for the thermal degradation of *n*-alkane molecules. *J. Phys. Chem. Ref. Data*, 9(3): 523–559.

Al-Sheikhly, M. and Silverman, J. 2008. Radiation processing of heavy oils. Patent Application No. PCT/US2007/088578.

Arakawa, K., Hayakawa, N., Yoshida, K., Tamura, N., Nakanishi, H., Yagi, T., and Kuroiwa, S. 1987. Lubricating oil blend resistant to ionizing radiation. US Patent 4,664,829.

Ayukawa, Y. and Ono, M. 2004. High-energy beam irradiating desulfurization device. US Patent 6,284,746B2.

Bakirov, M.Y., Yakubov, K.M., Mustafaev, I.I., and Gajiev, K.M. 1984. A method for hydrogen production. Author certificate SU 1181242.

Barker, R. 1968. Dose-rate effects in the radiolysis of liquid n-hexane. *Trans. Faraday Soc.*, 64: 430–439.

Basfar, A.A. and Khaled, A.-A.M. 2012. Method of removing sulfur from crude oil by ionizing irradiation. Publication No. US 2012/0138449 A1.

Batzle, M. and Zhijing, W. 1992. Seismic properties of pore fluids. *Geophysics* 57(11): 1396.

Bazyleva, A.B., Hasan, A., Michal, F.M., Becerra, M., and Shaw, J.M. 2010. Bitumen and heavy oil rheological properties: Reconciliation with viscosity measurements. *J. Chem. Eng. Data*, 55: 1389–1397.

Berejka, A.J. 2003. Reactor design concepts for radiation processing. *Radiat. Phys. Chem.*, 71: 311–316.

Berg, G.A. and Habibullin, S.G. 1986. *Catalytic Hydrorefining of Oil Residua*. Leningrad, Russia: Chimiya Publications, 192pp.

Bludenko, F.V., Ponomarev, A.V., Chulkov, A.V., Yakushev, I.A., and Yarullin, R.S. 2007. Electron-beam decomposition of bitumen-gas mixtures at high dose rates. *Mendeleev Commun.*, 17: 227–229.

Boyd, A.W. and Tomlinson, M. 1968. Radiolysis of ortho- and meta-terphenyl. Atomic Energy of Canada, Report AECL-2730, 20pp.

Brodskiy, A.M. and Lavrovskiy, K.P. 1963. On the temperature limit of radiation effect on the rate of chemical conversion. *Kinet. Catal.*, 4: 652–653.

Brodskiy, A.M., Lavrovskiy, K.P., and Titov, V.B. 1963. On temperature dependence of the rate of hydrocarbon radiation conversion. *Kinetika I Kataliz (Kinet. Catal.)*, 4: 337–342.

Brodskiy, A.M., Lavrovskiy, K.P., Zvonov, N.V., and Titov, V.B. 1961. Radiation-thermal conversion of oil fractions. *Neftekhimiya (Oil Chem.)*, 3(3): 370–383.

Bugaenko, L.P., Kuzmin, M.G., and Polak, L.S. 1988. *Chemistry of High Energies*. Moscow, Russia: Khimiya.

Burns, W.G., Holroyd, R.A., and Klein, G.V. 1966. Radical yields in the radiolysis of cyclohexane with different kinds of radiation. *J. Phys. Chem.*, 70: 910–924.

Burns, W.G. and Reed, C.R.V. 1968. Effect of dose-rate in the radiolysis of liquid cyclohexane. *Chem. Commun. (Lond.)*, 1968: 1468–1469.

Burrous, M.L. and Bolt, R.O. 1963. Petroleum refinery stream as nuclear reactor coolants—Radiolytic product investigations. Richmond, VA, California Research Corp., Report No. 21, 19pp.

Chaychian, M., Jones, C.A., Poster, D.L., Silverman, J., Neta, P., Huie, R.E., and Al-Sheikhly, M. 2002. Radiolytic dechlorination of polychlorinated biphenyls in transformer oil and in marine sediment. *Radiat. Phys. Chem.*, 65(4–5): 473–478.

Chester, L.R. and Lucchesi, P.J. 1962. Producing lubricating oils by irradiation. US Patent 3,043,759.

Comberg, H.J. 1988. Extraction and liquefaction of fossil fuels using gamma irradiation and solvents. US Patent 4,772,379.

Cserep, D., György, I., Roder, M., and Wojnárovits, L. 1985. *Radiation Chemistry of Hydrocarbons*. Moscow, Russia: Energoatomizdat, 304pp.

Denisov, E.T. 1971. *Rate Constants of Homolytical Liquid Phase Reactions*. Moscow, Russia: Nauka Publications.

Denisov, E.T. 1987. *Kinetics of Homogeneous Chemical Reactions*. Moscow, Russia: Vysshaya Shkola.

Dolgachev, G.I., Maslennikov, D.D. et al. 2008. Experimental plant ECHO for organic media modification by combined action of electric discharge and electron beam. *Voprosy Atomnoi Nauki I Tekhniki (Prob. Atom. Sci. Technol.) Ser. Nucl. Fusion*, 1: 57–68.

Dolivo, G., Gaumann, T., and Ruf, A. 1986. Photoinduced isomerization and fragmentation of the pentane radical cation in condensed phase. *Radiat. Phys. Chem.*, 28(2): 195–200.

Drelich, J., Krishna, B., Miller, J.D., and Francis, V.H. 1994. Surface tension of toluene-extracted bitumen from Utah oil sands. *Energy Fuels*, 8: 700–704.

Egiazarov, Y.G., Savchitz, M.F., and Ustilovskaya, E.Y. 1989. *Heterogeneous Catalytic Isomerization of Hydrocarbons*. Moscow, Russia: Nauka i Technika.

Egorov, G.F., Terekhov, G.A., and Medvedovsky, V.I. 1972. *Khimiya Vysokikh Energiy (High Energy Chem).*, Vol. 6: 425–429.

Esso Research and Engineering Co. 1960. Desulfurization of petroleum oils. US Patent 826,693.

Evans, R.C. 1964. *Introduction to Crystal Chemistry*. Cambridge, U.K.: University Press, 360pp.

Evdokimov, A.Y., Dzhamalov, A.A., and Lachshi, V.L. 1992. Used lubricants and environmental problems. *Khim. Tekhnol. Topl. Masel (Chem. Technol. Oils Fuels)*, 11: 26–30.

Farkhadova, G.T., Rustamov, M.I. et al. 1987. Influence of temperature on contents of *n*-heptane conversion products on zeolite-containing catalysts. *Neftekhimiya (Oil Chem.)*, 3: 386.

Fellows, A.T. 1966. Chemical conversion in presence of nuclear fission fragments. US Patent 3,228,850.

Földiàk, G. (Ed.) 1981. *Radiation Chemistry of Hydrocarbons*. Budapest, Hungary: Akademiai Kiado.

Folkins, H.O. 1962. Naphta hydroforming by irradiation. US Patent 3,055,814.

Ford, T.J. 1986. Liquid phase thermal decomposition of hexadecane: Reaction mechanisms. *Ind. Eng. Chem. Fundam.*, 25: 240–243.

Frenkel, Y.I. 1945. *The Kinetic Theory of Liquids*. Moscow, Russia: Acad. Sci. USSR

Frenkel, Y.I. 1972. *Introduction to the Theory of Metals*. Leningrad, Russia: Nauka Publications.

Gabsatarova, S.A. and Kabakchi, A.M. 1969. Effect of gamma-radiation dose rate on formation of unsaturated compounds in radiation-thermal cracking of n-heptane. *Khimiya Vysokikh Energiy (High-Energy. Chem.)*, 3: 126–129.

Gafiatullin, R.R., Makarov, I.E., Ponomarev, A.V., Pokhilo, S.B., Rygalov, V.A., Syrtlanov, A.S., and Khusainov, B.K. 1997a. The method for normal alkane processing. Patent of Russia 2099317.

Gafiatullin, R.R., Makarov, I.E., Ponomarev, A.V., Pokhillo, S.B., Rygalov, V.A., Syrtlanov, F.S., and Khusainov, B.K. 1997b. A method for processing condensed hydrocarbons. Patent of Russia RU 2087519.

Gaisin, M.F., Cherninov, C.C., Remnev, G.E., Pushkarev, A.I., Goncharov, D.V., and Urazbakhtina, L.V. 2003. A study of oil radiation cracking with a pulse electron beam. *Proceedings of the 5th International Conf. "Oil and Gas Chemistry,"* Tomsk, Russia, pp. 493–395.

Gäuman, T. and Hoigne, T. (Eds.) 1968. *Aspects of Hydrocarbon Radiolysis.* London, U.K.: Academic Press, 273pp.

Grigoryev, B.A., Bogatov, G.F., and Gerasimov, A.A. 1999. *Thermophysical Properties of Oil, Oil Products, Gas Condensates and Their Fractions.* Moscow, Russia: Moscow Institute of Energy, 372pp.

Haley, F.A. 1961. Effects of mixed-field radiation on lubricating oils. Falls Church, General Dynamics Corp., Report AD0267084, 69pp.

Hartzband, H.M., Tarmy, B.L., and Long, R.B. 1960. Preparing lubricating oils using radiation. US Patent 2,951,022.

Henley, J.B. and Repetti, R.V. 1959. Effect of gamma radiation upon the hydrocracking of a heavy paraffin. Presented before the *Division of Gas and Fuel Chemistry ACS,* Atlantic City, NJ, pp. 161–168.

Hoare, M.F., Garbett, T.A., and Pegg, R.E. 1959. The effect of gamma radiation on the thermal decomposition of light paraffinic hydrocarbons. *Proceedings of the 5th World Petroleum Congress,* New York, Paper 8912.

Hoehlein, G. and Freeman, G.R. 1970. Radiation-sensitized pyrolysis of diethyl ether. Free-radical reaction rate parameters. *J. Am. Chem. Soc.,* 92: 6118–6125.

Humphrey, E.L. and McGrath, J.J. 1964. Irradiation of lubricating oils. US Patent 3,153,622.

Ivanov, V.S. 1988. *Radiation Chemistry of Polymers.* Leningrad, Russia: Khimiya.

Jewell, B.M., Ruberto, R.G. et al. 1976. Structure of aromatic compounds in heavy oil. *Ind. Eng. Chem.,* 15(3): 206–211.

Jones, C.G., Silverman, J., Al-Sheikhly, M., Neta, P., and Poster, D.L. 2003. Dechlorination of polychlorinated biphenyls in industrial transformer oil by radiolytic and photolytic methods. *Environ. Sci. Technol.,* 37(24): 5773–5777.

Kamyanov, V.F. 2005. Ozonolysis in oil processing. *Technol. TEK,* 1(20): 32–45.

Kamyanov, V.F., Lebedev, F.K., and Sviridov, P.P. 1997. *Ozonolysis of Petroleum Feedstock.* Tomsk, Russia: Pasko, 271pp.

Kamyanov, V.F., Lebedev, A.K., Sivirilov, P.P., and Filimonova, T.A. 1991. The process and the products of heavy oil feedstock ozonizing. *Neftekhimiya (Petrol. Chem.),* 31(20): 255–263.

Kamyanov, V.F., Sivirilov, P.P., Lebedev, A.K., and Shabotkin, I.G. 1994. Production of motor fuels by iniated cracking of natural bitumen. *Oil Bitumen Kazan,* 5: 1750–1754.

Khorasheh, F. and Gray, M.R. 1993. High-pressure thermal cracking of *n*-hexadecane. *Ind. Eng. Chem. Res.,* 32: 1853–1863.

Kobzova, R.I., Shulzhenko, I.V., and Chepurova, M.B. 1978. Radiation-induced changes in high-temperature radiation greases. *Chem. Technol. Fuels Oils,* 14: 434–435.

Kochi, J.K. (Ed.) 1973. *Free Radicals,* Vol. 1. New York: Wiley & Sons.

Kossiakoff, A. and Rice, F.O. 1943. Thermal decomposition of hydrocarbons. Resonance stabilization and isomerization of free radicals. *J. Am. Chem. Soc.,* 65: 590–595.

Kurdumov, S.S. 1999. Concentration of heavy metals and demetallization of oil residua under thermal processing in presence of carriers of different nature and polymer wastes. *Neftekhimiya (Oil Chem.),* 39(4): 260–264.

Kuztetsov, P.N., Patrakov, Y.F., Torgashin, A.S., Kusnetsova, L.I., Semenova, S.A., Kurksanov, N.K., and Fadeev, S.N. 2005. Effect of accelerated electron beam processing on composition and molecular structure of brown and black coals of the metamorphic series. *Khimiya v interesakh ustouchivogo rasvitiya (Chem. Sustain. Dev.)*, 13: 71–77.

Lavrovskiy, K.P. 1976. *Catalytic, Thermal, and Radiation-Chemical Conversion in Hydrocarbons*. Moscow, Russia: Nauka Publications.

Lavrovskiy, K.P., Vinnitskiy, O.M., Rumyantsev, A.N., and Musaev, I.A. 1973. On the mechanism of high-molecular alkane cracking. *Neftekhimiya (Oil Chem.)*, 13: 422–430.

Likhterova, N.M., Lunin, V.V., Kukulin, V.I., Knipovich, O.M., and Torkhovskiy, V.N. 1998. A method for petroleum feedstock processing. Patent of Russia RU 212040.

Likhterova, N.M., Torkhovskiy, V.N., Lunin, V.V., Fionov, A.V., and Tretyakov, V.F. 2005. The role of ozone in heavy oil feedstock processing by ionizing radiation. *Proc. 1st All-Russia Conf. "Ozone and Other Environment Friendly Oxidizers. Science and Technology,"* Moscow, Russia, pp. 94–103.

Long, R.B., Hibshman, H.J., and Longwell, J.P. 1959b. Conversion of hydrocarbons in the presence of neutron irradiation and a cracking catalyst. US Patent 2,905,607, Patented September 22.

Long, R.B., Hibshman, H.J., Longwell, J.P., and Houston, R.W. 1959a. Conversion of hydrocarbons in the presence of neutron irradiation and a hydrogenation catalyst. US Patent 2,905,606.

Lucchese, P.J., Baeder, D.L., and Longwell, J.P. 1959. Radiation promoted hydrocarbon reactions. *Proc. 5th World Petroleum Congress*, New York, Paper 8909.

Lucchese, P.J., Tarmy, B.L., Long, R.B., Baeder, D.L., and Longwell, J.P. 1958. High temperature radiation chemistry of hydrocarbon. *Ind. Eng. Chem.*, 50(6): 879–884.

Lucchesi, P.J. and Long, R.B. 1961. Lubricants. US Patent 3,003,937.

Lunin, V.V., Frantsuzov, V.K., and Likhterova, N.M. 2002. Deepening of crude oil processing using ozonation and high energy impacts. *17th World Petroleum Congress*, Rio de Janeiro, Brazil, Paper 32215.

Lunin, V.V., Frantsuzov, V.K., and Likhterova, N.M. 2012. Desulfurization and demetallization of heavy oil fractions by ozonolysis and radiolysis. *Neftekhimiya (Petrol. Chem.)*, 42(36): 195–202.

Lunin, V.V., Likhterova, N.M., and Tarkhovskiy, V.N. 1999. Conversion of petroleum hydrocarbons under the action of electron beam and ozone. *Khimiya i Tekhnologiya Topliv i Masel (Chem. Technol. Fuels Oils)*, 4: 38–43.

Lunin, V.V., Riabchenko, P.V., and Solovetskii, Y.I. 1993. Influence of high temperature radiation-thermal treatments by electron accelerated beam on creaking of metaloxides catalytic systems. *J. Phys. Chem.*, 67: 444.

Lunin, V.V. and Solovetskii, Y.I. 1996. Environmentally friendly radiation-thermal processing of coal with the beams of accelerated 2 MeV electrons. *Proceedings of the VII Conference on Chemistry and Technology of Solid Fuels*, Moscow, Russia, p. 261.

Makarov, I.E., Ponomarev, A.V., and Ershov, B.G. 2007. Radiation synthesis of liquid branched alkanes. Khimiya Vysokikh Energiy (High Energy Chemistry), 41/2, 55–60.

Malkin, A.I. and Isayev, A.I. 2007. *Rheology: Conceptions, Methods, Applications*. St. Petersburg, FL: Professia Publications.

Massaut, B., Klassen, N., and Tilquin, B. 1991. Evidence for a dose rate effect in gamma-radiolysis—I. Methylmethhoxyacetate in aliphatic linear hydrocarbons. *Radiat. Phys. Chem.*, 38(6): 593–599.

Mathews, J. and Walker, R.L. 1964. *Mathematical Methods of Physics*. New York: W.A. Benjamin, Inc.

Matsuoka, S., Tamura, T., and Oshima, K. 1975. Radiation-sensitized thermal cracking of n-butane. 11. Ionic reactions. *Can. J. Chem.*, 56: 92–97.

Matsuoka, S., Tamura, T., Oshima, K., and Oshima, Y. 1974. Radiation-sensitized thermal cracking of n-butane. I. Radical reactions. *Can. J. Chem.*, 5(2): 2579–2589.

McDaniel, R.H. 1964. Materials—Lubricating oils—GTO-915—Irradiation under static and Dynamic conditions. General Dynamics Corp., Report AD0436701, 25pp.

Melikzade, M.M., Bakirov, M.Y., Yakubov, K.M., and Mustafaev, I.I. 1982. A method for hydrogen production. Author Certificate SU 1007315.

Melikzade, M.M., Garibov, A.A., Velibekova, G.Z., Salikhanov, M.S., and Mustafaev, I.I. 1980. The method for production of high-molecular mono-olefin hydrocarbons. Author Certificate SU 780422.

Meraliev, S.A., Gafner, V.V., Izteleuova, M.V., and Stekhun, A.I. 1996. Liquid products of heavy oil residua coking. *Oil Gas (Kazakhstan)*, l(1): 78–80.

Mezger, T.M. 2006. *The Rheology Handbook.* Hannover, Germany: Vincentz Network.

Mikhailov, N.V. and Lichtheim, A.M. 1955. Study of full rheological curves and formulas for calculation of the effective viscosity of the structured liquids with molecular-kinetic characteristics of the members. *Kolloidniy Zhurnal (Colloid J.)*, 17(5): 364–378.

Miley, G.H. and Martin, J.J. 1961. High temperature pile irradiation of the *n*-heptane-hydrogen system. *AIChE J.*, 7(4): 593–598.

Mirkin, G., Zaikin, Y.A., and Zaikina, R.F. 2003. Radiation methods for upgrading and refining of feedstock for oil chemistry. *Radiat. Phys. Chem.*, 67(3–4): 311–314.

Mishra, M.K. and Yaga, Y. 1998. *Handbook of Radical Vynil Polymerization.* New York: CRC Press, 424pp.

Mitsui, H. and Shimizu, Y. 1981. Radiation-thermal cracking of coal. *Radiat. Phys. Chem.*, 18(3–4): 817–826.

Musaev, G.A., Mamonova, T.B., Malibov, M.S., and Musaeva, Z.G. 1994. A study of Kazakhstan oil-bitumen properties by the thermocatalytic method. *Energy. Fuel Resour. Kazakhstan*, 4: 51–56.

Mustafaev, I.I. 1990. Radiation-thermal conversion of heavy oil fraction and organic part of oil bitumen rocks. *Khimiya Vysokikh Energiy (High Energy Chem.)*, 24: 22–26.

Mustafaev, I.I. 1996. Coal desulfurization under the Beam of Accelerated Electrons. *Proceedings of the VII Conference on Chemistry and Technology of Solid Fuels*, Moscow, Russia, p. 265.

Mustafaev, I.I., Bakirov, M.Y., Yakubov, K.M., and Gadzhiev, C.M. 1985. Kinetics of gas formation in radiolysis of brown coals and oil residues. *Khimiya Vysokikh Energiy (High Energy Chem.)*, 19(2): 184–185.

Mustafaev, I. and Gulieva, N. 1995. The principles of radiation-chemical technology of refining the petroleum residues. *Radiat. Phys. Chem.*, 46(4–6): 1313–1316.

Mustafaev, I.I., Gulieva, N.K., Yakubov, K.M., and Gadzhieva, N.N. 1990. Radiation-thermal conversion of pentadecane. *Proc. of the 2nd All-Union Conf. on Theoretical and Applied Chemistry*, Moscow, Russia, Institute of Technical and Economic Studies, p. 169.

Mustafaev, I., Jabbarova, L., Yagubov, R., and Gulieva, N. 2004. Radiation-thermal refining of oil-bituminous rocks. *J. Radioanalyt. Nucl. Chem.*, 262(2): 509–511.

Mustafaev, I.I., Yakubov, K.M., Mamedov, F.A., Gulieva, N.K., and Makhmudov, S.M. 1987. Production of gas fuel through radiation-chemical conversion of fuel oil and black coal. *Voprosy Atomnoi Nauki I Tekhniki (Prob. Atomic Sci. Technol.): Series Atomic-Hydrogen Eng. Technol.*, 2(8): 37–40.

Mustafaev, I.I., Yakubov, K.M., Mamedov, F.A., Gulieva, N.G., and Makhmudov, O.M. 1989. Thermoradiation processes of gaseous and liquid fuel production from bitumen rocks. *Voprosy Atomnoi Nauki I Tekhniki (Prob. Atomic Sci. Technol.): Atomic-Hydrogen Power Eng. Technol.*, 2: 23–26.

Nadirov, N.K. 1995. *Oil and Gas of Kazakhstan*, Vol. 2, Part 2. Almaty, Kazakhstan: Gylym Publications, 400pp.

Nadirov, N.K., Bakirova, S.F., Buyanova, N.S., and Ospanova, N.S. 1997a. Composition and physico-chemical properties of crude oils from Kumkol field. *Oil Gas (Kazakhstan)*, 3: 97–101.

Nadirov, N.K., Rudenko, N.V., and Bychkova, L.V. 1991. The method for sulfoxide production. Patent Application 4899327.04, Kazakhstan.

Nadirov, N.K., Zaikin, Y.A., and Zaikina, R.F. 1997d. Radiation technologies of for oil products refining. *Oil Gas Kazakstan*, 3: 43–45.

Nadirov, N.K., Zaikin, Y.A., Zaikina, R.F., Makulbekov, E.A., Petukhov, V.K., and Panin, Y.A. 1994b. Method for processing of heavy oil and oil residua. Patent of Kazakhstan 4676.

Nadirov, N.K., Zaykina, R.F., and Mamonova, T. 1997b. Prospects of ozone use to rise efficiency of high viscous oil transportation and refining. *Oil Gas Kazakstan*, 3: 159–164.

Nadirov, N.K., Zaikina, R.F., and Zaikin, Y.A. 1994a. Progress in high viscosity oil and natural bitumen refining by ionizing irradiation. *Oil Bitumen Kazan*, 4: 1638–1642.

Nadirov, N.K., Zaikina, R.F., and Zaikin, Y.A. 1995. New high-efficient technologies for heavy oil and oil residue refining. *Energy Fuel Resour. Kazakhstan* 1: 65–69.

Nadirov, N.K., Zaikina, R.F., Zaikin, Y.A., Mamonova, T.B., and Bakirova, S.F. 1997c. Radiation methods for sulfur conversion in oil products. *Oil Gas Kazakstan*, 3: 129–134.

Natkin, A.V., Kolfenbach, J.J., and Forster, E.O. 1961. Radiation of lubricating oils. US Patent 2,990,350.

Nevitt, T.D. and Wilson, W.A. 1961. Dose rate effects in liquid hydrocarbon radiolysis. Defense Technical Information Center, Vol. 1. Report number: ASD TR 61–169, Contract: AF 33(616)-7089, Project: 7360, Task: 73607, AD number: AD0267149, 27 pp.

Nevitt, T.D., Wilson, W.A., and Seelig, H.S. 1959. Radiation-induced reaction of hydrocarbons with *n*-butyl mercaptan. *Ind. Eng. Chem.*, 51: 311–312.

Noddings, C.R., Miller, W.E., and Engelder, T.C. 1959. Treatment of catalyst materials with high energy radiations. US Patent 2,905,608.

Odian, G., Sobel, M., Rossi, R.A., and Klein, R. 1961. Radiation-induced graft polymerization: The Tromsdorff effect of methanol. *J. Polym. Sci.*, 95: 663–673.

Ormiston, R.M., Kerber, Y.L., and Mazgarova, A. 1997. Demercaptanization pf the Tengiz oil field crude. *Oil Gas Kazakhstan*, 2: 71–83.

Panchenkov, G.M., Putilov, A.V., and Zhuravlev, G.I. 1981. Study of basic regularities of *n*-decane radiation-thermal cracking. *Khimiya Vysokikh Energiy (High Energy Chem.)*, 15: 426–430.

Pereverzev, A.E., Polak, L.S., and Chernyak, N.Y. 1968. Radiation-thermal cracking of propane. *Neftekhimiya (Petrol. Chem.)*, 8: 537–542.

Pierre, C., Barré, L., Pina, A., and Moan, M. 2004. Composition and heavy oil rheology. *Oil Gas Sci. Technol. Rev. IFP*, 59(5): 489–501.

Pikaev, A.K. 1987. *Modern Radiation Chemistry: Solids and Polymers: Applied Aspects*. Moscow, Russia: Nauka Publications.

Pokacalov, G.M. 1998. Method of treating heavy hydrocarbon raw material, particularly heavy fractions of crude oil and apparatus for performing said method. Patent of Czech Republic WO9804653.

Ponomarev, A.V. 2010. Major results of completed basic research in 2010. Institute of Chemistry of Solids and Mechanochemistry. www.solid.nsc.ru/rus/INST/activity/res2010.htm (accessed February 19, 2013).

Potanina, V.A., Gorbach, V.A., Siryuk, A.G., Zaslavskii, Y.S., and Ponomareva, T.P. 1977. Effect of ionizing radiation on aromatic hydrocarbons in lubricating oils. *Chem. Technol. Fuels Oils*, 13(8): 558–562.

Potanina, V.A., Zherdeva, L.G., Gorbatch, V.A., Siryuk, A.G., Zaslavskii, Y.S., Ponomareva, T.P., and Smirnyagina, N.A. 1976. Effect of ionizing radiation on naphthenic hydrocarbon in lubricating oils. *Chem. Technol. Fuels Oils*, 13(8): 558–562.

Pruss, G. 1999. Process for working up crude oil. Patent of Germany DE19826553.

Putilov, F.V., Zhuravlev, G.I., Breger, A.K., and Saltsov, S.V. 1981. Prospects of high-temperature nuclear reactor applications in radiation-thermal processes of oil chemistry. *Voprosy Atomnoi Nauki I Tekhniki (Prob. Atom. Sci. Technol.): Series Atomic-Hydrogen Power Eng. Technol.*, 2(8): 85–87.

Qu, Z., Yan, N., Jia, J., and Wu, D. 2006. Removal of thiophene-type sulfide by gamma radiation assisted with hydrogen peroxide. *Energy Fuels*, 20: 142–147.

Qu, Z., Yan, N., Jia, J., and Wu, D. 2007. Removal of dibenzothiophene from simulated petroleum by integrated γ-irradiation and Zr/alumina catalyst. *Appl. Catal. B: Environ.*, 71(1–2): 108–115.

Ray, A.B. and Feldman, P.L. 1983. Electron beam coal desulfurization. US Patent 4,406,762.

Razumovskiy, S.D. and Zaikov, G.E. 1974. *Ozone and Its Reactions with Organic Compounds.* Moscow, Russia: Nauka, Publications, 322pp.

Rebinder, P.A. 1978, 1979. *Selected Works*, Vols. 1 and 2. Moscow, Russia: Nauka Publications.

Remnev, G.E., Pushkarev, A.I., and Concharov, D.V. 2005. A study of the processes of hydrocarbon feedstock decomposition under a pulse electron beam. *Proc. 2nd International Conf. "Non-Traditional Methods for Oil Exploration, Preparation and Refining,"* Almaty, Kazakhstan, 11pp.

Rojas, M.A., Castagna, J., Krishnamoorti, R., Han, D., and Tutuncu, A. 2008. Shear thinning behavior of heavy oil samples: Laboratory measurements and modeling. *SEG Las Vegas 2008 Annual Meeting*, Las Vegas, NV, pp. 1714–1718.

Ruskin, S.L. 1959. Process for recovery of petroleum. US Patent 2,906,680.

Ruskin, S.L. 1961. Methods for hydrocarbon reforming and cracking. US Patent 2,992,173.

Saraeva, V.V. 1986. *Radiolysis of Hydrocarbons in Liquid Phase.* Moscow, Russia: State University Press.

Scapin, M.A., Duarte, C.L., Bustillos, J.O.W.M., and Sato, I.M. 2009. Assessment of gamma radiolytic degradation in waste lubricating oil by GC/MS and UV/VIS. *Radiat. Phys. Chem.*, 78: 733–735.

Scapin, M.A., Duarte, C., Sampa, M.H.O., and Sato, I.M. 2007. Recycling of the used automotive lubricating oil by ionizing irradiation. *Radiat. Phys. Chem.*, 76: 1899–1902.

Scapin, M.A., Duarte, C.L., and Sato, I.M. 2010. Sulphur removal from used automotive lubricating oil by ionizing irradiation. *Atom Peace*, 3(1): 50–55.

Scarborough, J.M. and Ingalls, R.B. 1966. Pyrolysis and radiolysis of *o*-terphenyl. *J. Phys. Chem.*, 71: 486–492.

Schmelling, D.C., Poster, D.L., Chaichian, M., Neta, P., Silverman, J., and Al-Sheikhly, M. 1998. Degradation of polychlorinated biphenyls induced by ionizing radiation in aqueous micellar solutions. *Environ. Sci. Technol.*, 32: 270–275.

Schultze, H.G. and Suttle, F.D. 1959. Conversion of organic compounds by radiation. US Patent 2,914, 452.

Sharipov, A. 1997. Production of oil sulfur-containing agents for hydrometallurgy. *Chem. Technol. Fuel Oil*, 3: 9–17.

Skripchenko, G.B., Sekrieru, D.B., Larina, N.K., Smutkina, Z.S., Miesserova, O.R., and Rudoi, V.A. 1986. The action of irradiation on heavy oil and coal products. *Khimiya Tverdogo Tela (Chem. Solids)*, 4: 55–59.

Solovetskii, Y.I. 1989. Petroleum- and gas-processing catalyst inactivation and new ways of their regeneration. In *Problems of Catalysts Inactivation.* Novosibirsk, Russia, Institute of Catalysis, pp. 92–110.

Solovetskii, Y.I., Chernavskii, P.A., Ryabchenko, P.V., Lunin, V.V., and Panteleev, D.M. 1998. Effect of electron beam irradiation on structure and properties of iron-containing Claus catalysts. *J. Chem. Phys.*, 72: 2268–2271.

Solovetskii, Y.I., Lunin, V.V., and Ryabchenko, P.V. 1995. Thermal irradiation of deactivated heterogeneous catalysts. *React. Kinet. Catalysis Lett.*, 55(2): 463–478.

Song, C., Lai, W.-C., and Schobert, H.H. 1994. Condensed phase pyrolysis of *n*-tetradecane at elevated temperatures for long duration. Product distribution and reaction mechanisms. *Ind. Eng. Chem. Res.*, 33: 534–537.

Sorokin, V. 2002. Method for treatment of heavy hydrocarbons and equipment thereto. Patent of Czech Republic WO0200811.

Stepukhovich, A.D. and Ulitsky, V.A. 1975. *Kinetics and Thermodynamics of Radical Cracking Reactions*. Moscow, Russia: Khimiya Publications, 256pp.

Stoops, C.E. and Day, J.M. 1960. *High VI Oil and Process for Preparing Same*. US Patent 2,954,334.

Sutherland, J.W. and Allen, A.O. 1961. Radiolysis of organic compounds in the adsorbed state. US Patent 3,002,911.

Talrose, V.L. 1974. To the theory of radiation-chemical initiation of chain reactions. *Khimiya Vysokikh Energiy (High-Energy Chem.)*, 8: 519–527.

Tarmy, B.L. and Long, R.B. 1959. Radiochemical treatment of heavy oil. US Patent 2,904,485.

Tian, Y., Yan, N., Li, D., Yao, S., and Wang, W. 2003. Removal of dibenzothiophene from simulated petroleum by integrated γ-irradiation and Zr/alumina catalyst. *J. Chem. Ind. Eng. (China)*, 54: 1279–1283.

Tomlinson, M. 1966. Radiation and thermal decomposition of terphenyls. Atomic Energy of Canada, Report AECL-2641, 20pp.

Tomlinson, M., Boyd, A.W., and Hatcher, S.R. 1966a. The radiation and thermal decomposition of terphenyls and Hydroterphenyls (A summary of Canadian experience). *Panel of the Use of Organic Liquids as Reactor Coolants and Moderators*, Vienna, Austria, 22pp.

Tomlinson, M., Smee, J.L., Winters, E.B., and Arneson, M.C. 1966b. Reactor organic coolants: I. Characteristics of irradiated hydrogenated terphenyls. *Nucl. Sci. Eng.*, 26: 547–558.

Tomlinson, M., Tymko, R.R., and Wuschke, D. 1967. Reactor organic coolants: II. Electron-irradiation studies of hydrogenated terphenyls. *Nucl. Sci. Eng.*, 30: 14–19.

Topchiev, A.V. 1964. *Radiolysis of Hydrocarbons*. Amsterdam, the Netherlands: El. Publ. Co.

Topchiev, A.V. and Polak, L.S. 1962. *Radiolysis of Hydrocarbons*. Moscow, Russia: Acad. Sci. USSR.

Topchiev, A.V., Polak, L.S., Chernyak, N.Y., Glushnev, V.E., Vereshchinskiy, I.V., and Glazunov, P.Y. 1959. Possibility of commercial application of radiation-thermal cracking of normal hydrocarbons. *Int. J. Appl. Radiat. Isotopes*, 7(2): 164–165.

Topchiev, A.V., Polak, L.S., Chernyak, N.Y., Glushnev, V.E., Vereschinskiy, I.V., and Glazunov, N.Y. 1960. Radiation-thermal cracking of hydrocarbons. *Reports Acad. Sci. USSR*, 130: 789–792.

Torgashin, A.S. 2009. Effects of mineral components and modifying processing on permolecular organization and reaction ability of brown coals. PhD dissertation, Krasnoyarsk, Russia, Institute of Chemistry and Chemical Technology.

Trutnev, Y.A., Mufazalov, R.S., Mukhortov, N.Y., Mitenkov, F.M., Zaripov, R.K., Pevnitskiy, A.V., Solovyev, V.P., and Tyupanov, A.A. 1998. The nuclear power plant for distillation and radiation-thermal cracking. Patent of Russia 2116330.

Uryev, N.B. 1980. *Highly Concentrated Disperse Systems*. Moscow, Russia: Khimiya.

Voge, H.H. and Good, G.M. 1949. Thermal cracking of higher paraffins. *J. Am. Chem. Soc.*, 71: 593–597.

Wentworth, R.L. and Canfield, M.P. 1964. *Patent Literature in Radiation Chemistry*. Cambridge, MA: Dynatech Corporation.

Wilson, A.T. and McCauley, D.A. 1965. *Irradiation Conversion of Paraffins*. US Patent 3,177,132.

Wojnárovits, L. and Schuler, R.H. 2000. Bond rupture in the radiolysis of *n*-alkanes: An application of gel permeation chromatography to studies of radical scavenging by iodine. *J. Phys. Chem. A*, 104: 1346–1358.

Woods, R.J. and Pikaev, A.K. 1994. *Applied Radiation Chemistry: Radiation Processing*. New York: Wiley-Interscience.

Wu, G., Katsamura, Y., Matsuura, C., and Ishigure, K. 1997b. Radiation effect in the thermal cracking of n-hexadecane. 2. A kinetic approach to chain reaction. *Ind. Eng. Chem. Res.*, 36: 3498–3504.

Wu, G., Katsamura, Y., Matsuura, C., Ishigure, K., and Kubo, J. 1997a. Radiation effect on the thermal cracking of *n*-hexadecane. 1. Products from radiation-thermal cracking. *Ind. Eng. Chem. Res.*, 36: 1973–1978.

Wuschke, D. and Tomlinson, M. 1968. Reactor organic coolants III: Temperature, intensity and postirradiation effects in the electron radiolysis of meta-terphenyl. *Nucl. Sci. Eng.*, 31: 521–530.

Yakubov, K.M., Bakirov, M.Y., Mustafaev, I.I., and Melikzade, M.M. 1983. A method for hydrogen production. Author Certificate SU 1096900.

Yan, N.Q., Zhao, Y.F., Wu, W.D., Jia, J.P., Yao, S.D., and Wang, W.F. 2004. Investigation on the removal of dibutyl sulfide by radiation of gamma rays. *Fuel Process. Technol.,* 85: 1393–1420.

Yang, D. 2009. Heavy oil upgrading from electron beam irradiation. MS thesis, College Station, TX, Texas A&M University, 143pp.

Yang, D., Kim, J., Silva, P., Barrufet, M.A., and Moreira, R. 2010. Electron beam (e–beam) irradiation improves conventional heavy-oil upgrading. *Proc. SPE Annual Technical Conference and Exhibition*, Florence, Italy, Vol. 2, pp. 1467–1482.

Yang, D., Kim, J., Silva, P., Barrufet, M.A., Moreira, R., and Sosa, J. 2009. Laboratory investigation of e-beam heavy oil upgrading. *Proc. Latin American and Caribbean Petroleum Engineering Conference*, Cartagena de Indias, Colombia, Paper 121911-MS, Vol. 2, pp. 616–623.

Yen, T.F. 1998. The realms and definitions of asphaltenes, Chapter 2. *Asphaltenes and Asphalts,* eds. T.F. Yen and G.V. Chilingar, Vol. 2. Amsterdam, the Netherlands: Elsevier Science.

Zaikin, Y.A. 2005. New technological approaches to cleaning, upgrading and desulfurization of oil wastes and low-grade oil products. *Proc. 4th International Conf. "Oils and Environment."* Gdansk, Poland, University of Technology, pp. 275–282.

Zaikin, Y.A. 2008. Low-temperature radiation-induced cracking of liquid hydrocarbons. *Radiat. Phys. Chem.* 77: 1069–1673.

Zaikin, Y.A. 2013a. Radiation-induced cracking of hydrocarbons. *Radiation Synthesis of Materials and Compounds*, eds. B.I. Kharissov, O.V. Kharissova, and U.O. Mendez. Boca Raton, FL: Taylor & Francis, pp. 355–379.

Zaikin, Y.A. 2013b. On the nature of radiation-excited unstable states of hydrocarbon molecules in heavy oil and bitumen. *Radiat. Phys. Chem.*, 84: 6–9.

Zaikin, Y.A., Ismailova, G.A., and Al-Sheikhly, M. 2007a. Effect of pulse electron beam characteristics on internal friction and structural alterations in epoxy. *Radiat. Phys. Chem.*, 76: 1404–1408.

Zaikin, Y.A. and Zaikina, R.F. 2002. Processing of oil products using complex radiation-thermal treatment and radiation ozonolysis. *Proc. of the 2nd Eurasian Conference "Nuclear Science and its Applications."* Almaty, Kazakhstan, Institute of Nuclear Physics, Vol. 3, pp. 164–169.

Zaikin, Y.A. and Zaikina, R.F. 2004a. Development of the methods for processing of oil products using complex radiation-thermal treatment and radiation ozonolysis. Final Report on IAEA Project (Research Contract # 11837/RO), Almaty, Kazakhstan, 34pp.

Zaikin, Y.A. and Zaikina, R.F. 2004b. Bitumen radiation processing. *Radiat. Phys. Chem.*, 71: 471–474.

Zaikin, Y.A. and Zaikina, R.F. 2004c. Criteria of synergetic effects in radiation-induced transformations of complex hydrocarbon mixtures. *Neft i Gas (Oil Gas Kazakhstan)* 2: 64–73.

Zaikin, Y.A. and Zaikina, R.F. 2004d. Stimulation of radiation-thermal cracking of oil products by reactive ozone-containing mixtures. *Radiat. Phys. Chem.*, 71: 475–478.

Zaikin, Y.A. and Zaikina, R.F. 2004e. Radiation technologies for priority branches of Kazakhstan industry. *News Kazakhstan Sci.*, 2(81): 40–44.

Zaikin, Y.A. and Zaikina, R.F. 2006. Development of experimental facility for processing hydrocarbon components of oil bitumen. Technical Report on ISTC project K-930. Almaty, Kazakhstan, 44pp.

Zaikin, Y.A. and Zaikina, R.F. 2007. Effect of radiation-induced isomerization on gasoline upgrading. *Proc. of the 8th Topical Meeting on Nuclear Applications and Utilization of Accelerators AcAPP'07*, Pocatello, ID, pp. 993–998.

Zaikin, Y.A. and Zaikina, R.F. 2008a. New trends in the radiation processing of petroleum. In: *Radiation Research Progress.* New York: Nova Science Publishers, pp. 17–103.

Zaikin, Y.A. and Zaikina, R.F. 2008b. PetroBeam process for heavy oil upgrading. In *Proceedings of the World Heavy Oil Congress*, Edmonton, Alberta, Canada, American Nuclear Society, Paper 2008-461.

Zaikin, Y.A. and Zaikina, R.F. 2012. Self-sustaining cold cracking of hydrocarbons. US patent 8,192,591, Eurasian patent 016698.

Zaikin, Y.A. and Zaikina, R.F. 2013. Polymerization as a limiting factor for light product yields in radiation cracking of heavy oil and bitumen. *Radiat. Phys. Chem.*, 84: 2–5.

Zaikin, Y.A., Zaikina, R.F., and Chappas, W.J. 2007b. Effect of electron beam characteristics on the rate of radiation-thermal cracking of petroleum feedstock. *Proc. of the 8th International Topical Meeting on Nuclear Applications and Utilization of Accelerators*, Pocatello, ID, American Nuclear Society, pp. 701–707.

Zaikin, Y.A., Zaikina, R.F., Mamonova, T.B., and Nadirov, N.K. 2001. Radiation-thermal processing of high-viscous oil from Karazhanbas field. *Radiat. Phys. Chem.*, 60: 211–221.

Zaikin, Y.A., Zaikina, R.F., and Mirkin, G. 2003. On Energetics of hydrocarbon chemical reactions by ionizing irradiation. *Radiat. Phys. Chem.*, 67(3–4): 305–309.

Zaikin, Y.A., Zaikina, R.F., Mirkin, G., and Nadirov, N.K. 1999c. Radiation technology as a real base for the new generation of oil refineries. *Oil Gas (Kazakhstan)*, 3(7): 93–99.

Zaikin, Y.A., Zaikina, R.F., and Nadirov, N.K. 1999a. Radiation processing of the wastes of high-paraffinic oil extraction. *Oil Gas (Kazakhstan)*, 1(5): 36–42.

Zaikin, Y.A., Zaikina, R.F., and Nadirov, N.K. 1999b. Radiation technologies for oil industry. *Proc. 2nd International Conf. "Nuclear and Radiation Physics,"* Almaty, Kazakhstan, Institute of Nuclear Physics, Vol. 1, pp. 209–214.

Zaikin, Y.A., Zaikina, R.F., and Nadirov, N.K. 2004b. Radiation-thermal cracking of hydrocarbons and its application for deep conversion of oil feedstock. *Oil Gas (Kazakstan)*, 4(24): 47–54.

Zaikin, Y.A., Zaikina, R.F., Nadirov, N.K., and Mamonova, T.B. 1998. On the new approaches to development of effective methods for desulfurization of heavy petroleum feedstock. *Oil Gas Kazakhstan*, 4: 91–96.

Zaikin, Y.A., Zaikina, R.F., and Silverman, J. 2004a. Radiation-thermal conversion of paraffinic oil. *Radiat. Phys. Chem.*, 69(3): 229–238.

Zaikina, R.F. and Aliyev, B.A. 2001. Synergetic effects in high-viscous and paraffinic oil. In: *Problems of Open System Evolution.* Almaty, Kazakhstan: Complex Publications, pp. 114–120.

Zaikina, R.F. and Mamonova, T.B. 1999. Effect of storage conditions on fuel oil properties. *Neft' i Gas (Oil Gas Kazakhstan)*, 1(5): 77–81.

Zaikina, R.F. and Nadirov, N.K. 2001. Effect of the conditions of oil radiation-thermal processing on transformation efficiency of metal-containing compounds. *Oil Gas (Kazakhstan)*, 2: 40–45.

Zaikina, R.F., Pivovarov, S.P., and Akhemetkalieva, G.B. 1998. Structure and properties of the lubricant product obtained by fuel oil radiation and thermal processing. *Oil Gas (Kazakhstan)*, 4: 87–91.

Zaikina, R.F. and Zaikin, Y.A. 2003. Radiation technologies for production and regeneration of motor fuels and lubricants. *Radiat. Phys. Chem.*, 65: 169–172.

Zaikina, R.F., Zaikin, Y.A., and Nadirov, N.K. 1996. Radiation methods for used oil products regeneration and refining. *Oil Gas (Kazakhstan)*, 1: 81–85.

Zaikina, R.F., Zaikin, Y.A., and Nadirov, N.K. 1997. Mechanisms and kinetics of radiation-induced thermal cracking of heavy oil fractions. *Oil Gas (Kazakhstan)*, 2: 83–89.

Zaikina, R.F., Zaikin, Y.A., Nadirov, N.K., and Mamonova, T.B. 1999a. A method for hydro-carbon feedstock cleaning with extraction of sulfur compounds. Patent of Kazakhstan 11995.

Zaikina, R.F., Zaikin, Y.A., Nadirov, N.K., and Sarsembinov, S.S. 1999b. Method and apparatus for reprocessing of used and residual mixtures of oil products. Patent of Kazakhstan 8142.

Zaikina, R.F., Zaikin, Y.A., and Mamonova, T.B. 2002a. Radiation methods for demercaptani-zation and desulfurization of oil products. *Radiat. Phys. Chem.*, 63(2): 617–619.

Zaikina, R.F., Zaikin, Y.A., and Mamonova, T.B. 2002c. The method for cleaning hydrocarbon feedstock with separation of sulfur compounds. Patent of Kazakhstan 11995, 2002.

Zaikina, R.F., Zaikin, Y.A., Mirkin, G., and Nadirov, N.K. 2002b. Prospects of radiation tech-nology application in oil industry. *Radiat. Phys. Chem.*, 63(2): 621–624.

Zaikina, R., Zaikin, Y., Silverman, J., and Al-Sheikhly, M. 2005. Potentialities of hydrocar-bon radiation processing for reduction of environmental pollution by petroleum prod-ucts. *Proc. 4th International Conf. "Oils and Environment,"* Gdansk, Poland, Gdansk University of Technology, pp. 296–303.

Zaikina, R.F., Zaikin, Y.A., Yagudin, S.G., and Fahruddinov, I.M. 2004. Specific approaches to radiation processing of high-sulfuric oil. *Radiat. Phys. Chem.*, 71: 467–470.

Zhao, Y.F., Yan, N.G., Wu, D., Jia, J.P., and Qu, Z. 2004. Investigation on removal of thiophene-type sulfide by gamma rays radiation assisted with hydrogen peroxide. *J. Shanghai Jiatong Univ.*, 38(10):1719–1723.

Zhuravlev, G.I., Voznesenskaya, S.V., Borisenko, I.V., and Bilan, L.A. 1991. Radiation-thermal processing of heavy oil residua. *Khimiya Vysokikh Energiy (High Energy. Chem.)*, 25(1): 27–31.

Zhussupov, D. 2006. Assessing the potential and limitations of heavy oil upgrading by electron beam irradiation. MS thesis, Texas A&M University, College Station, TX, 134pp.

Index

Printed and bound by CPI Group (UK) Ltd, Croydon, CR0 4YY

23/10/2024

01778242-0016